A·N·N·U·A·L EDITIONS

Environment

Twenty-third Edition

04/05

EDITOR

John L. Allen
University of Wyoming

John L. Allen is professor of geography at the University of Wyoming. He received his bachelor's degree in 1963 and his M.A. in 1964 from the University of Wyoming, and in 1969 he received his Ph.D. from Clark University. His special area of interest is the impact of contemporary human societies on environmental systems.

McGraw-Hill/Dushkin

2460 Kerper Blvd., Dubuque, IA 52001

Visit us on the Internet
http://www.dushkin.com

Credits

1. **The Global Environment: An Emerging World View**
   Unit photo—2004 by PhotoDisc, Inc.
2. **Population, Policy, and Economy**
   Unit photo—United Nations photo.
3. **Energy: Present and Future Problems**
   Unit photo—2004 by Sweet By
4. **Biosphere: Endangered Species**
   Unit photo—United Nations photo.
5. **Resources: Land and Water**
   Unit photo—United Nations photo.
6. **The Hazards of Growth: Pollution and Climate Change**
   Unit photo—AP/Wide World photo.
7. **Perspectives and Trends**
   Unit photo—United Nations photo by Rick Grunbaum

Copyright

Cataloging in Publication Data
Main entry under title: Annual Editions: Environment 2004/2005.
1. Environment—Periodicals. I. Allen, John L., *comp*. II. Title: Environment.
ISBN 0–07–286147–9   658'.05   ISSN 0272–9008

Twenty-Third Edition

Cover image © 2004 PhotoDisc, Inc.
Printed in the United States of America   1234567890QPDQPD0987654

Editors/Advisory Board

Members of the Advisory Board are instrumental in the final selection of articles for each edition of ANNUAL EDITIONS. Their review of articles for content, level, currentness, and appropriateness provides critical direction to the editor and staff. We think that you will find their careful consideration well reflected in this volume. ·

To the Reader

In publishing ANNUAL EDITIONS we recognize the enormous role played by the magazines, newspapers, and journals of the public press in providing current, first-rate educational information in a broad spectrum of interest areas. Many of these articles are appropriate for students, researchers, and professionals seeking accurate, current material to help bridge the gap between principles and theories and the real world. These articles, however, become more useful for study when those of lasting value are carefully collected, organized, indexed, and reproduced in a low-cost format, which provides easy and permanent access when the material is needed. That is the role played by ANNUAL EDITIONS.

At the beginning of our new millennium, environmental dilemmas long foreseen by natural and social scientists began to emerge in a number of guises: regional imbalances in numbers of people and the food required to feed them, international environmental crime, energy scarcity, acid rain, buildup of toxic and hazardous wastes, ozone depletion, water shortages, massive soil erosion, global atmospheric pollution and possible climate change, forest dieback and tropical deforestation, and the highest rates of plant and animal extinction the world has known in 65 million years.

These and other environmental problems continue to worsen despite an increasing amount of national and international attention to the issues surrounding them and increased environmental awareness and legislation at both global and national levels. The problems have resulted from centuries of exploitation and unwise use of resources, accelerated recently by the shortsighted public policies that have favored the short-term, expedient approach to problem solving over longer-term economic and ecological good sense. In Africa, for example, the drive to produce enough food to support a growing population has caused the use of increasingly fragile and marginal resources, resulting in the dryland deterioration that brings famine to that troubled continent. Similar social and economic problems have contributed to massive deforestation in middle and South America and in Southeast Asia.

Part of the problem is that efforts to deal with environmental issues have been intermittent. During the decade of the 1980s, economic problems generated by resource scarcity caused the relaxation of environmental quality standards and contributed to the refusal of many of the world's governments and international organizations to develop environmentally sound protective measures, which were viewed as too costly. More recently, in the late twentieth and early twenty-first centuries, as environmental protection policies were adopted, they were often cosmetic, designed for good press and TV sound bites but little else. Even with these public relations policies, governments often lacked either the will or the means to implement them properly. The absence of effective environmental policy has been particularly apparent in those countries that are striving to become economically developed. But even in the more highly developed nations, economic concerns tend to favor a loosening of environmental controls. In the United States, for example, the interests of maintaining jobs for the timber industry imperil many of the last areas of old-growth forests, and the desire to maintain agricultural productivity at all costs causes the continued use of destructive and toxic chemicals on the nation's farmlands. In addition, concerns over energy availability have created the need for foreign policy and military action to protect the developed nations' access to cheap oil and have prompted increasing reliance on technological quick fixes, as well as the development of environmentally sensitive areas to new energy resource exploration and exploitation.

Despite the recent tendency of the U.S. government to turn its back on environmental issues and refuse to participate in international environmental accords, particularly those related to global warming, there is some reason to hope that, globally, a new environmental consciousness is awakening. Unfortunately, increasing globalization of the economy has meant globalization of other things as well, such as internal conflict and disease transmission. The emergence of terrorism as an instrument of national or quasi-national policy—particularly where terrorism may employ environmental contamination as a weapon—has the potential to produce future environmental problems that are almost too frightening to think about.

In *Annual Editions: Environment 04/05* every effort has been made to choose articles that encourage an understanding of the nature of the environmental problems that beset us and how, with wisdom and knowledge and the proper perspective, they can be solved or at least mitigated. Accordingly, the selections in this book have been chosen more for their intellectual content than for their emotional tone. They have been arranged into an order of topics—the global environment; population, policy, and economy; energy; the biosphere; land and water resources; and pollution—that lends itself to a progressive understanding of the causes and effects of human modifications of Earth's environmental systems.

Readers can have input into the next edition of *Annual Editions: Environment* by completing and returning the postpaid *article rating form* at the back of the book.

John L. Allen
Editor

Contents

UNIT 1
The Global Environment: An Emerging World View

Five selections provide information on the current state of Earth and the changes we will face.

The concepts in bold italics are developed in the article. For further expansion, please refer to the Topic Guide and the Index.

UNIT 2
Population, Policy, and Economy

Five unit selections examine the problems that the world will have in feeding and caring for its ever-increasing population.

The concepts in bold italics are developed in the article. For further expansion, please refer to the Topic Guide and the Index.

UNIT 3
Energy: Present and Future Problems

Five articles in this unit consider the problems of meeting present and future energy needs. Alternative energy sources are also examined.

UNIT 4
Biosphere: Endangered Species

Four unit articles examine the problems in the world's biosphere that include economic issues, natural ecosystems, and bioinvasion.

The concepts in bold italics are developed in the article. For further expansion, please refer to the Topic Guide and the Index.

UNIT 5
Resources: Land and Water

In this unit, five selections discuss the environment problems that affect our land and water resources.

The concepts in bold italics are developed in the article. For further expansion, please refer to the Topic Guide and the Index.

UNIT 6
The Hazards of Growth: Polution and Climate Change

The four selections in this unit weigh the environmental impacts of the growth of human population.

The concepts in bold italics are developed in the article. For further expansion, please refer to the Topic Guide and the Index.

Topic Guide

This topic guide suggests how the selections in this book relate to the subjects covered in your course. You may want to use the topics listed on these pages to search the Web more easily.

On the following pages a number of Web sites have been gathered specifically for this book. They are arranged to reflect the units of this *Annual Edition.* You can link to these sites by going to the DUSHKIN ONLINE support site at *http://www.dushkin.com/online/.*

ALL THE ARTICLES THAT RELATE TO EACH TOPIC ARE LISTED BELOW THE BOLD-FACED TERM.

Aerosols
27. Solving Hazy Mysteries

Agribusiness
20. Where Have All the Farmers Gone?

Agriculture
20. Where Have All the Farmers Gone?

Air pollution
25. Three Pollutants and an Emission

Alternative energy
11. Beyond Oil: The Future of Energy

Aquifers
23. Our Perilous Dependence on Groundwater

Atmosphere
28. Feeling the Heat: Life in the Greenhouse

Biodiversity
16. What Is Nature Worth?

Bioinvasion
18. Invasive Species: Pathogens of Globalization

Biosphere
16. What Is Nature Worth?

Carbon dioxide emissions
25. Three Pollutants and an Emission

Clean Water Act
26. The Quest for Clean Water

Climate
28. Feeling the Heat: Life in the Greenhouse

Climate change
1. How Many Planets? A Survey of the Global Environment
24. Oceans Are on the Critical List
28. Feeling the Heat: Life in the Greenhouse

Community
28. Feeling the Heat: Life in the Greenhouse

Conservation
17. Where Wildlife Rules
18. Invasive Species: Pathogens of Globalization
20. Where Have All the Farmers Gone?

Conservation of energy
15. Fossil Fuels and Energy Independence

Conservation strategies
15. Fossil Fuels and Energy Independence

Cultural customs
20. Where Have All the Farmers Gone?

Cultural values
20. Where Have All the Farmers Gone?

Development, economic
20. Where Have All the Farmers Gone?

Development, social
20. Where Have All the Farmers Gone?

Ecology
16. What Is Nature Worth?
18. Invasive Species: Pathogens of Globalization
20. Where Have All the Farmers Gone?

Ecology, human influences on
28. Feeling the Heat: Life in the Greenhouse

Ecology and environment
28. Feeling the Heat: Life in the Greenhouse

Economics
20. Where Have All the Farmers Gone?

Economic systems
16. What Is Nature Worth?

Economic theories
7. An Economy for the Earth

Ecosystem
2. Forget Nature. Even Eden Is Engineered
16. What Is Nature Worth?

Endangered species
3. Crimes of (a) Global Nature

Energy alternatives
13. Living Without Oil

Energy development
12. Powder Keg

Energy issues
15. Fossil Fuels and Energy Independence

Energy production
15. Fossil Fuels and Energy Independence

World Wide Web Sites

The following World Wide Web sites have been carefully researched and selected to support the articles found in this reader. The easiest way to access these selected sites is to go to our DUSHKIN ONLINE support site at *http://www.dushkin.com/online/*.

AE: Environment 04/05

The following sites were available at the time of publication. Visit our Web site—we update DUSHKIN ONLINE regularly to reflect any changes.

General Sources

Britannica's Internet Guide
http://www.britannica.com

This site presents extensive links to material on world geography and culture, encompassing material on wildlife, human lifestyles, and the environment.

EnviroLink
http://www.envirolink.org/

One of the world's largest environmental information clearinghouses, EnviroLink is a grassroots nonprofit organization that unites organizations and volunteers around the world and provides up-to-date information and resources.

Library of Congress
http://www.loc.gov

Examine this extensive Web site to learn about resource tools, library services/resources, exhibitions, and databases in many different subfields of environmental studies.

The New York Times
http://www.nytimes.com

Browsing through the archives of the *New York Times* will provide a wide array of articles and information related to the different subfields of the environment.

SocioSite: Sociological Subject Areas
http://www.pscw.uva.nl/sociosite/TOPICS/

This huge sociological site from the University of Amsterdam provides many discussions and references of interest to students of the environment, such as the links to information on ecology and consumerism.

U.S. Geological Survey
http://www.usgs.gov

This site and its many links are replete with information and resources in environmental studies, from explanations of El Niño to discussion of concerns about water resources.

UNIT 1: The Global Environment: An Emerging World View

Alternative Energy Institute (AEI)
http://www.altenergy.org

The AEI will continue to monitor the transition from today's energy forms to the future in a "surprising journey of twists and turns." This site is the beginning of an incredible journey.

Earth Science Enterprise
http://www.earth.nasa.gov

Information about NASA's Mission to Planet Earth program and its Science of the Earth System can be found here. Surf to learn about satellites, El Niño, and even "strategic visions" of interest to environmentalists.

IISDnet
http://iisd.ca

The International Institute for Sustainable Development, a Canadian organization, presents information through gateways entitled Business, Climate Change, Measurement and Assessment, and Natural Resources. IISD Linkages is its multimedia resource for environment and development policy makers.

National Geographic Society
http://www.nationalgeographic.com

Links to *National Geographic*'s huge archive are provided here. There is a great deal of material related to the atmosphere, the oceans, and other environmental topics.

Research and Reference (Library of Congress)
http://lcweb.loc.gov/rr/

This research and reference site of the Library of Congress will lead to invaluable information on different countries. It provides links to numerous publications, bibliographies, and guides in area studies that can be of great help to environmentalists.

Santa Fe Institute
http://acoma.santafe.edu

This home page of the Santa Fe Institute—a nonprofit, multidisciplinary research and education center—will lead to many interesting links related to its primary goal: to create a new kind of scientific research community, pursuing emerging science.

Solstice: Documents and Databases
http://solstice.crest.org/index.html

In this online source for sustainable energy information, the Center for Renewable Energy and Sustainable Technology (CREST) offers documents and databases on renewable energy, energy efficiency, and sustainable living. The site also offers related Web sites, case studies, and policy issues.

United Nations
http://www.unsystem.org

Visit this official Web site Locator for the United Nations System of Organizations to get a sense of the scope of international environmental inquiry today. Various UN organizations concern themselves with everything from maritime law to habitat protection to agriculture.

United Nations Environment Programme (UNEP)
http://www.unep.ch

Consult this home page of UNEP for links to critical topics of concern to environmentalists, including desertification, migratory species, and the impact of trade on the environment. The site will direct you to useful databases and global resource information.

UNIT 2: Population, Policy, and Economy

The Hunger Project
http://www.thp.org

Browse through this nonprofit organization's site to explore the ways in which it attempts to achieve its goal: the sustainable end to global hunger through leadership at all levels of society. The Hunger Project contends that the persistence of hunger is at the heart of the major security issues that are threatening our planet.

Poverty Mapping

http://www.povertymap.net

Poverty maps can quickly provide information on the spatial distribution of poverty. This site provides maps, graphics, data, publications, news, and links that provide the public with poverty mapping from the global to the subnational level.

World Health Organization

http://www.who.int

The home page of the World Health Organization provides links to a wealth of statistical and analytical information about health and the environment in the developing world.

World Population and Demographic Data

http://geography.about.com/cs/worldpopulation/

On this site, information about world population and additional demographic data for all the countries of the world are provided.

WWW Virtual Library: Demography & Population Studies

http://demography.anu.edu.au/VirtualLibrary/

This is a definitive guide to demography and population studies. A multitude of important links to information about global poverty and hunger can be found here.

UNIT 3: Energy: Present and Future Problems

Alliance for Global Sustainability (AGS)

http://globalsustainability.org/

The AGS is a cooperative venture seeking solutions to today's urgent and complex environmental problems. Research teams from four universities study large-scale, multidisciplinary environmental problems that are faced by the world's ecosystems, economies, and societies.

Alternative Energy Institute, Inc.

http://www.altenergy.org

On this site created by a nonprofit organization, discover how the use of conventional fuels affects the environment. Also learn about research work on new forms of energy.

Communications for a Sustainable Future

http://csf.colorado.edu

This site will lead to information on topics in international environmental sustainability. It pays particular attention to the political economics of protecting the environment.

Energy and the Environment: Resources for a Networked World

http://zebu.uoregon.edu/energy.html

An extensive array of materials having to do with energy sources—both renewable and nonrenewable—as well as other topics of interest to students of the environment is found on this site.

Institute for Global Communication/EcoNet

http://www.igc.org/

This environmentally friendly site provides links to dozens of governmental, organizational, and commercial sites having to do with energy sources. Resources address energy efficiency, renewable generating sources, global warming, and more.

Nuclear Power Introduction

http://library.thinkquest.org/17658/pdfs/nucintro.pdf

Information regarding alternative energy forms can be accessed here. There is a brief introduction to nuclear power and a link to maps that show where nuclear power plants exist.

U.S. Department of Energy

http://www.energy.gov

Scrolling through the links provided by this Department of Energy home page will lead to information about fossil fuels and a variety of sustainable/renewable energy sources.

UNIT 4: Biosphere: Endangered Species

Endangered Species

http://www.endangeredspecie.com/

This site provides a wealth of information on endangered species anywhere in the world. Links providing data on the causes, interesting facts, law issues, case studies, and other issues on endangered species are available.

Friends of the Earth

http://www.foe.co.uk/index.html

Friends of the Earth, a nonprofit organization based in the United Kingdom, pursues a number of campaigns to protect the Earth and its living creatures. This site has links to many important environmental sites, covering such broad topics as ozone depletion, soil erosion, and biodiversity.

Smithsonian Institution Web Site

http://www.si.edu

Looking through this site, which will provide access to many of the enormous resources of the Smithsonian, offers a sense of the biological diversity that is threatened by humans' unsound environmental policies and practices.

World Wildlife Federation (WWF)

http://www.wwf.org

This home page of the WWF leads to an extensive array of information links about endangered species, wildlife management and preservation, and more. It provides many suggestions for how to take an active part in protecting the biosphere.

UNIT 5: Resources: Land and Water

Agriculture Production Statistics

http://www.wri.org/statistics/fao-prd.html

The Food and Agriculture Organization of the UN (FAO) provides annual statistics, on a worldwide basis, on all important data of crop and livestock production. Coverage includes land use, irrigation, human population, index numbers of agriculture production, major crops, livestock numbers, livestock products, food supply, and means of production for individual countries, continents, and the world. Web links to the FAOSTAT Database Gateway are provided.

Global Climate Change

http://www.puc.state.oh.us/consumer/gcc/index.html

The goal of this PUCO (Public Utilities Commission of Ohio) site is to serve as a clearinghouse of information related to global climate change. Its extensive links provide an explanation of the science and chronology of global climate change, acronyms, definitions, and more.

National Oceanic and Atmospheric Administration (NOAA)

http://www.noaa.gov

Through this home page of NOAA, you can find information about coastal issues, fisheries, climate, and more.

National Operational Hydrologic Remote Sensing Center (NOHRSC)

http://www.nohrsc.nws.gov

Flood images are available at this site of the NOHRSC, which works with the U.S. National Weather Service to track weather-related information.

Virtual Seminar in Global Political Economy/Global Cities & Social Movements

http://csf.colorado.edu/gpe/gpe95b/resources.html

Links to subjects of interest in regional environmental studies, covering topics such as sustainable cities, megacities, and urban planning are available here. Many international nongovernmental organizations are included.

Terrestrial Sciences

http://www.cgd.ucar.edu/tss/

The Terrestrial Sciences Section (TSS) is part of the Climate and Global Dynamics (CGD) Division at the National Center for Atmospheric Research (NCAR) in Boulder, Colorado. Scientists in the section study land-atmosphere interactions, in particular surface forcing of the atmosphere, through model development, application, and observational analyses. Here, you'll find a link to VEMAP, The Vegetation/Ecosystem Modeling and Analysis Project.

UNIT 6: The Hazards of Growth: Polution and Climate Change

IISDnet

http://www.iisd.org/default.asp

The International Institute for Sustainable Development's site presents information through links on business, climate change, communities and livelihoods, trade, and more.

Persistent Organic Pollutants (POP)

http://www.chem.unep.ch/pops/

Visit this site to learn more about persistent organic pollutants (POPs) and the issues and concerns surrounding them.

School of Labor and Industrial Relations (SLIR): Hot Links

http://www.lir.msu.edu/hotlinks/

Michigan State University's SLIR page connects to industrial relations sites throughout the world. It has links to U.S. government statistics, newspapers and libraries, international intergovernmental organizations, and more.

Space Research Institute

http://arc.iki.rssi.ru/Welcome.html

For a change of pace, browse through this home page of Russia's Space Research Institute for information on its Environment Monitoring Information Systems, the IKI Satellite Situation Center, and its Data Archive.

Worldwatch Institute

http://www.worldwatch.org

The Worldwatch Institute, dedicated to fostering the evolution of an environmentally sustainable society, presents this site with access to *World Watch Magazine* and *State of the World 2000*. Click on In the News and Press Releases for discussions of current problems.

We highly recommend that you review our Web site for expanded information and our other product lines. We are continually updating and adding links to our Web site in order to offer you the most usable and useful information that will support and expand the value of your Annual Editions. You can reach us at: *http://www.dushkin.com/annualeditions/*.

UNIT 1

The Global Environment: An Emerging World View

Unit Selections

1. **How Many Planets? A Survey of the Global Environment**, The Economist
2. **Forget Nature. Even Eden Is Engineered**, Andrew C. Revkin
3. **Crimes of (a) Global Nature**, Lisa Mastny and Hilary French
4. **Toward a Sustainability Transition: The International Consensus**, Thomas M. Parris
5. **Making the Global Local: Responding to Climate Change Concerns From the Ground Up**, Robert W. Kates and Thomas J. Wilbanks

Key Points to Consider

- What are the connections between the attempts to develop sustainable systems and the quantity and quality of environmental data? Are there also relationships between data and the role of technology and economic systems in shaping the environmental future?

- In what ways are environmental changes engineered by humans, and how can human planning systems develop mechanisms to ensure that the human-designed environment will be a sustainable one?

- How well do international agreements work in controlling "environmental crimes" such as the taking of endangered species, hazardous waste dumping, or emissions of harmful pollutants? Are there ways in which international environmental accords could be made more enforceable?

- How can you link the scales of environmental impact from the global to the local and still make sense of how local actions fit into global change—and vice versa?

 **Links: www.dushkin.com/online/**
These sites are annotated in the World Wide Web pages.

Alternative Energy Institute (AEI)
http://www.altenergy.org

Earth Science Enterprise
http://www.earth.nasa.gov

IISDnet
http://iisd.ca

National Geographic Society
http://www.nationalgeographic.com

Research and Reference (Library of Congress)
http://lcweb.loc.gov/rr/

Santa Fe Institute
http://acoma.santafe.edu

Solstice: Documents and Databases
http://solstice.crest.org/index.html

United Nations
http://www.unsystem.org

United Nations Environment Programme (UNEP)
http://www.unep.ch

More than three decades after the celebration of the first Earth Day in 1970, public apprehension over the environmental future of the planet has reached levels unprecedented even during the late 1960's and early 1970's "Age of Aquarius." No longer are those concerned about the environment dismissed as "ecofreaks" and "tree-huggers." Most serious scientists have joined the rising clamor for environmental protection, as have the more traditional environmentally conscious public-interest groups. There are a number of reasons for this increased environmental awareness. Some of these reasons arise from environmental events; it is, for example, becoming increasingly difficult to deny the effects of global warming. But more arise simply from the process of globalization: the increasing unity of the world's economic, social, and information systems. Hailed by many as the salvation of the future, globalization has done little to make the world a better or safer place. Diseases once confined to specific regions now have the capacity for widespread dissemination. Increasing human mobility has allowed human-caused disruptions to political, cultural, and economic systems to spread, and acts of terrorism now take place in locations once thought safe from such manifestations of hatred and despair. On the more positive side, the expansion of global information systems has fostered a maturation of concepts about the global nature of environmental processes.

Much of what has been learned through this increased information flow, particularly by American observers, has been of the environmentally ravaged world behind the old Iron Curtain—a chilling forecast of what other industrialized regions as well as the developing countries can become in the near future unless strict international environmental measures are put in place. For perhaps the first time ever, countries are beginning to recognize that environmental problems have no boundaries and that international cooperation is the only way to solve them.

The subtitle of this first unit, "An Emerging World View," is an optimistic assessment of the future: a future in which less money is spent on defense and more on environmental protection and cleanup—a new world order in which political influence might be based more on leadership in environmental and economic issues than on military might. It is probably far too early to make such optimistic predictions, to conclude that the world's nations—developed and underdeveloped—will begin to recognize that Earth's environment is a single unit. Thus far those nations have shown no tendency to recognize that humankind is a single unit and that what harms one harms all. The recent emergence of wide-scale terrorism as an instrument of political and social policy is evidence of such a failure of recognition. Nevertheless, there is a growing international realization—aided by the information superhighway—that we are all, as environmental activists have been saying for decades, inhabitants of Spaceship Earth and will survive or succumb together.

The articles selected for this unit have been chosen to illustrate the increasingly global perspective on environmental problems and the degree to which their solutions must be linked to political, economic, and social problems and solutions. In the lead piece of the unit, "How Many Planets?" the editors of *The Economist* attempt an analysis of what they admit is a very slippery subject by beginning with the observation that "it comes as a shock to discover how little information there is on the environment." They note the lip service paid everywhere to the concept of sustainability and acknowledge that economic growth and environmental health are not mutually inconsistent, but that a great deal more work is necessary to make them compatible. They also conclude that governments, corporations, and individuals are more prepared now to think about how to use the planet than they were even 10 years ago.

Planning for the wise use of the planet is the subject of the next selection in the unit. In "Forget Nature. Even Eden Is Engineered," science writer Andrew Revkin directs his attention toward the interconnectedness of environmental systems and toward new concepts and ways of thinking about the environment and human impact. Science has finally recognized, says Revkin, the degree to which people have significantly altered global atmospheric systems and the biosphere. With that recognition comes the first real chance to begin balancing economic development with a sustainable future on Earth for human beings. Whatever we make of that chance, the world of the future will be one in which, through technology and sheer numbers, the forces of human society rather than the forces of nature will be the chief architect.

Some of the impediments in making the human-engineered Earth one in which we would want to live are discussed in the third article in this unit. Lisa Mastny and Hilary French of the Worldwatch Institute describe the difficulties of developing and enforcing international environmental treaties and other agreements. In "Crimes of (a) Global Nature," Mastny and French focus on three types of international environmental agreements: those related to the taking and sale of endangered terrestrial and aquatic species; those governing the disposal of toxic and hazardous waste materials; and those mandating restrictions on the manufacture and use of certain chemicals that damage portions of the atmosphere. It is easy enough for countries to agree to international environmental accords. It is much more difficult for officials to enforce those same accords, and some of the authors' statistics are staggering—for example, illegally cut wood accounted for 65 percent of the world's supply in 2000. The future of the environment depends not just upon international agreement but upon enforcement.

The final two selections in the opening section deal with issues of global change. In "Toward a Sustainability Transition" research scientist and contributing editor of *Environment* magazine, Thomas Parris, suggests that part of the difficulty in developing sustainable systems to reduce the pace of global environmental change is how we define what we mean by "sustainability" and how that definition changes in both time and place. Geographers Robert Kates and Thomas Wilbanks also deal with the issue of definition in "Making the Global Local." They attempt to arrive at an understanding of how locally-based activities influence such global phenomena as climate change. In a time of disappointing progress in global initiatives to curb greenhouse gas emissions, the authors suggest that the most promising path may be that of localized action.

How many planets?

A survey of the global environment

The great race

Growth need not be the enemy of greenery. But much more effort is required to make the two compatible, says Vijay Vaitheeswaran

Sustainable development is a dangerously slippery concept. Who could possibly be against something that invokes such alluring images of untouched wildernesses and happy creatures? The difficulty comes in trying to reconcile the "development" with the "sustainable" bit: look more closely, and you will notice that there are no people in the picture.

That seems unlikely to stop a contingent of some of 60,000 world leaders, businessmen, activists, bureaucrats and journalists from travelling to South Africa next month for the UN-sponsored World Summit on Sustainable Development in Johannesburg. Whether the summit achieves anything remains to be seen, but at least it is asking the right questions. This survey will argue that sustainable development cuts to the heart of mankind's relationship with nature—or, as Paul Portney of Resources for the Future, an American think-tank, puts it, "the great race between development and degradation". It will also explain why there is reason for hope about the planet's future.

The best way known to help the poor today—economic growth—has to be handled with care, or it can leave a degraded or even devastated natural environment for the future. That explains why ecologists and economists have long held diametrically opposed views on development. The difficult part is to work out what we owe future generations, and how to reconcile that moral obligation with what we owe the poorest among us today.

It is worth recalling some of the arguments fielded in the run-up to the big Earth Summit in Rio de Janeiro a decade ago. A publication from UNESCO, a United Nations agency, offered the following vision of the future: "Every generation should leave water, air and soil resources as pure and unpolluted as when it came on earth. Each generation should leave undiminished all the species of ani-

mals it found existing on earth." Man, that suggests, is but a strand in the web of life, and the natural order is fixed and supreme. Put earth first, it seems to say.

Robert Solow, an economist at the Massachusetts Institute of Technology, replied at the time that this was "fundamentally the wrong way to go", arguing that the obligation to the future is "not to leave the world as we found it in detail, but rather to leave the option or the capacity to be as well off as we are." Implicit in that argument is the seemingly hard-hearted notion of "fungibility": that natural resources, whether petroleum or giant pandas, are substitutable.

Rio's fatal flaw

Champions of development and defenders of the environment have been locked in battle ever since a UN summit in Stockholm launched the sustainable-development debate three decades ago. Over the years, this debate often pitted indignant politicians and social activists from the poor world against equally indignant politicians and greens from the rich world. But by the time the Rio summit came along, it seemed they had reached a truce. With the help of a committee of grandees led by Gro Harlem Brundtland, a former Norwegian prime minister, the interested parties struck a deal in 1987: development and the environment, they declared, were inextricably linked. That compromise generated a good deal of euphoria. Green groups grew concerned over poverty, and development charities waxed lyrical about greenery. Even the World Bank joined in. Its World Development Report in 1992 gushed about "win-win" strategies, such as ending environmentally harmful subsidies, that would help both the economy and the environment.

By nearly universal agreement, those grand aspirations have fallen flat in the decade since that summit. Little headway has been made with environmental problems such as climate change and loss of biodiversity. Such progress as has been achieved has been largely due to three factors that this survey will explore in later sections: more decision-making at local level, technological innovation, and the rise of market forces in environmental matters.

The main explanation for the disappointment—and the chief lesson for those about to gather in South Africa—is that Rio overreached itself. Its participants were so anxious to reach a political consensus that they agreed to the Brundtland definition of sustainable development, which Daniel Esty of Yale University thinks has turned into "a buzz-word largely devoid of content". The biggest mistake, he reckons, is that it slides over the difficult trade-offs between environment and development in the real world. He is careful to note that there are plenty of cases where those goals are linked—but also many where they are not: "Environmental and economic policy goals are distinct, and the actions needed to achieve them are not the same."

No such thing as win-win

To insist that the two are "impossible to separate", as the Brundtland commission claimed, is nonsense. Even the World Bank now accepts that its much-trumpeted 1992 report was much too optimistic. Kristalina Georgieva, the Bank's director for the environment, echoes comments from various colleagues when she says: "I've never seen a real win-win in my life. There's always somebody, usually an elite group grabbing rents, that loses. And we've learned in the past decade that those losers fight hard to make sure that technically elegant win-win policies do not get very fat."

So would it be better to ditch the concept of sustainable development altogether? Probably not. Even people with their feet firmly planted on the ground think one aspect of it is worth salvaging: the emphasis on the future.

Nobody would accuse John Graham of jumping on green bandwagons. As an official in President George Bush's Office of Management and Budget, and previously as head of Harvard University's Centre for Risk Analysis, he has built a reputation for evidence-based policymaking. Yet he insists sustainable development is a worthwhile concept: "It's good therapy for the tunnel vision common in government ministries, as it forces integrated policymaking. In practical terms, it means that you have to take economic cost-benefit trade-offs into account in environmental laws, and keep environmental trade-offs in mind with economic development."

Jose Maria Figueres, a former president of Costa Rica, takes a similar view. "As a politician, I saw at first hand how often policies were dictated by short-term considerations such as elections or partisan pressure. Sustainability is a useful template to align short-term policies with medium- to long-term goals."

It is not only politicians who see value in saving the sensible aspects of sustainable development. Achim Steiner,

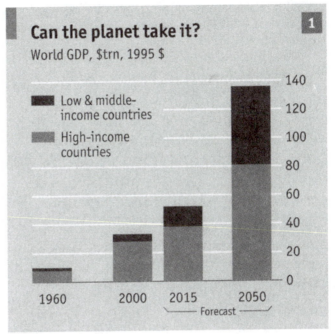

Can the planet take it?
World GDP, $trn, 1995 $

■ Low & middle-income countries
■ High-income countries

Source: World Bank

head of the International Union for the Conservation of Nature, the world's biggest conservation group, puts it this way: "Let's be honest: greens and businesses do not have the same objective, but they can find common ground. We look for pragmatic ways to save species. From our own work on the ground on poverty, our members—be they bird watchers or passionate ecologists—have learned that 'sustainable use' is a better way to conserve."

Sir Robert Wilson, boss of Rio Tinto, a mining giant, agrees. He and other business leaders say it forces hard choices about the future out into the open: "I like this concept because it frames the trade-offs inherent in a business like ours. It means that single-issue activism is simply not as viable."

Kenneth Arrow and Larry Goulder, two economists at Stanford University, suggest that the old ideological enemies are converging: "Many economists now accept the idea that natural capital has to be valued, and that we need to account for ecosystem services. Many ecologists now accept that prohibiting everything in the name of protecting nature is not useful, and so are being selective." They think the debate is narrowing to the more empirical question of how far it is possible to substitute natural capital with the man-made sort, and specific forms of natural capital for one another.

The job for Johannesburg

So what can the Johannesburg summit contribute? The prospects are limited. There are no big, set-piece political treaties to be signed as there were at Rio. America's acrimonious departure from the Kyoto Protocol, a UN treaty on climate change, has left a bitter taste in many mouths. And the final pre-summit gathering, held in early June in

Indonesia, broke up in disarray. Still, the gathered worthies could usefully concentrate on a handful of areas where international co-operation can help deal with environmental problems. Those include improving access for the poor to cleaner energy and to safe drinking water, two areas where concerns about human health and the environmental overlap. If rich countries want to make progress, they must agree on firm targets and offer the money needed to meet them. Only if they do so will poor countries be willing to cooperate on problems such as global warming that rich countries care about.

That seems like a modest goal, but it just might get the world thinking seriously about sustainability once again. If the Johannesburg summit helps rebuild a bit of faith in international environmental cooperation, then it will have been worthwhile. Minimising the harm that future economic growth does to the environment will require the rich world to work hand in glove with the poor world—which seems nearly unimaginable in today's atmosphere poisoned by the shortcomings of Rio and Kyoto.

To understand why this matters, recall that great race between development and degradation. Mankind has stayed comfortably ahead in that race so far, but can it go on doing so? The sheer magnitude of the economic growth that is hoped for in the coming decades (see chart 1) makes it seem inevitable that the clashes between mankind and nature will grow worse. Some are now asking whether all this economic growth is really necessary or useful in the first place, citing past advocates of the simple life.

"God forbid that India should ever take to industrialism after the manner of the West… It took Britain half the resources of the planet to achieve this prosperity. How many planets will a country like India require?", Mahatma Gandhi asked half a century ago. That question encapsulated the bundle of worries that haunts the sustainable-development debate to this day. Today, the vast majority of Gandhi's countrymen are still living the simple life—full of simple misery, malnourishment and material want. Grinding poverty, it turns out, is pretty sustainable.

If Gandhi were alive today, he might look at China next door and find that the country, once as poor as India, has been transformed beyond recognition by two decades of roaring economic growth. Vast numbers of people have been lifted out of poverty and into middle-class comfort. That could prompt him to reframe his question: how many planets will it take to satisfy China's needs if it ever achieves profligate America's affluence? One green group reckons the answer is three. The next section looks at the environmental data that might underpin such claims. It makes for alarming reading—though not for the reason that first springs to mind.

Flying blind

It comes as a shock to discover how little information there is on the environment

WHAT is the true state of the planet? It depends from which side you are peering at it. "Things are really looking up," comes the cry from one corner (usually overflowing with economists and technologists), pointing to a set of rosy statistics. "Disaster is nigh," shouts the other corner (usually full of ecologists and environmental lobbyists), holding up a rival set of troubling indicators.

According to the optimists, the 20th century marked a period of unprecedented economic growth that lifted masses of people out of abject poverty. It also brought technological innovations such as vaccines and other advances in public health that tackled many preventable diseases. The result has been a breath-taking enhancement of human welfare and longer, better lives for people everywhere on earth (see chart 2).

At this point, the pessimists interject: "Ah, but at what ecological cost?" They note that the economic growth which made all these gains possible sprang from the rapid spread of industrialisation and its resource-guzzling cousins, urbanisation, motorisation and electrification. The earth provided the necessary raw materials, ranging from coal to pulp to iron. Its ecosystems—rivers, seas, the atmosphere—also absorbed much of the noxious fallout from that process. The sheer magnitude of ecological change resulting directly from the past century's economic activity is remarkable (see table 3).

To answer that Gandhian question about how many planets it would take if everybody lived like the West, we need to know how much—or how little—damage the West's transformation from poverty to plenty has done to the planet to date. Economists point to the remarkable improvement in local air and water pollution in the rich world in recent decades. "It's Getting Better All the Time", a cheerful tract co-written by the late Julian Simon, insists that: "One of the greatest trends of the past 100 years has been the astonishing rate of progress in reducing almost every form of pollution." The conclusion seems unavoidable: "Relax! If we keep growing as usual, we'll inevitably grow greener."

The ecologically minded crowd takes a different view. "GEO3", a new report from the United Nations Environment Programme, looks back at the past few decades and sees much reason for concern. Its thoughtful boss, Klaus Töpfer (a former German environment minister),

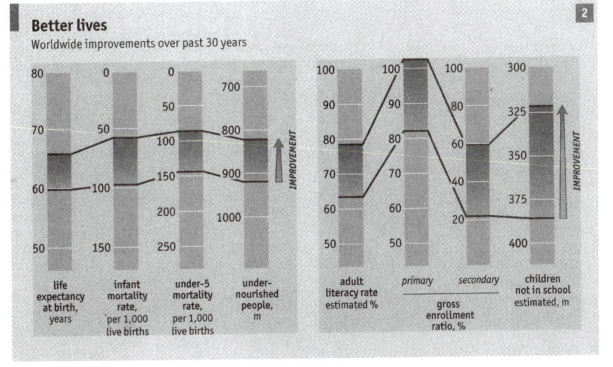

Better lives
Worldwide improvements over past 30 years

Source: UNEP, GEO3

insists that his report is not "a document of doom and gloom". Yet, in summing it up, UNEP decries "the declining environmental quality of planet earth", and wags a finger at economic prosperity: "Currently, one-fifth of the world's population enjoys high, some would say excessive, levels of affluence." The conclusion seems unavoidable: "Panic! If we keep growing as usual, we'll inevitably choke the planet to death."

"People and Ecosystems", a collaboration between the World Resources Institute, the World Bank and the United Nations, tried to gauge the condition of ecosystems by examining the goods and services they produce—food, fibre, clean water, carbon storage and so on—and their capacity to continue producing them. The authors explain why ecosystems matter: half of all jobs worldwide are in agriculture, forestry and fishing, and the output from those three commodity businesses still dominates the economies of a quarter of the world's countries.

The report reached two chief conclusions after surveying the best available environmental data. First, a number of ecosystems are "fraying" under the impact of human activity. Second, ecosystems in future will be less able than in the past to deliver the goods and services human life depends upon, which points to unsustainability. But it took care to say: "It's hard, of course, to know what will be truly sustainable." The reason this collection of leading experts could not reach a firm conclusion was that, remarkably, much of the information they needed was in-

complete or missing altogether: "Our knowledge of ecosystems has increased dramatically, but it simply has not kept pace with our ability to alter them."

Another group of experts, this time organised by the World Economic Forum, found itself similarly frustrated. The leader of that project, Daniel Esty of Yale, exclaims, throwing his arms in the air: "Why hasn't anyone done careful environmental measurement before? Businessmen always say, 'what matters gets measured.' Social scientists started quantitative measurement 30 years ago, and even political science turned to hard numbers 15 years ago. Yet look at environmental policy, and the data are lousy."

Gaping holes

At long last, efforts are under way to improve environmental data collection. The most ambitious of these is the Millennium Ecosystem Assessment, a joint effort among leading development agencies and environmental groups. This four-year effort is billed as an attempt to establish systematic data sets on all environmental matters across the world. But one of the researchers involved grouses that it "has very, very little new money to collect or analyse new data". It seems astonishing that governments have been making sweeping decisions on environmental policy for decades without such a baseline in the first place.

One positive sign is the growing interest of the private sector in collecting environmental data. It seems plain that leaving the task to the public sector has not worked.

A century that changed the world

Change between 1890 (=1) and 1990s

Industrial output	40
Marine fish catch	35
Carbon dioxide emissions	17
Energy use	16
World economy	14
World urban population	13
Coal production	7
Air pollution	5
Irrigated area	5
World population	**4**
Horse population	1.1
Bird and mammal species	0.99
Forest area	0.8
Blue-whale population	0.0025

Source: "Something New Under the Sun" by John McNeill

Information on the environment comes far lower on the bureaucratic pecking order than data on education or social affairs, which tend to be overseen by ministries with bigger budgets and more political clout. A number of countries, ranging from New Zealand to Austria, are now looking to the private sector to help collect and manage data in areas such as climate. Development banks are also considering using private contractors to monitor urban air quality, in part to get around the corruption and apathy in some city governments.

"I see a revolution in environmental data collection coming because of computing power, satellite mapping, remote sensing and other such information technologies," says Mr Esty. The arrival of hard data in this notoriously fuzzy area could cut down on environmental disputes by reducing uncertainty. One example is the long-running squabble between America's mid-western states, which rely heavily on coal, and the north-eastern states, which suffer from acid rain. Technology helped disprove claims by the mid-western states that New York's problems all resulted from home-grown pollution.

The arrival of good data would have other benefits as well, such as helping markets to work more robustly: witness America's pioneering scheme to trade emissions of sulphur dioxide, made possible by fancy equipment capable of monitoring emissions in real time. Mr Esty raises an even more intriguing possibility: "Like in the American West a hundred years ago, when barbed wire helped establish rights and prevent overgrazing, information technology can help establish 'virtual barbed wire' that secures property rights and so prevents over-exploitation of the commons." He points to fishing in the waters between Australia and New Zealand, where tracking and monitoring devices have reduced over-exploitation.

Best of all, there are signs that the use of such fancy technology will not be confined to rich countries. Calestous Juma of Harvard University shares Mr Esty's excitement about the possibility of such a technology-driven revolution even in Africa: "In the past, the only environmental 'database' we had in Africa was our grandmothers. Now, with global information systems and such, the potential is enormous." Conservationists in Namibia, for example, already use satellite tracking to keep count of their elephants. Farmers in Mali receive satellite updates about impending storms on hand-wound radios. Mr Juma thinks the day is not far off when such technology, combined with ground-based monitoring, will help Africans measure trends in deforestation, soil erosion and climate change, and assess the effects on their local environment.

Make a start

That is at once a sweeping vision and a modest one. Sweeping, because it will require heavy investment in both sophisticated hardware and nuts-and-bolts information infrastructure on the ground to make sense of all these new data. As the poor world clearly cannot afford to pay for all this, the rich world must help—partly for altruistic reasons, partly with the selfish aim of discovering in good time whether any global environmental calamities are in the making. A number of multilateral agencies now say they are willing to invest in this area as a "neglected global public good"—neglected especially by those agencies themselves. Even President Bush's administration has recently indicated that it will give environmental satellite data free to poor countries.

But that vision is also quite a modest one. Assuming that this data "revolution" does take place, all it will deliver is a reliable assessment of the health of the planet today. We will still not be able to answer the broader question of whether current trends are sustainable or not.

To do that, we need to look more closely at two very different sorts of environmental problems: global crises and local troubles. The global sort is hard to pin down, but can involve irreversible changes. The local kind is common and can have a big effect on the qualify of life, but is usually reversible. Data on both are predictably inadequate. We turn first to the most elusive environmental problem of all, global warming.

Blowing hot and cold

Climate change may be slow and uncertain, but that is no excuse for inaction

WHAT would Winston Churchill have done about climate change? Imagine that Britain's visionary wartime leader had been presented with a potential time bomb capable of wreaking global havoc, although not certain to do so. Warding it off would require concerted global action and economic sacrifice on the home front. Would he have done nothing?

Not if you put it that way. After all, Churchill did not dismiss the Nazi threat for lack of conclusive evidence of Hitler's evil intentions. But the answer might be less straightforward if the following provisos had been added: evidence of this problem would remain cloudy for decades; the worst effects might not be felt for a century; but the costs of tackling the problem would start biting immediately. That, in a nutshell, is the dilemma of climate change. It is asking a great deal of politicians to take action on behalf of voters who have not even been born yet.

One reason why uncertainty over climate looks to be with us for a long time is that the oceans, which absorb carbon from the atmosphere, act as a time-delay mechanism. Their massive thermal inertia means that the climate system responds only very slowly to changes in the composition of the atmosphere. Another complication arises from the relationship between carbon dioxide (CO_2), the principal greenhouse gas (GHG), and sulphur dioxide (SO_2), a common pollutant. Efforts to reduce man-made emissions of GHGs by cutting down on fossil-fuel use will reduce emissions of both gases. The reduction in CO_2 will cut warming, but the concurrent SO_2 cut may mask that effect by contributing to the warming.

There are so many such fuzzy factors—ranging from aerosol particles to clouds to cosmic radiation—that we are likely to see disruptions to familiar climate patterns for many years without knowing why they are happening or what to do about them. Tom Wigley, a leading climate scientist and member of the UN's Intergovernmental Panel on Climate Change (IPCC), goes further. He argues in an excellent book published by the Aspen Institute, "US Policies on Climate Change: What Next?", that whatever policy changes governments pursue, scientific uncertainties will "make it difficult to detect the effects of such changes, probably for many decades."

As evidence, he points to the negligible short- to medium-term difference in temperature resulting from an array of emissions "pathways" on which the world could choose to embark if it decided to tackle climate change (see chart 4). He plots various strategies for reducing GHGs (including the Kyoto one) that will lead in the next century to the stabilisation of atmospheric concentrations of CO_2 at 550 parts per million (ppm). That is roughly double the level which prevailed in pre-industrial times, and

is often mooted by climate scientists as a reasonable target. But even by 2040, the temperature differences between the various options will still be tiny—and certainly within the magnitude of natural climatic variance. In short, in another four decades we will probably still not know if we have over- or undershot.

Ignorance is not bliss

However, that does not mean we know nothing. We do know, for a start, that the "greenhouse effect" is real: without the heat-trapping effect of water vapour, CO_2, methane and other naturally occurring GHGs, our planet would be a lifeless 30°C or so colder. Some of these GHG emissions are captured and stored by "sinks", such as the oceans, forests and agricultural land, as part of nature's carbon cycle.

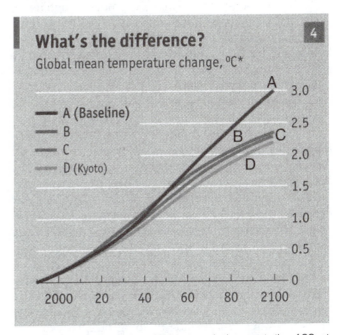

What's the difference?
Global mean temperature change, °C*

- A (Baseline)
- B
- C
- D (Kyoto)

*Various plausible scenarios for stabilising atmospheric concentration of CO_2 at 550 parts per million

Source: Tom Wigley

We also know that since the industrial revolution began, mankind's actions have contributed significantly to that greenhouse effect. Atmospheric concentrations of GHGs have risen from around 280ppm two centuries ago to around 370ppm today, thanks chiefly to mankind's use of fossil fuels and, to a lesser degree, to deforestation and other land-use changes. Both surface temperatures and sea levels have been rising for some time.

There are good reasons to think temperatures will continue rising. The IPCC has estimated a likely range for that increase of 1.4°C–5.8°C over the next century, although the lower end of that range is more likely. Since what matters is not just the absolute temperature level but the rate of change as well, it makes sense to try to slow down the increase.

The worry is that a rapid rise in temperatures would lead to climate changes that could be devastating for many (though not all) parts of the world. Central America, most of Africa, much of south Asia and northern China could all be hit by droughts, storms and floods and otherwise made miserable. Because they are poor and have the misfortune to live near the tropics, those most likely to be affected will be least able to adapt.

The colder parts of the world may benefit from warming, but they too face perils. One is the conceivable collapse of the Atlantic "conveyor belt", a system of currents that gives much of Europe its relatively mild climate; if temperatures climb too high, say scientists, the system may undergo radical changes that damage both Europe and America. That points to the biggest fear: warming may trigger irreversible changes that transform the earth into a largely uninhabitable environment.

Given that possibility, extremely remote though it is, it is no comfort to know that any attempts to stabilise atmospheric concentrations of GHGs at a particular level will take a very long time. Because of the oceans' thermal inertia, explains Mr Wigley, even once atmospheric concentrations of GHGs are stabilised, it will take decades or centuries for the climate to follow suit. And even then the sea level will continue to rise, perhaps for millennia.

This is a vast challenge, and it is worth bearing in mind that mankind's contribution to warming is the only factor that can be controlled. So the sooner we start drawing up a long-term strategy for climate change, the better.

What should such a grand plan look like? First and foremost, it must be global. Since CO_2 lingers in the atmosphere for a century or more, any plan must also extend across several generations.

The plan must recognise, too, that climate change is nothing new: the climate has fluctuated through history, and mankind has adapted to those changes—and must continue doing so. In the rich world, some of the more obvious measures will include building bigger dykes and flood defences. But since the most vulnerable people are those in poor countries, they too have to be helped to adapt to rising seas and unpredictable storms. Infrastructure improvements will be useful, but the best investment will probably be to help the developing world get wealthier.

It is essential to be clear about the plan's long-term objective. A growing chorus of scientists now argues that we need to keep temperatures from rising by much more than 2–3°C in all. That will require the stabilisation of atmospheric concentrations of GHGs. James Edmonds of the University of Maryland points out that because of the long life of CO_2, stabilisation of CO_2 concentrations is not at all the same thing as stabilisation of CO_2 emissions. That, says Mr Edmonds, points to an unavoidable conclusion: "In the

very long term, global net CO_2 emissions must eventually peak and gradually decline toward zero, regardless of whether we go for a target of 350ppm or 1,000ppm."

A low-carbon world

That is why the long-term objective for climate policy must be a transition to a low-carbon energy system. Such a transition can be very gradual and need not necessarily lead to a world powered only by bicycles and windmills, for two reasons that are often overlooked.

One involves the precise form in which the carbon in the ground is distributed. According to Michael Grubb of the Carbon Trust, a British quasi-governmental body, the long-term problem is coal. In theory, we can burn all of the conventional oil and natural gas in the ground and still meet the most ambitious goals for tackling climate change. If we do that, we must ensure that the far greater amounts of carbon trapped as coal (and unconventional resources like tar sands) never enter the atmosphere.

The snag is that poor countries are likely to continue burning cheap domestic reserves of coal for decades. That suggests the rich world should speed the development and diffusion of "low carbon" technologies using the energy content of coal without releasing its carbon into the atmosphere. This could be far off, so it still makes sense to keep a watchful eye on the soaring carbon emissions from oil and gas.

The other reason, as Mr Edmonds took care to point out, is that it is net emissions of CO_2 that need to peak and decline. That leaves scope for the continued use of fossil fuels as the main source of modern energy if only some magical way can be found to capture and dispose of the associated CO_2. Happily, scientists already have some magic in the works.

One option is the biological "sequestration" of carbon in forests and agricultural land. Another promising idea is capturing and storing CO_2—underground, as a solid or even at the bottom of the ocean. Planting "energy crops" such as switch-grass and using them in conjunction with sequestration techniques could even result in negative net CO_2 emissions, because such plants use carbon from the atmosphere. If sequestration is combined with techniques for stripping the hydrogen out of this hydrocarbon, then coal could even offer a way to sustainable hydrogen energy.

But is anyone going to pay attention to these long-term principles? After all, over the past couple of years all participants in the Kyoto debate have excelled at producing short-sighted, selfish and disingenuous arguments. And the political rift continues: the EU and Japan pushed ahead with ratification of the Kyoto treaty a month ago, whereas President Bush reaffirmed his opposition.

However, go back a decade and you will find precisely those principles enshrined in a treaty approved by the elder George Bush and since reaffirmed by his son: the UN Framework Convention on Climate Change (FCCC). This treaty was perhaps the most important outcome of the Rio summit, and it remains the basis for the international climate-policy regime, including Kyoto.

The treaty is global in nature and long-term in perspective. It commits signatories to pursuing "the stabilisation of GHG concentrations in the atmosphere at a level that would prevent dangerous interference with the climate system." Note that the agreement covers GHG concentrations, not merely emissions. In effect, this commits even gas-guzzling America to the goal of declining emissions.

Better than Kyoto

Crucially, the FCCC treaty not only lays down the ends but also specifies the means: any strategy to achieve stabilisation of GHG concentrations, it insists, "must not be disruptive of the global economy". That was the stumbling block for the Kyoto treaty, which is built upon the FCCC agreement: its targets and timetables proved unrealistic.

Any revised Kyoto treaty or follow-up accord (which must include the United States and the big developing countries) should rest on the three basic pillars. First, governments everywhere (but especially in Europe) must understand that a reduction in emissions has to start modestly. That is because the capital stock involved in the global energy system is vast and long-lived, so a dash to scrap fossil-fuel production would be hugely expensive. However, as Mr Grubb points out, that pragmatism must be flanked by policies that encourage a switch to low-carbon technologies when replacing existing plants.

Second, governments everywhere (but especially in America) must send a powerful signal that carbon is going out of fashion. The best way to do this is to levy a carbon tax. However, whether it is done through taxes, mandated restrictions on GHG emissions or market mechanisms is less important than that the signal is sent clearly, forcefully and unambiguously. This is where President Bush's mixed signals have done a lot of harm: America's industry, unlike Europe's, has little incentive to invest in low-carbon technology. The irony is that even some coal-fired utilities in America are now clamouring for CO_2 regulation so that they can invest in new plants with confidence.

The third pillar is to promote science and technology. That means encouraging basic climate and energy research, and giving incentives for spreading the results.

Rich countries and aid agencies must also find ways to help the poor world adapt to climate change. This is especially important if the world starts off with small cuts in emissions, leaving deeper cuts for later. That, observes Mr Wigley, means that by mid-century "very large investments would have to have been made—and yet the 'return' on these investments would not be visible. Continued investment is going to require more faith in climate science than currently appears to be the case."

Even a visionary like Churchill might have lost heart in the face of all this uncertainty. Nevertheless, there is a glimmer of hope that today's peacetime politicians may rise to the occasion.

Miracles sometimes happen

Two decades ago, the world faced a similar dilemma: evidence of a hole in the ozone layer. Some inconclusive signs suggested that it was man-made, caused by the use of chlorofluorocarbons (CFCs). There was the distant threat of disaster, and the knowledge about a concerted global response was required. Industry was reluctant at first, yet with leadership from Britain and America the Montreal Protocol was signed in 1987. That deal has proved surprisingly successful. The manufacture of CFCs is nearly phased out, and there are already signs that the ozone layer is on the way to recovery.

This story holds several lessons for the admittedly far more complex climate problem. First, it is the rich world which has caused the problem and which must lead the way in solving it. Second, the poor world must agree to help, but is right to insist on being given time—as well as money and technology—to help it adjust. Third, industry holds the key: in the ozone-depletion story, it was only after DuPont and ICI broke ranks with the rest of the CFC manufacturers that a deal became possible. On the climate issue, BP and Shell have similarly broken ranks with Big Oil, but the American energy industry—especially the coal sector—remains hostile.

The final lesson is the most important: that the uncertainty surrounding a threat such as climate change is no excuse for inaction. New scientific evidence shows that the threat from ozone depletion had been much deadlier than was thought at the time when the world decided to act. Churchill would surely have approved.

Local difficulties

Greenery is for the poor too, particularly on their own doorstep

WHY should we care about the environment? Ask a European, and he will probably point to global warming. Ask the two little boys playing outside a newsstand in Da Shilan, a shabby neighbourhood in the heart of Beijing,

and they will tell you about the city's notoriously foul air: "It's bad—like a virus!"

Given all the media coverage in the rich world, people there might believe that global scares are the chief envi-

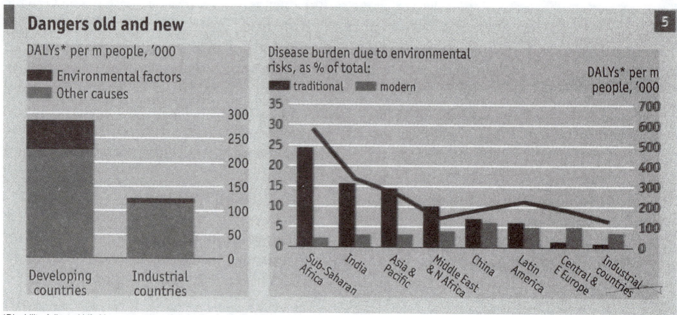

Dangers old and new

DALYs* per m people, '000
- Environmental factors
- Other causes

300
250
200
150
100
50
0

Developing countries
Industrial countries

Disease burden due to environmental risks, as % of total:
- traditional
- modern

DALYs* per m people, '000

35 — 700
30 — 600
25 — 500
20 — 400
15 — 300
10 — 200
5 — 100
0 — 0

Sub-Saharan Africa
India
Asia & Pacific
Middle East & N Africa
China
Latin America
Central & E Europe
Industrial countries

*Disability Adjusted Life Years—number of years lived with disability and years lost to premature death
Source: World Bank

ronmental problems facing humanity today. They would be wrong. Partha Dasgupta, an economics professor at Cambridge University, thinks the current interest in global, future-oriented problems has "drawn attention away from the economic misery and ecological degradation endemic in large parts of the world today. Disaster is not something for which the poorest have to wait; it is a frequent occurrence."

Every year in developing countries, a million people die from urban air pollution and twice that number from exposure to stove smoke inside their homes. Another 3m unfortunates die prematurely every year from water-related diseases. All told, premature deaths and illnesses arising from environmental factors account for about a fifth of all diseases in poor countries, bigger than any other preventable factor, including malnutrition. The problem is so serious that Ian Johnson, the World Bank's vice-president for the environment, tells his colleagues, with a touch of irony, that he is really the bank's vice-president for health: "I say tackling the underlying environmental causes of health problems will do a lot more good than just more hospitals and drugs."

The link between environment and poverty is central to that great race for sustainability. It is a pity, then, that several powerful fallacies keep getting in the way of sensible debate. One popular myth is that trade and economic growth make poor countries' environmental problems worse. Growth, it is said, brings with it urbanisation, higher energy consumption and industrialisation—all factors that contribute to pollution and pose health risks.

In a static world, that would be true, because every new factory causes extra pollution. But in the real world, economic growth unleashes many dynamic forces that, in the longer run, more than offset that extra pollution. As

chart 5 makes clear, traditional environmental risks (such as water-borne diseases) cause far more health problems in poor countries than modern environmental risks (such as industrial pollution).

Rigged rules

However, this is not to say that trade and economic growth will solve all environmental problems. Among the reasons for doubt are the "perverse" conditions under which world trade is carried on, argues Oxfam. The British charity thinks the rules of trade are "unfairly rigged against the poor", and cites in evidence the enormous subsidies lavished by rich countries on industries such as agriculture, as well as trade protection offered to manufacturing industries such as textiles. These measurements hurt the environment because they force the world's poorest countries to rely heavily on commodities—a particularly energy-intensive and ungreen sector.

Mr Dasgupta argues that this distortion of trade amounts to a massive subsidy of rich-world consumption paid by the world's poorest people. The most persuasive critique of all goes as follows: "Economic growth is not sufficient for turning environmental degradation around. If economic incentives facing producers and consumers do not change with higher incomes, pollution will continue to grow unabated with the growing scale of economic activity." Those words come not from some anti-globalist green group, but from the World Trade Organisation.

Another common view is that poor countries, being unable to afford greenery, should pollute now and clean up later. Certainly poor countries should not be made to adopt American or European environmental standards. But there is evidence to suggest that poor countries can and should try to tackle some environmental problems now, rather than wait till they have become richer.

This so-called "smart growth" strategy contradicts conventional wisdom. For many years, economists have observed that as agrarian societies industrialised, pollution increased at first, but as the societies grew wealthier it declined again. The trouble is that this applies only to some pollutants, such as sulphur dioxide, but not to others, such as carbon dioxide. Even more troublesome, those smooth curves going up, then down, turn out to be misleading. They are what you get when you plot data for poor and rich countries together at a given moment in time, but actual levels of various pollutants in any individual country plotted over time wiggle around a lot more. This suggests that the familiar bell-shaped curve reflects no immutable law, and that intelligent government policies might well help to reduce pollution levels even while countries are still relatively poor.

Developing countries are getting the message. From Mexico to the Philippines, they are now trying to curb the worst of the air and water pollution that typically accompanies industrialisation. China, for example, was persuaded by outside experts that it was losing so much potential economic output through health troubles caused by pollution (according to one World Bank study, somewhere between 3.5% and 7.7% of GDP) that tackling it was cheaper than ignoring it.

One powerful—and until recently ignored—weapon in the fight for a better environment is local people. Old-fashioned paternalists in the capitals of developing countries used to argue that poor villagers could not be relied on to look after natural resources. In fact, much academic research has shown that the poor are more often victims than perpetrators of resource depletion: it tends to be rich locals or outsiders who are responsible for the worst exploitation.

Local people usually have a better knowledge of local ecological conditions than experts in faraway capitals, as well as a direct interest in improving the quality of life in their village. A good example of this comes from the bone-dry state of Rajasthan in India, where local activism and indigenous know-how about rainwater "harvesting" provided the people with reliable water supplies—something the government had failed to do. In Bangladesh, villages with active community groups or concerned mullahs proved greener than less active neighbouring villages.

Community-based forestry initiatives from Bolivia to Nepal have shown that local people can be good custodians of nature. Several hundred million of the world's poorest people live in and around forests. Giving those villagers an incentive to preserve forests by allowing sustainable levels of harvesting, it turns out, is a far better way to save those forests than erecting tall fences around them.

To harness local energies effectively, it is particularly important to give local people secure property rights, argues Mr Dasgupta. In most parts of the developing world, control over resources at the village level is ill-defined. This often means that local elites usurp a disproportionate share of those resources, and that individuals have little incentive to maintain and upgrade forests or agricultural land. Authorities in Thailand tried to remedy this problem by distributing 5.5m land titles over a 20-year period. Agricultural output increased, access to credit improved and the value of the land shot up.

Name and shame

Another powerful tool for improving the local environment is the free flow of information. As local democracy flourishes, ordinary people are pressing for greater environmental disclosure by companies. In some countries, such as Indonesia, governments have adopted a "sunshine" policy that involves naming and shaming companies that do not meet environmental regulations. It seems to achieve results.

Bringing greenery to the grass roots is good, but on its own it will not avert perceived threats to global "public goods" such as the climate or biodiversity. Paul Portney of Resources for the Future explains: "Brazilian villagers may think very carefully and unselfishly about their future descendants, but there's no reason for them to care about and protect species or habitats that no future generation of Brazilians will care about."

That is why rich countries must do more than make pious noises about global threats to the environment. If they believe that scientific evidence suggests a credible threat, they must be willing to pay poor countries to protect such things as their tropical forests. Rather than thinking of this as charity, they should see it as payment for environmental services (say, for carbon storage) or as a form of insurance.

In the case of biodiversity, such payments could even be seen as a trade in luxury goods: rich countries would pay poor countries to look after creatures that only the rich care about. Indeed, private green groups are already buying up biodiversity "hot spots" to protect them. One such initiative, led by Conservation International and the International Union for the Conservation of Nature (IUCN), put the cost of buying and preserving 25 hot spots exceptionally rich in species diversity at less than $30 billion. Sceptics say it will cost more, as hot spots will need buffer zones of "sustainable harvesting" around them. Whatever the right figure, such creative approaches are more likely to achieve results than bullying the poor into conservation.

It is not that the poor do not have green concerns, but that those concerns are very different from those of the rich. In Beijing's Da Shilan, for instance, the air is full of soot from the many tiny coal boilers. Unlike most of the neighbouring districts, which have recently converted from coal to natural gas, this area has been considered too poor to make the transition. Yet ask Liu Shihua, a shopkeeper who has lived in the same spot for over 20 years,

11

and he insists he would readily pay a bit more for the cleaner air that would come from using natural gas. So would his neighbours.

To discover the best reason why poor countries should not ignore pollution, ask those two little boys outside Mr Liu's shop what colour the sky is. "Grey!" says one tyke, as if it were the most obvious thing in the world. "No, stupid, it's blue!" retorts the other. The children deserve blue skies and clean air. And now there is reason to think they will see them in their lifetime.

Working miracles

Can technology save the planet?

"Nothing endures but change." That observation by Heraclitus often seems lost on modern environmental thinkers. Many invoke scary scenarios assuming that resources—both natural ones, like oil, and man-made ones, like knowledge—are fixed. Yet in real life man and nature are entwined in a dynamic dance of development, scarcity, degradation, innovation and substitution.

The nightmare about China turning into a resource-guzzling America raises two questions: will the world run out of resources? And even if it does not, could the growing affluence of developing nations lead to global environmental disaster?

The first fear is the easier to refute; indeed, history has done so time and again. Malthus, Ricardo and Mill all worried that scarcity of resources would snuff out growth. It did not. A few decades ago, the limits-to-growth camp raised worries that the world might soon run out of oil, and that it might not be able to feed the world's exploding population. Yet there are now more proven reserves of petroleum than three decades ago; there is more food produced than ever; and the past decade has seen history's greatest economic boom.

What made these miracles possible? Fears of oil scarcity prompted investment that led to better ways of producing oil, and to more efficient engines. In food production, technological advances have sharply reduced the amount of land required to feed a person in the past 50 years. Jesse Ausubel of Rockefeller University calculates that if in the next 60 to 70 years the world's average farmer reaches the yield of today's average (not best) American maize grower, then feeding 10 billion people will require just half of today's cropland. All farmers need to do is maintain the 2%-a-year productivity gain that has been the global norm since 1960.

"Scarcity and Growth", a book published by Resources for the Future, sums it up brilliantly: "Decades ago Vermont granite was only building and tombstone material; now it is a potential fuel, each ton of which has a usable energy content (uranium) equal to 150 tons of coal. The notion of an absolute limit to natural resource availability is untenable when the definition of resources changes drastically and unpredictably over time." Those words

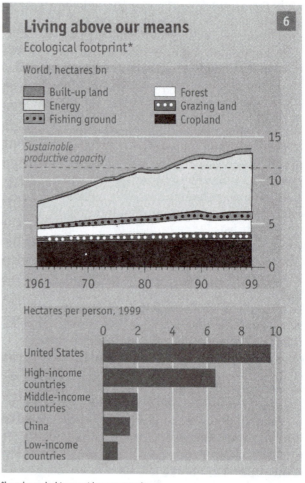

*Land needed to meet human needs

Source: WWF, Living Planet Report 2002

were written by Harold Barnett and Chandler Morse in 1963, long before the limits-to-growth bandwagon got rolling.

Giant footprint

Not so fast, argue greens. Even if we are not going to run out of resources, guzzling ever more resources could still do irreversible damage to fragile ecosystems.

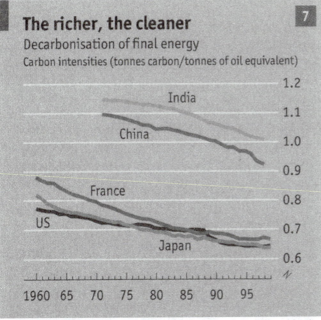

The richer, the cleaner
Decarbonisation of final energy
Carbon intensities (tonnes carbon/tonnes of oil equivalent)

Source: Nebojsa Nakicenovic and Arnulf Gruebler, International Institute for Applied Systems Analysis

WWF, an environmental group, regularly calculates mankind's "ecological footprint", which it defines as the "biologically productive land and water areas required to produce the resources consumed and assimilate the wastes generated by a given population using prevailing technology." The group reckons the planet has around 11.4 billion "biologically productive" hectares of land available to meet continuing human needs. As chart 6 overleaf shows, WWF thinks mankind has recently been using more than that. This is possible because a forest harvested at twice its regeneration rate, for example, appears in the footprint accounts at twice its area—an unsustainable practice which the group calls "ecological overshoot."

Any analysis of this sort must be viewed with scepticism. Everyone knows that environmental data are incomplete. What is more, the biggest factor by far is the land required to absorb CO_2 emissions of fossil fuels. If that problem could be managed some other way, then mankind's ecological footprint would look much more sustainable.

Even so, the WWF analysis makes an important point: if China's economy were transformed overnight into a clone of America's, an ecological nightmare could ensue. If a billion eager new consumers were suddenly to produce CO_2 emissions at American rates, they would be bound to accelerate global warming. And if the whole of the developing world were to adopt an American lifestyle tomorrow, local environmental crises such as desertification, aquifer depletion and topsoil loss could make humans miserable.

So is this cause for concern? Yes, but not for panic. The global ecological footprint is determined by three factors: population size, average consumption per person and technology. Fortunately, global population growth now appears to be moderating. Consumption per person in poor countries is rising as they become better off, but there are signs that the rich world is reducing the footprint of its consumption (as this survey's final section explains). The most powerful reason for hope—innovation—was foreshadowed by WWF's own definition. Today's "prevailing technologies" will, in time, be displaced by tomorrow's greener ones.

"The rest of the world will not live like America," insists Mr Ausubel. Of course poor people around the world covet the creature comforts that Americans enjoy, but they know full well that the economic growth needed to improve their lot will take time. Ask Wu Chengjian, an environmental official in booming Shanghai, what he thinks of the popular notion that his city might become as rich as today's Hong Kong by 2020: "Impossible—that's just not enough time." And that is Shanghai, not the impoverished countryside.

Leaps of faith

This extra time will allow poor countries to embrace new technologies that are more efficient and less environmentally damaging. That still does not guarantee a smaller ecological footprint for China in a few decades' time than for America now, but it greatly improves the chances. To see why, consider the history of "dematerialisation" and "decarbonisation" (see chart 7). Viewed across very long spans of time, productivity improvements allow economies to use ever fewer material inputs—and to emit ever fewer pollutants—per unit of economic output. Mr Ausubel concludes: "When China has today's American mobility, it will not have today's American cars," but the cleaner and more efficient cars of tomorrow.

The snag is that consumers in developing countries want to drive cars not tomorrow but today. The resulting emissions have led many to despair that technology (in the form of vehicles) is making matters worse, not better.

Can they really hope to "leapfrog" ahead to cleaner air? The evidence from Los Angeles—a pioneer in the fight against air pollution—suggests the answer is yes. "When I moved to Los Angeles in the 1960s, there was so much soot in the air that it felt like there was a man standing on your chest most of the time," says Ron Loveridge, the mayor of Riverside, a city to the east of LA that suffers the worst of the region's pollution. But, he says, "We have come an extraordinary distance in LA."

Four decades ago, the city had the worst air quality in America. The main problem was the city's infamous "smog" (an amalgam of "smoke" and "fog"). It took a while to figure out that this unhealthy ozone soup developed as a result of complex chemical reactions between nitrogen oxides and volatile organic compounds that need sunlight to trigger them off.

Arthur Winer, an atmospheric chemist at the University of California at Los Angeles, explains that tackling smog required tremendous perseverance and political will. Early regulatory efforts met stiff resistance from business interests, and began to falter when they failed to show dramatic results.

Clean-air advocates like Mr Loveridge began to despair: "We used to say that we needed a 'London fog' [a reference to an air-pollution episode in 1952 that may have killed 12,000 people in that city] here to force change." Even so, Californian officials forged ahead with an ambitious plan that combined regional regulation with stiff mandates for cleaner air. Despite uncertainties about the cause of the problem, the authorities introduced a sequence of controversial measures: unleaded and low-sulphur petrol, on-board diagnostics for cars to minimise emissions, three-way catalytic converters, vapour-recovery attachments for petrol nozzles and so on.

As a result, the city that two decades ago hardly ever met federal ozone standards has not had to issue a single alert in the past three years. Peak ozone levels are down by 50% since the 1960s. Though the population has shot up in recent years, and the vehicle-miles driven by car-crazy Angelenos have tripled, ozone levels have fallen by two-thirds. The city's air is much cleaner than it was two decades ago.

"California, in solving its air-quality problem, has solved it for the rest of the United States and the world—but it doesn't get credit for it," says Joe Norbeck of the University of California at Riverside. He is adamant that the poor world's cities can indeed leapfrog ahead by embracing some of the cleaner technologies developed specifically for the Californian market. He points to China's vehicle fleet as an example: "China's typical car has the emissions of a 1974 Ford Pinto, but the new Buicks sold there use 1990s emissions technology." The typical car sold today produces less than a tenth of the local pollution of a comparable model from the 1970s.

That suggests one lesson for poor cities such as Beijing that are keen to clean up: they can order polluters to meet high emissions standards. Indeed, from Beijing to Mexico city, regulators are now imposing rich-world rules, mandating new, cleaner technologies. In China's cities, where pollution from sooty coal fires in homes and industrial boilers had been a particular hazard, officials are keen to switch to natural-gas furnaces.

However, there are several reasons why such mandates—which worked wonders in LA—may be trickier to achieve in impoverished or politically weak cities. For a start, city officials must be willing to pay the political price of reforms that raise prices for voters. Besides, higher standards for new cars, useful though they are, cannot do the trick on their own. Often, clean technologies such as catalytic converters will require cleaner grades of petrol too. Introducing cleaner fuels, say ex-

perts, is an essential lesson from LA for poor countries. This will not come free either.

There is another reason why merely ordering cleaner new cars is inadequate: it does nothing about the vast stock of dirty old ones already on the streets. In most cities of the developing world, the oldest fifth of the vehicles on the road is likely to produce over half of the total pollution caused by all vehicles taken together. Policies that encourage a speedier turnover of the fleet therefore make more sense than "zero emissions" mandates.

Policy matters

In sum, there is hope that the poor can leapfrog at least some environmental problems, but they need more than just technology. Luisa and Mario Molina of the Massachusetts Institute of Technology, who have studied such questions closely, reckon that technology is less important than the institutional capacity, legal safeguards and financial resources to back it up: "The most important underlying factor is political will." And even a techno-optimist such as Mr Ausubel accepts that: "There is nothing automatic about technological innovation and adoption; in fact, at the micro level, it's bloody."

Clearly innovation is a powerful force, but government policy still matters. That suggests two rules for policymakers. First, don't do stupid things that inhibit innovation. Second, do sensible things that reward the development and adoption of technologies that enhance, rather than degrade, the environment.

The greatest threat to sustainability may well be the rejection of science. Consider Britain's hysterical reaction to genetically modified crops, and the European Commission's recent embrace of a woolly "precautionary principle". Precaution applied case-by-case is undoubtedly a good thing, but applying any such principle across the board could prove disastrous.

Explaining how not to stifle innovation that could help the environment is a lot easier than finding ways to encourage it. Technological change often goes hand-in-hand with greenery by saving resources, as the long history of dematerialisation shows—but not always. Sports utility vehicles, for instance, are technologically innovative, but hardly green. Yet if those SUVs were to come with hydrogen-powered fuel cells that emit little pollution, the picture would be transformed.

The best way to encourage such green innovations is to send powerful signals to the market that the environment matters. And there is no more powerful signal than price, as the next section explains.

The invisible green hand

Markets could be a potent force for greenery—if only greens could learn to love them

"MANDATE, regulate and litigate." That has been the environmentalists' rallying cry for ages. Nowhere in the green manifesto has there been much mention of the market. And, oddly, it was market-minded America that led the dirigiste trend. Three decades ago, Congress passed a sequence of laws, including the Clean Air Act, which set lofty goals and generally set rigid technological standards. Much of the world followed America's lead.

This top-down approach to greenery has long been a point of pride for groups such as the Natural Resources Defence Council (NRDC), one of America's most influential environmental outfits. And with some reason, for it has had its successes: the air and water in the developed world is undoubtedly cleaner than it was three decades ago, even though the rich world's economies have grown by leaps and bounds. This has convinced such groups stoutly to defend the green status quo.

But times may be changing. Gus Speth, now head of Yale University's environment school and formerly head of the World Resources Institute and the UNDP, as well as one of the founders of the NRDC, recently explained how he was converted to market economics: "Thirty years ago, the economists at Resources for the Future were pushing the idea of pollution taxes. We lawyers at NRDC thought they were nuts, and feared that they would derail command-and-control measures like the Clean Air Act, so we opposed them. Looking back, I'd have to say this was the single biggest failure in environmental management—not getting the prices right."

A remarkable mea culpa; but in truth, the command-and-control approach was never as successful as its advocates claimed. For example, although it has cleaned up the air and water in rich countries, it has notably failed in dealing with waste management, hazardous emissions and fisheries depletion. Also, the gains achieved have come at a needlessly high price. That is because technology mandates and bureaucratic edicts stifle innovation and ignore local realities, such as varying costs of abatement. They also fail to use cost-benefit analysis to judge trade-offs.

Command-and-control methods will also be ill-suited to the problems of the future, which are getting trickier. One reason is that the obvious issues—like dirty air and water—have been tackled already. Another is increasing technological complexity: future problems are more likely to involve subtle linkages—like those involved in ozone depletion and global warming—that will require sophisticated responses. The most important factor may be society's ever-rising expectations; as countries grow wealthier, their people start clamouring for an ever-cleaner environment. But because the cheap and simple things have been done, that is proving increasingly expensive. Hence the greens' new interest in the market.

Carrots, not just sticks

In recent years, market-based greenery has taken off in several ways. With emissions trading, officials decide on a pollution target and then allocate tradable credits to companies based on that target. Those that find it expensive to cut emissions can buy credits from those that find it cheaper, so the target is achieved at the minimum cost and disruption.

The greatest green success story of the past decade is probably America's innovative scheme to cut emissions of sulphur dioxide (SO_2). Dan Dudek of Environmental Defence, a most unusual green group, and his market-minded colleagues persuaded the elder George Bush to agree to an amendment to the sacred Clean Air Act that would introduce an emissions-trading system to achieve sharp cuts in SO_2. At the time, this was hugely controversial: America's power industry insisted the cuts were prohibitively costly, while nearly every other green group decried the measure as a sham. In the event, ED has been vindicated. America's scheme has surpassed its initial objectives, and at far lower cost than expected. So great is the interest worldwide in trading that ED is now advising groups ranging from hard-nosed oilmen at BP to bureaucrats in China and Russia.

Europe, meanwhile, is forging ahead with another sort of market-based instrument: pollution taxes. The idea is to levy charges on goods and services so that their price reflects their "externalities"—jargon for how much harm they do to the environment and human health. Sweden introduced a sulphur tax a decade ago, and found that the sulphur content of fuels dropped 50% below legal requirements.

Though "tax" still remains a dirty word in America, other parts of the world are beginning to embrace green tax reform by shifting taxes from employment to pollution. Robert Williams of Princeton University has looked at energy use (especially the terrible effects on health of particulate pollution) and concluded that such externalities are comparable in size to the direct economic costs of producing that energy.

Externalities are only half the battle in fixing market distortions. The other half involves scrapping environmentally harmful subsidies. These range from prices below market levels for electricity and water to shameless

15

cash handouts for industries such as coal. The boffins at the OECD reckon that stripping away harmful subsidies, along with introducing taxes on carbon-based fuels and chemicals use, would result in dramatically lower emissions by 2020 than current policies would be able to achieve. If the revenues raised were then used to reduce other taxes, the cost of these virtuous policies would be less than 1% of the OECD's economic output in 2020.

Such subsidies are nothing short of perverse, in the words of Norman Myers of Oxford University. They do double damage, by distorting markets and by encouraging behaviour that harms the environment. Development banks say such subsidies add up to $700 billion a year, but Mr Myers reckons the true sum is closer to $2 trillion a year. Moreover, the numbers do not fully reflect the harm done. For example, EU countries subsidise their fishing fleets to the tune of $1 billion a year, but that has encouraged enough overfishing to drive many North Atlantic fishing grounds to near-collapse.

Fishing is an example of the "tragedy of the commons", which pops up frequently in the environmental debate. A resource such as the ocean is common to many, but an individual "free rider" can benefit from plundering that commons or dumping waste into it, knowing that the costs of his actions will probably be distributed among many neighbours. In the case of shared fishing grounds, the absence of individual ownership drives each fisherman to snatch as many fish as he can—to the detriment of all.

Of rights and wrongs

Assigning property rights can help, because providing secure rights (set at a sustainable level) aligns the interests of the individual with the wider good of preserving nature. This is what sceptical conservationists have observed in New Zealand and Iceland, where schemes for tradable quotas have helped revive fishing stocks. Similar rights-based approaches have led to revivals in stocks of African elephants in southern Africa, for example, where the authorities stress property rights and private conservation.

All this talk of property rights and markets makes many mainstream environmentalists nervous. Carl Pope, the boss of the Sierra Club, one of America's biggest green groups, does not reject market forces out of hand, but expresses deep scepticism about their scope. Pointing to the difficult problem of climate change, he asks: "Who has property rights over the commons?"

Even so, some greens have become converts. Achim Steiner of the IUCN reckons that the only way forward is rights-based conservation, allowing poor people "sustainable use" of their local environment. Paul Faeth of the World Resources Institute goes further. He says he is convinced that market forces could deliver that holy grail of environmentalism, sustainability—"but only if we get prices right."

The limits to markets

Economic liberals argue that the market itself is the greatest price-discovery mechanism known to man. Allow it to function freely and without government meddling, goes the argument, and prices are discovered and internalised automatically. Jerry Taylor of the Cato Institute, a libertarian think-tank, insists that "The world today is already sustainable—except those parts where western capitalism doesn't exist." He notes that countries that have relied on central planning, such as the Soviet Union, China and India, have invariably misallocated investment, stifled innovation and fouled their environment far more than the prosperous market economies of the world have done.

All true. Even so, markets are currently not very good at valuing environmental goods. Noble attempts are under way to help them do better. For example, the Katoomba Group, a collection of financial and energy companies that have linked up with environmental outfits, is trying to speed the development of markets for some of forestry's ignored "co-benefits" such as carbon storage and watershed management, thereby producing new revenue flows for forest owners. This approach shows promise: water consumers ranging from officials in New York City to private hydro-electric operators in Costa Rica are now paying people upstream to manage their forests and agricultural land better. Paying for greenery upstream turns out to be cheaper than cleaning up water downstream after it has been fouled.

Economists too are getting into the game of helping capitalism "get prices right." The World Bank's Ian Johnson argues that conventional economic measures such as gross domestic product are not measuring wealth creation properly because they ignore the effects of environmental degradation. He points to the positive contribution to China's GDP from the logging industry, arguing that such a calculation completely ignores the billions of dollars-worth of damage from devastating floods caused by over-logging. He advocates a more comprehensive measure the Bank is working on, dubbed "genuine GDP", that tries (imperfectly, he accepts) to measure depletion of natural resources.

That could make a dramatic difference to how the welfare of the poor is assessed. Using conventional market measures, nearly the whole of the developing world save Africa has grown wealthier in the past couple of decades. But when the degradation of nature is properly accounted for, argues Mr Dasgupta at Cambridge, the countries of Africa and south Asia are actually much worse off today than they were a few decades ago—and even China, whose economic "miracle" has been much trumpeted, comes out barely ahead.

The explanation, he reckons, lies in a particularly perverse form of market distortion: "Countries that are exporting resource-based products (often among the poorest) may be subsidising the consumption of countries that are doing the importing (often among the richest)." As evidence, he points to the common practice in poor countries of encouraging resource extraction. Whether through licenses granted at below-market rates, heavily subsidised exports or corrupt officials tolerating illegal exploitation, he reckons the result is the same: "The cruel paradox we face may well be that contemporary economic development is unsustainable in poor countries because it is sustainable in rich countries."

One does not have to agree with Mr Dasgupta's conclusion to acknowledge that markets have their limits. That should not dissuade the world from attempting to get prices right—or at least to stop getting them so wrong. For grotesque subsidies, the direction of change should be obvious. In other areas, the market itself may not provide enough information to value nature adequately. This is true of threats to essential assets, such as nature's ability to absorb and "recycle" CO_2, that have no substitute at any price. That is when governments must step in, ensuring that an informed public debate takes place.

Robert Stavins of Harvard University argues that the thorny notion of sustainable development can be reduced to two simple ideas: efficiency and intergenerational equity. The first is about making the economic pie as large as possible; he reckons that economists are well equipped to handle it, and that market-based policies can be used to achieve it. On the second (the subject of the next section), he is convinced that markets must yield to public discourse and government policy: "Markets can be efficient, but nobody ever said they're fair. The question is, what do we owe the future?"

Insuring a brighter future

How to hedge against tomorrow's environmental risks

So WHAT do we owe the future? A precise definition for sustainable development is likely to remain elusive but, as this survey has argued, the hazy outline of a useful one is emerging from the experience of the past decade.

For a start, we cannot hope to turn back the clock and return nature to a pristine state. Nor must we freeze nature in the state it is today, for that gift to the future would impose an unacceptable burden on the poorest alive today. Besides, we cannot forecast the tastes, demands or concerns of future generations. Recall that the overwhelming pollution problem a century ago was horse manure clogging up city streets: a century hence, many of today's problems will surely seem equally irrelevant. We should therefore think of our debt to the future as including not just natural resources but also technology, institutions and especially the capacity to innovate. Robert Solow got it mostly right a decade ago: the most important thing to leave future generations, he said, is the capacity to live as well as we do today.

However, as the past decade has made clear, there is a limit to that argument. If we really care about the "sustainable" part of sustainable development, we must be much more watchful about environmental problems with critical thresholds. Most local problems are reversible and hence no cause for alarm. Not all, however: the depletion of aquifers and the loss of topsoil could trigger irreversible changes that would leave future generations worse off. And global or long-term threats, where victims are far removed in time and space, are easy to brush aside.

In areas such as biodiversity, where there is little evidence of a sustainability problem, a voluntary approach is best. Those in the rich world who wish to preserve pandas, or hunt for miracle drugs in the rainforest, should pay for their predilections. However, where there are strong scientific indications of unsustainability, we must act on behalf of the future—even at the price of today's development. That may be expensive, so it is prudent to try to minimise those risks in the first place.

A riskier world

Human ingenuity and a bit of luck have helped mankind stay a few steps ahead of the forces degrading the environment this past century, the first full one in which the planet has been exposed to industrialisation. In the century ahead, the great race between development and degradation could well become a closer call.

On one hand, the demands of development seem sure to grow at a cracking pace in the next few decades as the Chinas, Indias and Brazils of this world grow wealthy enough to start enjoying not only the necessities but also some of the luxuries of life. On the other hand, we seem to be entering a period of huge technological advances in emerging fields such as biotechnology that could greatly increase resource productivity and more than offset the effect of growth on the environment. The trouble is, nobody knows for sure.

Since uncertainty will define the coming era, it makes sense to invest in ways that reduce that risk at relatively low cost. Governments must think seriously about the future implications of today's policies. Their best bet is to encourage the three powerful forces for sustainability outlined in this survey: the empowerment of local people to manage local resources and adapt to environmental change; the encouragement of science and technology, especially innovations that reduce the ecological footprint of consumption; and the greening of markets to get prices right.

To advocate these interventions is not to call for a return to the hubris of yesteryear's central planners. These

Out of the blue

8

Financial costs of weather-related great natural disasters, 2001 prices, $bn

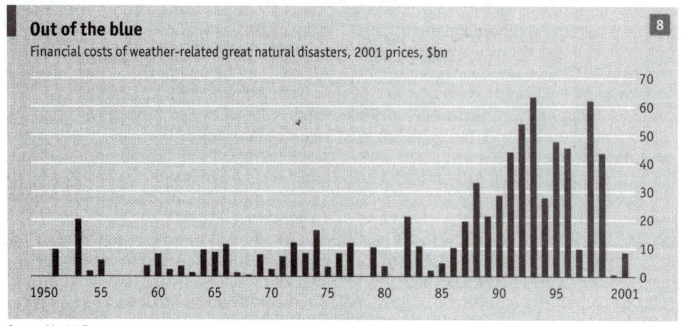

Source: Munich Re

measures would merely give individuals the power to make greener choices if they care to. In practice, argues Chris Heady of the OECD, this may still not add up to sustainability "because we might still decide to be greedy, and leave less for our children."

Happily, there are signs of an emerging bottom-up push for greenery. Even such icons of western consumerism as Unilever and Procter & Gamble now sing the virtues of "sustainable consumption." Unilever has vowed that by 2005 it will be buying fish only from sustainable sources, and P&G is coming up with innovative products such as detergents that require less water, heat and packaging. It would be naive to label such actions as expressions of "corporate social responsibility": in the long run, firms will embrace greenery only if they see profit in it. And that, in turn, will depend on choices made by individuals.

Such interventions should really be thought of as a kind of insurance that tilts the odds of winning that great race just a little in humanity's favour. Indeed, even some of the world's most conservative insurance firms increasingly see things this way. As losses from weather-related disasters have risen of late (see chart 8), the industry is getting more involved in policy debates on long-term environmental issues such as climate change.

Bruno Porro, chief risk officer at Swiss Re, argues that: "The world is entering a future in which risks are more concentrated and more complex. That is why we are pressing for policies that reduce those risks through preparation, adaptation and mitigation. That will be cheaper than covering tomorrow's losses after disaster strikes."

Jeffrey Sachs of Columbia University agrees: "When you think about the scale of risk that the world faces, it is clear that we grossly underinvest in knowledge… we have enough income to live very comfortably in the developed world and to prevent dire need in the developing world. So we should have the confidence to invest in longer-term issues like the environment. Let's help insure the sustainability of this wonderful situation."

He is right. After all, we have only one planet, now and in the future. We need to think harder about how to use it wisely.

Acknowledgements

In addition to those cited in the text, the author would like to thank Robert Socolow, David Victor, Geoffrey Heal, and experts at Tsinghua University, Friends of the Earth, the European Commission, the World Business Council for Sustainable Development, the International Energy Agency, the OECD and the UN for sharing their ideas with him. A list of sources can be found on *The Economist's* website.

MANAGING PLANET EARTH

Forget Nature. Even Eden Is Engineered

By ANDREW C. REVKIN

Nearly 70 years ago, a Soviet geochemist, reflecting on his world, made a startling observation: through technology and sheer numbers, he wrote, people were becoming a geological force, shaping the planet's future just as rivers and earthquakes had shaped its past.

Eventually, wrote the scientist, Vladimir I. Vernadsky, global society, guided by science, would soften the human environmental impact, and earth would become a "noosphere," a planet of the mind, "life's domain ruled by reason."

Today, a broad range of scientists say, part of Vernadsky's thinking has already been proved right: people have significantly altered the atmosphere and are the dominant influence on ecosystems and natural selection. The question now, scientists say, is whether the rest of his vision will come to pass. Choices made in the next few years will determine the answer.

Aided by satellites and supercomputers, and mobilized by the evident environmental damage of the last century, humans have a real chance to begin balancing economic development with sustaining earth's ecological webs, said Dr. William C. Clark, a biologist at Harvard who heads an international effort to build a scientific foundation for such a shift.

"We've come through a period of finally understanding the nature and magnitude of humanity's transformation of the earth," Dr. Clark said. "Having realized it, can we become clever enough at a big enough scale to be able to maintain the rates of progress? I think we can."

Some scientists say it is anthropocentric hubris to think people understand the living planet well enough to know how to manage it. But that prospect is attracting more than 100 world leaders and thousands of other participants to the United Nations' World Summit on Sustainable Development, which starts on Monday in Johannesburg.

No matter what they come up with, ice ages, volcanoes and shifting tectonic plates will dwarf human activities in the long run. But communities and countries face concrete choices in the next decade that are likely to determine the quality of human life and the environment well into the 22nd century.

Human activity is such a pervasive influence on the planet's ecological framework that it is no longer possible to separate people and nature.

Emissions of heat-trapping carbon dioxide, whether from an Ohio power plant or a Bangkok taxicab, contribute to global warming.

Seafood lovers dining in Manhattan bistros prompt fishing vessels to sweep Antarctic waters for slow-growing Chilean sea bass. Shoppers in Tokyo seeking inexpensive picture frames send loggers deep into Indonesian forests.

In a new book "Great Transition: The Promise and Lure of the Times Ahead," published by the Stockholm Environment Institute, a group of top geographers, economists, engineers and other experts concludes that the same inventiveness that accelerated the human ascent can be harnessed to soften human impact.

The need for a new approach is urgent, the researchers say, because a surge of growth in quickly industrializing regions of Asia and the Americas could have environmental effects that exceed those of the industrialization of the West. More pressure for change comes from southern Africa and other pockets of extreme poverty where the brutal calculus of Malthus still holds sway.

Even in the industrialized north, after generations of prosperity, people are hemmed in by concrete, seeing commuting times grow and starting to question their definition of progress.

As a result, countless communities, from the charred fringes of the Amazon to the spreading suburbs of Seattle, are balancing growing needs and limited resources.

If development does not change course, the new book concludes, "the nightmare of an impoverished, mean, destructive future looms."

The World Remade By Human Hands

When delegates convene in Johannesburg next week for a United Nations environment summit, they will confront the problems human interference has brought to the world's ecosystems. But some trends once seen as problems may not be so bad after all. And some seemingly beneficial trends have serious unexpected side effects. Here are some examples.

MEGACITIES

By 2015, demographers predict, the number of cities with more than 10 million people—so-called megacities— will rise to 36, from about 20 today. Megacities are notorious for pollution and sprawling slums. But people who live in them tend to have smaller families and greater opportunities for jobs and educations. And environmental improvements can come quickly to cities. For example, in Mexico City, the percentage of days violating pollution standards has fallen to less than 20 percent, from 50 percent. Now the city is considering a mass transit program that would bring further improvements.

FOOD SUPPLIES

Demographers are lowering their estimates for how much the population of the world will increase before it stabilizes, but even so they predict 50 percent growth by the end of the century. Demand for food will grow even faster, as people in developing countries start consuming more meat and fish. Aquaculture, once thought of as a solution, has produced problems of its own. In Thailand, and elsewhere, shrimp farms have replaced valuable mangrove ecosystems. In China, however, a recent experiment in mixing rice varieties in a single field greatly increased yields, without added chemicals.

FALLING FORESTS

In the tropics, forests continue to fall before farmers, loggers and timber processors like the saw mill, on a branch of the Amazon River in Brazil. But new surveys show that in the 1990's the rate of cutting was 23 percent less than earlier estimates. Some timber companies have gained certification from environmental groups as managing forests with care. Companies that buy their wood, like the Gibson guitar company, promote their products as environmentally friendly. Environmental groups are working with growers of coca, bananas, coffee and other commodities to adopt similar programs. And some conservationists now say that judicious logging can actually benefit the plants and animals that live in the forest.

Signs of Improvement

Unexpected Cleanups Generate Optimism

Over the past 30 years, "sustainability" has become the mantra of many private groups, government officials, scientists and, even, a growing number of businesses. Most define the notion as advancing human endeavors without diminishing prospects for future generations. The Johannesburg summit will be the third global conclave in that span chasing this elusive goal.

But movement toward concrete action has been slow. The first meeting, in Stockholm in 1972, rang an alarm about despoiling the earth. Wealthy nations began cleaning air and water, but continued to assault forests and other resources elsewhere to fuel growth.

In 1992 came the second meeting, in Rio de Janeiro, called the Earth Summit. There, diplomats forged ambitious agreements aimed at holding back deserts and protecting the atmosphere, forests and pockets of biological richness.

But the agreements were vague, relying more on good will than on concrete obligations. Developing countries refused to take on obligations, saying the north should step first.

After Rio, population continued to grow, poverty persisted, forests retreated, soils eroded, fish stocks shrank, and concentrations of heat-trapping greenhouse gases rose, despite a treaty in which industrialized countries pledged to "strive" to reduce them.

Now, a host of satellites provides streams of data that powerful computers sift and disseminate on the Web. Communities can track forest loss in Indonesia, sprawl in Indiana and the flow of pollution from state to state, country to country.

After disasters like the chemical release in Bhopal, India, in 1984 and the grounding of the Exxon Valdez in Alaska in 1989, many companies have shifted practices to avoid environmental damage, shareholder wrath and consumer boycotts.

Fast-growing developing countries including China and Mexico are rapidly cutting urban air pollution..

They have been spurred both by community pressure and awareness of the high costs of treating illnesses caused by pollution.

Indonesia, China and other countries are posting factories' chemical emissions on the Web. The technique, pioneered in the United States, is prompting cleanups.

No one expects that people will be able to manage the planet like some giant corporation—the real Big Blue.

"If you mean making the thousands of little decisions that need to be made, we can no more effectively manage the world than the Soviet Union could manage its centrally planned economy," said Dr. Robert W. Kates, a geographer who headed a National Academy of Sciences committee on sustainable development and is an author of "Great Transition."

But Dr. Kates says the potential exists to make informed choices that spread the benefits of development to an impoverished majority while not depleting vital assets.

One impediment to such a transition is the change itself, the environmental and societal turbulence created by explosive

Look What We've Done...

Experts from the World Bank and elsewhere say the availability of environmental information can, in itself, lead to environmental improvements. Satellites are producing a wealth of data.

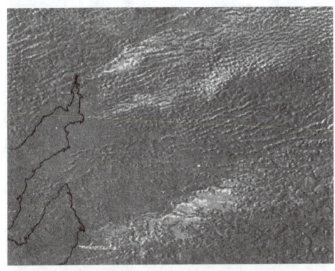

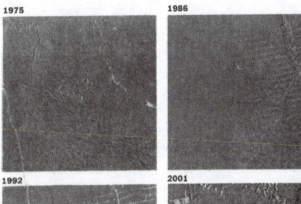

Top **POLLUTION PLUMES OVER AUSTRALIA** In this image, the clouds near Adelaide on the south coast of Australia are color coded to show their density. The light streaks indicate pollution emanating from smokestacks.

Right **DEFORESTATION IN RONDÔNIA, BRAZIL** Brazil completed the Cuiaba-Porto Velho highway in 1960 but growth in the region was slow until after the World Bank paid to pave the highway in 1980. The region, in northwest Brazil, has since grown to become a substantial farming community.

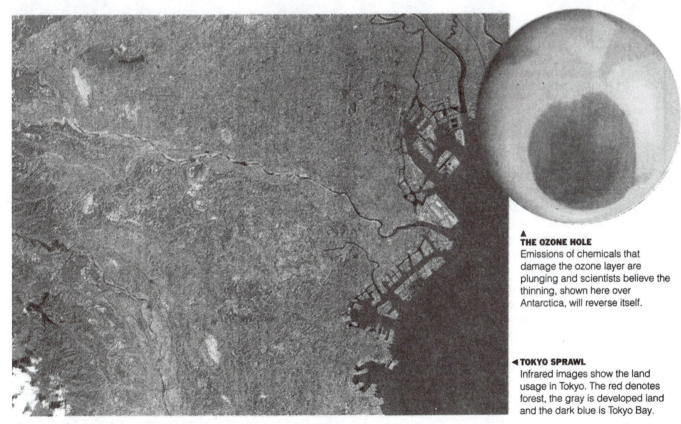

▲
THE OZONE HOLE
Emissions of chemicals that damage the ozone layer are plunging and scientists believe the thinning, shown here over Antarctica, will reverse itself.

◄ **TOKYO SPRAWL**
Infrared images show the land usage in Tokyo. The red denotes forest, the gray is developed land and the dark blue is Tokyo Bay.

continued on next page

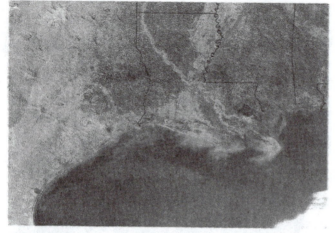

GULF SEDIMENT
- Sediment
- Phytoplankton

Agricultural runoff and other pollutants flow into the Mississippi River and into the Gulf of Mexico where they nourish phytoplankton, creating, in effect, a dead zone for fish.

INDONESIA SMOG
- Smoke
- Smog

Wildfires in Indonesia in 1997 left a plume of smoke and smog over Southeast Asia. Researchers are using satellite data to track how different types of pollution move through the atmosphere.

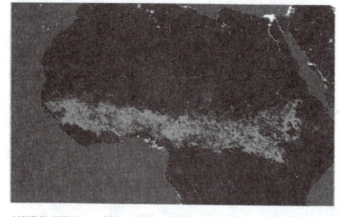

LIGHT IN AFRICA
- Gas flares
- City lights
- Fires

Using a light-sensing satellite, scientists tracked light sources over the course of six months to study land usage. The data will help researchers gauge global change in the environment and population.

SQUID FISHING
- Fishing lights
- Land
- Ocean

Fishing boats use lights to attract squid. This image shows fishing activity near Japan and Korea.

Sources: NASA; National Oceanic & Atmospheric Administration; U.S. Geological Survey; Daniel Rosenfeld, The Hebrew University of Jerusalem; Michael King, Goddard Space Flight Center; University of Colorado; Colorado State University. Images by Daniel Rosenfeld, The Hebrew University of Jerusalem (pollution over Australia); Advanced Spaceborne Thermal Emission and Reflection Radiometer (Tokyo sprawl) Landsat (deforestation in Rondônia, Brazil); Terra satellite (Gulf sediment); Total Ozone Mapping Spectrometer (ozone hole, Indonesia smog); Defense Meterological Satellite Program (lights in Africa, squid fishing)

Brett Taylor/The New York Times

human growth, technological advance and the planetwide linkup of disparate cultures, Dr. Kates and other experts say.

Another barrier, they add, is the enormous growth of population and consumption. Although global population appears headed for a 50 percent increase in the next 50 years, for example, demand for food will likely double, as prosperity raises the per capita consumption of calories.

There is another roadblock. Not every problem of consequence comes with a Bhopal-style wake-up call. Global warming and species extinction are examples of potential catastrophes that are hiding in plain sight, experts say.

Scientists are helping identify problems and opportunities. But communities will make choices guided only in part by what makes sense for the long haul.

For one thing, big gaps persist in the basic information needed to measure progress. When a team from Yale and Columbia studied dozens of trends in 142 countries to rank their sustainability, the members had to leave 40 percent of their spreadsheet blank, said Daniel C. Esty, a Yale law professor, who was a leader of the project.

Nonetheless, optimists say they see signs of hope. Not the least of them is the intensifying dialogue on the problem, which includes parties as disparate as multinational companies and tribal bands.

In essence, the human capacity for understanding the world is catching up with the human capacity to change it, Dr. Clark at Harvard said. "It really is a plausible case that we're coming on a key stage now, with the cold war under control, with globalization happening," he said.

The hard part, he said, will be for societies to overcome a habit of focusing on present needs. "Do we move beyond simply being a big bull in a china shop, having impacts, to becoming a reflective capacity on the planet?" Dr. Clark asked. "Or do we simply bungle ahead?"

An Altered World

Human Imprints From Pole to Pole

Evidence abounds now that the world is a human-dominated place.

By flooding the atmosphere with synthetic chemicals and heat-trapping carbon dioxide and other greenhouse gases, for example, people damaged the protective ozone layer and contributed to a warming climate, scientists have said. The ozone depletion became vividly and unexpectedly evident in the 1980's, when a gaping hole was detected over Antarctica.

The hole will shrink in the next 50 years because of a ban on ozone-eating chlorofluorocarbons. Other damage will not be so easy to repair.

Long before they are cataloged, thousands of plant and animal species are likely to be driven to extinction as forests, wetlands, mountain slopes and other habitats are exploited or harmed by climate change.

Satellites that map vegetation and the nighttime signature of human activity—fire and light—show that people have altered more than one-third of the terrestrial landscape. Once it is changed, it is usually changed forever, Dr. G. David Tilman, an ecologist at the University of Minnesota, said. "When you add another 1,000 acres of shopping mall or another highway, far into the future those are probably close to permanent acts," Dr. Tilman said.

Where progress is seen, too often it is only in a slowing rate of destruction, ecologists say. For example, new satellite surveys show that forest loss in the tropics through the 1990's occurred at a rate 23 percent less than previous estimates. But losses still add up to some 14 million acres a year, with 5 million more acres visibly damaged.

The human imprint is evident almost everywhere. In the South Atlantic, fleets illuminate so much of the ocean to attract squid that the illuminated area dwarfs the megalopolis of São Paulo. The squid harvest has in part grown because commercial fish stocks have been overfished.

Altogether, scientists have found that two-thirds of commercial marine fish species are fully exploited or diminishing, prompting companies to move down the food chain.

Aquaculture is a fast-growing alternative, but often causes damage like the destruction of coastal mangroves in southern Asia to make way for shrimp farms.

Also, in many cases, farmed species are fed fish meal made of other fish. So the cultivation still indirectly depletes the oceans.

Hydrologists estimate that people appropriate half the world's flowing fresh water. Across the American West in the last 20 years, circular patches made by great rotating irrigation rigs have peppered the land like an expanding checkerboard, marking the draining of aquifers under the plains.

Scientists have concluded that humans not only now dominate the planet, but have also become the dominant driver of natural selection, the machinery of evolution.

The main influence, experts say, is the continuing chemical arms race against germs and pests, which kills most, but leaves a resistant minority behind.

Also, by wittingly and unwittingly moving myriad species around the globe, humans have become a biological blender, carrying West Nile virus to America and overrunning the Bordeaux countryside in France with American bullfrogs that residents say do not even taste good.

Troubling Trends

An S.U.V. Culture Shifts to Third World

Projections for the next two generations do not bode well for easing environmental problems. Even with the population bomb predicted in the 1960's substantially defused, the human population is likely headed for at least nine billion before leveling off.

Most of the growth will be in poor countries, and as people there pursue prosperity, consumption of natural resources will rapidly increase.

Half the world's 17,000 major wildlife refuges are already being heavily used for agriculture, recent studies showed.

Car companies are racing to build factories to assemble sport utility vehicles in India, even as its once-legendary rail system, plagued by mismanagement, is deteriorating and losing freight and passengers, said Dr. Rajendra K. Pachauri, director general of the Tata Energy Research Institute of India. The private organization assesses energy and environmental problems in India.

"In the last 10 years, every major manufacturer has set up facilities in India," Dr. Pachauri said. "They see it as a major market. They have the buzz of the people. This is something that should cause real concern."

Depending on how power is generated, bringing electricity to the two billion people in the world who still lack it could greatly increase emissions of greenhouse gases. But, experts note, the options available to those who remain off the grid also cause harm. The two billion people who cook on wood or dung fires live in acrid clouds of toxic smoke and deplete forests.

In India, Dr. Pachauri said, millions of people light their homes with kerosene, using government-subsidized fuel. Together, cooking fires and sputtering lanterns create indoor pollution that causes asthma and other ailments and that, in India alone, is estimated to kill 600,000 women a year.

Dr. Pachauri's group has experimented with distributing solar-powered lanterns to rural communities. Other projects push cleaner ovens that use less-polluting fuels.

But the question of how to take good ideas from pilot projects to the new norm remains largely unsolved.

Then there are the costs. For example, about a billion people have no clean water. More than twice that number live where raw sewage flows unchecked. So water is a prime focus of the delegates in Johannesburg.

When even countries like the United States lag tens of billions of dollars in improving their sewage systems, experts say the prospects of big investments elsewhere are dim.

New Strategies

Learning to Harvest More From Less

The hardest part of meshing economic and environmental progress, experts say, is that this shift cannot be engineered with top-down directives.

It will be a result of 10,000 decisions, large and small, by countries, communities, companies and individuals, said Dr. Kates, the geographer.

Action will have to be focused where the human imprint is most intense, in forests, on the farm and in the fast-expanding cities, experts say.

An analysis by the World Wildlife Fund found that 20 percent of the existing forest area could provide all the world's future needs for wood and pulp if it was all managed according to environmentally sound practices that a few big companies have already adopted.

Some forestry companies and wood and paper buyers, including Home Depot and Ikea, participate in a program in which wood is certified by the Forest Stewardship Council, a private group. The council monitors forest holdings and products to ensure that wood marketed as environmentally friendly is produced with limited damage.

Other organizations run similar certification programs to encourage growers of other crops like bananas and cocoa to preserve habitat or limit pesticide use.

Those efforts remain a tiny fringe of the markets for those commodities. The Forest Stewardship Council has certified 70 million acres of forests—just 4 percent of the total acreage controlled by timber companies.

If farming does not change drastically in the next few decades, enormous ecological damage will result, many scientists say.

Dr. Tilman of Minnesota notes that farming has already produced the biggest global imprint of humanity, affecting half the earth's habitable land. The challenge now, he said, will be to double agricultural productivity without using substantially more land.

There are signs this can be achieved. In an extraordinary experiment several years ago, in Yunnan Province, China, rice farmers were recruited to intersperse two varieties on 8,000 acres instead of planting one. The yield nearly doubled, and the occurrence of rice blast, the most harmful disease in the world's biggest crop, fell 94 percent.

Recognizing a good thing, China has expanded the work to 250,000 acres, said Dr. Christopher C. Mundt, a plant pathologist at Oregon State University who helped conduct the research.

"Perhaps more important," he added, "the Chinese are taking the general concept of diversification into related approaches" to other crops.

The key to the next green revolution, Dr. Mundt said, will be abandoning most of the industrial model of agriculture of the 20th century and shifting to a "biological model based on management of ecological processes" like applying fertilizer and water only as needed.

Such efforts are being made in agriculture, forestry, fisheries and other fields around the world. But, once again, experts say the challenge lies in moving to the scales needed to avert widespread harm.

The Role of the City

The Megalopolis As Eco-Strategy

Another focal point for experts who envision a managed earth is cities. In many ways, they are where the battle will be won or lost.

Cities are where almost all remaining population growth will occur, demographers say. The roster of megacities, those with populations exceeding 10 million, is widely expected to climb, from 20 today to 36 by 2015.

These vast metropolises have been widely characterized as a nightmarish element of the new century, sprawling and chaotic and spawning waste and illness.

But increasingly, demographers and other experts say that cities may actually be a critical means of limiting environmental damage. Most significantly, they say, family size drops sharply in urban areas.

"The city is perhaps the most effective device for reducing the birthrate," said Dr. George Bugliarello, chancellor of the Polytechnic University in Brooklyn and an expert on urban trends.

For the poor, access to health care, schools and other basic services is generally greater in the city than in the countryside. Energy is used more efficiently, and drinking and wastewater systems, although lacking now, can be built relatively easily.

And for every person who moves to a city, that is one person fewer chopping firewood or poaching game.

Still, many cities face decisions now that may permanently alter the quality of human lives and the environment.

Dr. Kates said the pivotal nature of these times is perfectly illustrated by Mexico City, which is just behind Tokyo atop the list of megacities. The sprawling megalopolis, where traffic is paralyzed, is about to choose in a referendum between double-decking its downtown highways or expanding its subway system.

One course could encourage sprawl and pollution; the other would conserve energy, experts say.

Chances of Change

'Pernicious Fad' Or Real Prospect?

Some environmentalists say the whole notion of sustainable development is an oxymoron, that the Western industrial model of endless growth, however packaged, cannot possibly persist without grievous environmental damage. At the same time, some business leaders still scoff at the effort.

In a new PricewaterhouseCoopers survey of Fortune 1000 executives' attitudes toward sustainable development, one corporate vice president for environmental affairs called the concept "a pernicious fad."

Some skeptics note that even if cleaner, less destructive industries take hold everywhere, the sheer volume of economic growth could still cause big problems.

But a durable line of pragmatic optimists from science, business and environmental groups, for now, holds center stage.

They say that cleaning the environment and reducing poverty are not only required from an ethical standpoint, but also because they are in humanity's self-interest.

Without improvements, said Nitin Desai, the United Nations official running the Johannesburg meeting, "you create societies that live in a state of perpetual hopelessness."

"That is obviously going to hit back at you at some stage," Mr. Desai said.

Hundreds of businesses, though still a minority, have added sustainability managers to their executive roster.

Prof. Jeffrey D. Sachs, director of the Columbia University Earth Institute, said the world was quickly shifting from a model in which wealth was derived mainly through exploiting resources.

"Most growth now comes from increased knowledge, not from the mining of nature," Professor Sachs said. "And knowledge isn't limited in the way that, say, soil fertility is."

Dr. Kates agrees that economic advancement is vital and says evidence is emerging in many places that it can occur without too much environmental harm. Despite federal inaction on climate change, Dr. Kates said, 129 American cities have programs to cut greenhouse-gas emissions, and California has moved to reduce car emissions.

"This fits with history," he said. "States and cities have always been a major set of social experimenters. This was true on disability insurance and child-labor laws, on antimonopoly laws."

It may well end up being the case that local communities, here and abroad, lead the way in harmonizing people and the planet, he said.

"That ferment," Dr. Kates said, "is the most encouraging sign."

Crimes of (a) Global Nature

**FORGING ENVIRONMENTAL TREATIES
IS DIFFICULT.
ENFORCING THEM IS EVEN TOUGHER.**

by Lisa Mastny and Hilary French

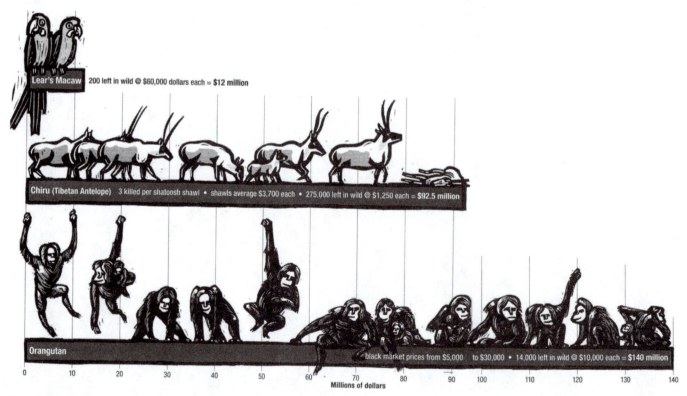

Lear's Macaw   200 left in wild @ $60,000 dollars each = $12 million

Chiru (Tibetan Antelope)   3 killed per shatoosh shawl • shawls average $3,700 each • 275,000 left in wild @ $1,250 each = $92.5 million

Orangutan   black market prices from $5,000   to $30,000 • 14,000 left in wild @ $10,000 each = $140 million

0   10   20   30   40   50   60   70   80   90   100   110   120   130   140
                                              Millions of dollars

Extinction's Payoff
Pets, aphrodisiacs, distinctive clothing: these are a few of our favorite things—even if having them drives a species or two to extinction. The staggering prices some threatened animals fetch on the black market create powerful incentives for illegal trafficking and help increase the risk of extinction. The prices in the graph are probably conservative, since prices would tend to rise with increasing scarcity.

Last February, armed troops and fisheries officials on two Australian navy ships and a helicopter boarded and seized the Volga and the Lena, Russian-flagged fishing vessels operating near Heard Island, some 2,200 nautical miles southwest of Perth. The two ships were found to be carrying about 200 tons of illegally caught Patagonian toothfish in their holds. This bounty, valued at an estimated $1.25 million, had been taken in violation of conservation agreements negotiated under the auspices of the Commission for the Conservation of Antarctic Marine Living Resources.

Few casual seafood lovers have heard of Patagonian toothfish, but many are familiar with Chilean sea bass, a different name for the same fish. Chilean sea bass began appearing on menus in the early 1990s, and consumption of the flaky, white fish took off fast, quickly endangering the health of the fishery. Large-scale commercial fishing of the species began only a decade ago, but scientists estimate that at current rates of plunder the fish could become commercially extinct in less than five years.

A few months after the drama in the Southern Ocean, a different front in the same battle opened up thousands of miles away in Washington, D.C., where nearly 60 restaurants and caterers pledged to keep the fish off their menus. More than 90 restaurants in the Los Angeles area did the same a few weeks later, following similar promises by chefs in Northern California, Chicago, and Houston. Thus the fight to save the Patagonian toothfish is beginning to hit close to home. But many Chilean sea bass fans remain unaware that they may be accomplices to a growing phenomenon known as international environmental crime.

Although variously defined, in this article international environmental crime means an activity that violates the letter or the spirit of an international environmental treaty and the implementing national legislation. Trade in endangered species, illegal fishing, CFC smuggling, and the illicit dumping of wastes are all cases in point that are explored below. Illegal logging is another major category of environmental crime, although environmental treaties currently impose few specific constraints on logging (see Box, *Logging Illogic*). The rapidly growing illegal trade in these environmentally sensitive products stems from strong demand, low risk, and other factors (see Box, *Variable Crimes, Constant Incentives*).

The number of international environmental accords has exploded as countries awaken to the seriousness of transboundary and global ecological threats. The UN Environment Programme (UNEP) estimates that there are now more than 500 international treaties and other agreements related to the environment, more than 300 of them negotiated in the last 30 years.

But reaching such agreements is only the first step. The larger challenge is seeing that the ideals expressed in them become reality. What is needed is not necessarily more agreements, but a commitment to breathe life into the hundreds of existing accords by implementing and enforcing them.

Here the genteel world of diplomacy often runs into hard-nosed domestic politics. Countries that ratify treaties are responsible for upholding them by enacting and enforcing the necessary domestic laws. This requires the backing of businesses, consumers, and other constituencies, which may not be easily secured. Countries with strong fossil fuel industries, for instance, may meet staunch resistance to international rules to mitigate climate change. And countries where natural resource industries are politically powerful will probably find it difficult to adequately enforce environmental treaties designed to regulate resource-related activity. The effect has been to expand trafficking in a number of restricted substances, an increasingly urgent problem that is beginning to stimulate a stronger international response.

Trading in Wildlife

Undercover Russian police officers in the port city of Vladivostok recently trailed two investigators from environmental groups posing as eager purchasers of Siberian tiger skins from a corrupt official. When the deal went down, the officers arrested the wildlife trader on the spot. Russian investigators earlier had infiltrated the wildlife trade crime ring and determined that it was raking in some $5 million a year from smuggling wild ginseng, tiger skins, and bear paws and gallbladders across the Russian border.

Trade in these wildlife products is restricted under the terms of the 1973 Convention on International Trade in Endangered Species of Wild Fauna and Flora (CITES), which bans international trade in some 900 animal and plant species in danger of extinction, including all tigers, great apes, and sea turtles, and many species of elephants, orchids, and crocodiles. CITES also restricts trade in some 29,000 additional species that are threatened by commerce; among them birdwing butterflies, parrots, black and stony corals, and some hummingbirds.

CITES has shrunk the trade in many threatened species, including cheetahs, chimpanzees, crocodiles, and elephants. But the trafficking in these and other species continues, earning smugglers profits of $8 billion to $12 billion annually. Among the most coveted black market items are tigers and other large cats, rhinos, reptiles, rare birds, and botanical specimens.

Most illegally traded wildlife originates in developing countries, home to most of the world's biological diversity. Brazil alone supplies some 10 percent of the global black market, and its nonprofit wildlife-trade monitoring body, RENCTAS, estimates that poachers steal some 38 million animals a year from the country's Amazon forests, Pantanal wetlands, and other important habitats, generating annual revenues of $1 billion. Southeast Asian wildlife has also been plundered: the Gibbon Foundation reports that in a single recent year, traders smuggled out some 2,000 orangutans from Indonesia—at an average street price of $10,000 apiece.

The demand for illegal wildlife—for food and medicine, as clothing and ornamentation, for display in zoo collections and horticulture, and as pets—comes primarily from wealthy collectors and other consumers in Europe, North America, Asia, and the Middle East. In the United States (the world's largest market for reptile trafficking) exotic pets such as the Komodo dragon of Indonesia, the plowshare tortoise of northeast Madagascar, and the tuatara (a small lizard-like reptile from New Zealand) reportedly sell for as much as $30,000 each on the black market. Wildlife smuggling is also a growing concern in the United Kingdom, where in 1999 alone, customs officials confiscated some 1,600 live animal and birds, 1,800 plants, 52,000 parts and derivatives of endangered species, and 388,000 grams of smuggled caviar.

The multibillion-dollar traditional Asian medicine industry has also been a strong source of demand for illegal wildlife, with adherents from Beijing to New York purchasing potions made from ground tiger bone, rhino horn, and other wildlife derivatives for their alleged effect on ailments from impotence to asthma. In parts of Asia, bile from bear gall bladders (used to

VARIABLE CRIMES, CONSTANT INCENTIVES

International environmental crime involves many different kinds of activities and contraband, ranging from illegal fishing and trading in endangered wildlife to smuggling ozone-depleting chlorofluorocarbons (CFCs) across borders. But in most cases there are several common elements:

• **Booming demand, minimal investment, high profits**. Such crime is generally very attractive to traders and smugglers. Demand for increasingly rare plants, fish, or CFCs is strong. Investment can be minimal and the rewards great, with prices climbing substantially as the items change hands. For instance, a Senegalese wholesaler may buy a grey parrot in Gabon for $16–20 and sell it to a European wholesaler for $300–360, who then gets $600–1,200 for it.

• **Low risk, low penalties**. Environmental crimes normally carry low risk, with domestic penalties often light or nonexistent. A 2000 study by the secretariat of the Convention on International Trade in Endangered Species of Wild Fauna and Flora found that roughly half of the treaty's 158 parties failed to implement the treaty adequately. A common deficiency is a lack of appropriate penalties to deter treaty violations.

• **Creative smuggling methods**. Smugglers have devised clever ways to evade controls on restricted items, from using unauthorized crossing points to shipping illegal items along with legal consignments.

• **Links to organized crime**. Environmental criminals sometimes launder or funnel the proceeds from their trafficking to other illicit activities, such as buying drugs or weapons. The National Network for Combating the Traffic of Wild Animals (RENCTAS), a Brazilian NGO, reports that as much as 40 percent of the 300 to 400 criminal groups that control that country's wildlife trade also have links with drug trafficking.

• **Weak domestic treaty enforcement**. Because treaty secretariats have little centralized authority, signatory countries are expected to designate their own permitting authorities and train local customs inspectors and police to detect and penalize illegal activity. But many countries simply don't have the resources to do this.

• **Porous borders**. Many countries lack the money, equipment, or political will to effectively monitor illegal activity at border crossings. In the developing world in particular, customs offices are chronically understaffed and underfunded. Agents are often untrained to spot crimes or overwhelmed by the sheer numbers of items they must track.

• **Permitting challenges**. Local authorities may unwittingly issue permits for animals, seafood, chemicals, or other items they don't know are restricted, or let consignments slip through because they can't identify a fake or altered permit. Corrupt local officials may issue false permits or overlook illicit consignments in return for bribes or kickbacks.

Gold
$4,672 per pound

Tuatara
$13,636 per pound

Scarcity=Value
The tuatara has a third eye on top of its skull and can live to be 100 years old. Its cousins in the reptilian order Rhynchocephalia all died out 139 million years ago. These oddities may help explain why collectors will pay up to $30,000 apiece. At least one species is nominally protected under CITES.

treat cancers, asthma, eye disease, and other afflictions) can be worth more than narcotics. The New York-based Wildlife Conservation Society reports that the illegal hunting and trafficking of animals for medicine, aphrodisiacs, and gourmet food is now the single greatest threat to endangered species in Asia.

The scale of the illegal wildlife trade reflects the serious obstacles to enforcement under CITES. Smugglers may conceal the items on their persons or in vehicles, baggage, or postal and courier shipments, resulting in fatality rates as high as 90 percent for many live species. They may also alter the required CITES permits to indicate a different quantity, type, or destination of species, or to change the appearance of items so they appear ordinary. In February 2000, the U.S. Fish and Wildlife Service arrested a Cote d'Ivoire man for smuggling 72 elephant ivory carvings, valued at $200,000, through New York's John F. Kennedy airport. Many of them had been painted to resemble stone.

As wildlife smuggling grows in sophistication, it accounts for a rapidly rising share of international criminal activity. RENCTAS now ranks the illicit wildlife trade as the third largest illegal cross-border activity, after the arms and drug trades. Wildlife smugglers commonly rely on the same international trafficking routes as dealers of other contraband goods, such as gems and drugs. In the United States, consignments of snakes have been found stuffed with cocaine, and illegally traded turtles have entered on the same boats as marijuana. The U.S.-Mexico border has become a significant transfer point for environmental contraband: in the first eight months of 2001, Mexican authorities reported seizing more than 50,000 smuggled animals en route to the U.S. border.

As the scope of international wildlife trafficking becomes clearer, authorities are beginning to take action on the domestic, regional, and global fronts. In Europe, officials are working together to improve regional cooperation in enforcement of CITES and to strengthen legal mechanisms for prosecuting violators. The United Kingdom has established a new national police unit to combat wildlife crime and announced tougher penalties for persistent offenders. And in the first program of its kind in East Asia, a team at South Korea's Seoul airport now uses specially trained dogs to detect tiger bone, musk, bear gall bladders, and other illegal wildlife derivatives smuggled in luggage and freight.

In general, developing countries have had greater difficulty in controlling wildlife smuggling and implementing CITES. Many lack the political will, money, or equipment to effectively monitor their wildlife populations, much less the wildlife trade. In Kenya, where elephant deaths declined dramatically following a CITES-imposed ban on the international ivory trade in the late 1980s, poaching is again surging as funds to hire additional game wardens have run dry. In two separate incidents in April 2002, poachers armed with automatic weapons slaughtered 25 elephants in the country's wildlife parks and removed the tusks for sale on the black market.

At the international level, one tool that has proved successful in enticing (some would say coercing) countries to uphold their obligations under CITES is the use of trade sanctions. CITES is empowered to recommend that its members temporarily suspend all wildlife trade with noncomplying countries. Within the past year, such sanctions have been levied against the United Arab Emirates for not taking strong enough measures to combat the illicit trade in falcons, against Russia for not cracking down on the illegal caviar trade, and against Fiji and Vietnam for not enacting adequate wildlife trade legislation by the required deadline. In most instances, the governments scrambled to strengthen their legislation and enforcement, and the sanctions were soon lifted.

High Seas, High Crimes

Although CITES focuses mainly on terrestrial animals and plants, it protects several highly endangered fish species as well. In February 2002, the Philippine navy arrested 95 Chinese fishermen near a national marine park in the Sulu Sea and charged them with multiple counts of poaching fish, harvesting endangered species, and using illegal fishing methods like poison and explosives. Among the species found aboard their four vessels were endangered sea turtles and giant clams, both of which are prohibited from trade under CITES.

Countries have negotiated a wide range of other agreements to oversee and regulate the world's fisheries. Like CITES, many of these are poorly enforced, resulting in illegal fishing in all types of fisheries and in all the world's oceans, including national waters, regionally managed fisheries, and the high seas. In some of the world's most important fisheries, as much as 30 percent of the catch is illegal, according to the UN Food and Agriculture Organization (FAO).

As with other forms of wildlife trade, booming consumer demand is an important driver behind the rise in this activity. Big profits can be made selling black market seafood to selective buyers, who are willing to pay a premium for increasingly rare items. In Japan, species such as the threatened Southern bluefin tuna now fetch up to $50,000 per fish.

One major form of illegal fishing occurs when foreign vessels fish without authorization in the waters of other countries, often developing nations unable to patrol their shores adequately. In early 2000, the Tanzanian government estimated that more than 70 vessels, most of them from Mediterranean countries and the Far East, were fishing illegally in its waters. Tanzania has few police boats of its own and has had to rely on assistance from France and the United Kingdom to crack down on offenders. Mozambique, Somalia, and other African countries also report increased illegal fishing, often by heavily armed foreign vessels, and worry that the continued poaching will damage national economies and deprive coastal villagers of the healthy fisheries they depend upon for their livelihoods.

Commercial fishers are also turning to the largely unmonitored high seas, including the Mediterranean Sea and the Indian, South Atlantic, and Southern Oceans. Their ships illegally penetrate the borders of regional fishing grounds that are closed or restricted under international law, and deliberately hide their flags or other markings to avoid recognition. Some offload their illicit catch to other vessels to further disguise its origin and to minimize the penalties if discovered. Often these fishers do not report their activity, or if they do, they may falsify the equipment used, the fishing area frequented, or the species or amount caught.

TRAFFIC, a UK-based nonprofit group that monitors the international wildlife trade, reports that illegal Russian trawlers are accelerating the collapse of once productive fisheries in the Bering Sea. Backed by the Russian mafia, the fishers remove billions of dollars worth of pollack, cod, herring, flounder, halibut, and other species from the ecosystem each year, often trawling in prohibited areas and using illegal nets and other gear. They then transfer the catches to ships bound for ports in the United States, Canada, and Asia, in particular the South Korean port of Pusan, where vessel inspections are rare.

One of the most serious challenges to adequate fisheries enforcement is the rapid rise in so-called flag-of-convenience (FOC) fishing. Increasingly, commercial fishing companies register their ships in countries known to be lax enforcers of international fisheries laws or that are not members of major maritime agreements. By transferring their allegiance to these new "flags," the companies can easily enter the waters of their adopted countries or operate undercover on the high seas where only the country of registry (the flag state) can make an official arrest. Some vessels change their flags frequently in order to hide their origins or identities, making it difficult for enforcement officials to track them down.

An estimated 5 to 10 percent of the vessels in the world fishing fleet now fly flags of convenience. This includes more than 1,300 large industrial fishing vessels registered in such countries as Belize, Honduras, Panama, St. Vincent, and Equatorial Guinea.

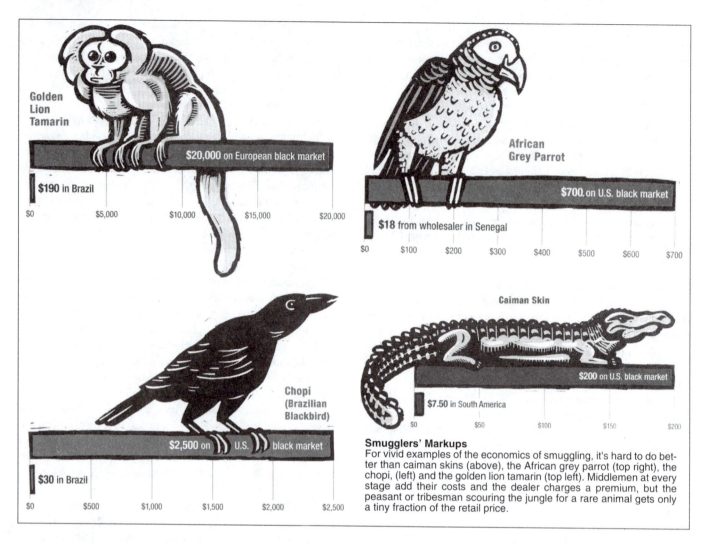

Golden Lion Tamarin
$20,000 on European black market
$190 in Brazil
$0 $5,000 $10,000 $15,000 $20,000

African Grey Parrot
$700 on U.S. black market
$18 from wholesaler in Senegal
$0 $100 $200 $300 $400 $500 $600 $700

Chopi (Brazilian Blackbird)
$2,500 on U.S. black market
$30 in Brazil
$0 $500 $1,000 $1,500 $2,000 $2,500

Caiman Skin
$200 on U.S. black market
$7.50 in South America
$0 $50 $100 $150 $200

Smugglers' Markups
For vivid examples of the economics of smuggling, it's hard to do better than caiman skins (above), the African grey parrot (top right), the chopi, (left) and the golden lion tamarin (top left). Middlemen at every stage add their costs and the dealer charges a premium, but the peasant or tribesman scouring the jungle for a rare animal gets only a tiny fraction of the retail price.

The countries and companies that own the reflagged vessels are also partially to blame for the problem. Fishing companies from Europe (primarily Spain) and Taiwan own the highest numbers of FOC vessels. Many of these firms receive special government subsidies to register their ships abroad.

FOC vessels also pose a serious problem in the cruise and shipping industries. Freight companies may re-flag their vessels in order to avoid higher taxes, labor, and operating costs at home, or to take advantage of the lower safety standards and weaker pollution laws in countries like Liberia, Panama, and Malta. The International Transport Workers' Federation reports that FOC ships accounted for nearly a quarter of all large freight vessels in 2000, and carried 53 percent of the world's gross tonnage. The re-flagged ships account for a disproportionate share of pollution and accidents at sea, as well as detentions in ports for violations of maritime laws.

Some of the most serious offenses have been breaches of MARPOL, the international treaty regulating the disposal of garbage and other pollutants at sea. Vessel owners may decide to dump their waste overboard illegally either because they lack the storage or incineration capacity onboard or because they wish to avoid the high costs of eventual disposal on land. Often this at-sea discharge occurs in the waters of countries where monitoring and enforcement are minimal and the activity can go undetected. In 1999, however, U.S. officials fined Royal Caribbean Cruises Ltd. a record $18 million for releasing oil waste from its ships into U.S. waters, among other violations. In April 2002, rival Carnival Corporation was also fined $18 million for similar offenses.

As the scope of maritime violations becomes increasingly evident, many countries are redoubling their efforts to combat illegal fishing, including by exchanging information about illicit activity, making vessel registries more transparent, increasing inspections at sea and in ports, denying landing and trans-shipment rights for illegally caught fish, and improving monitoring and surveillance of fleets. These steps appear to be working, at least in some places: in the North Pacific, Canadian officials attribute the recent drop in the number of vessels using illegal driftnets to stepped-up air patrols by Canada and its partners (the United States, Russia, and China) in regional enforcement.

Significant effort has also been made at the global level to crack down on illegal fishing. In March of last year, 114 countries agreed to a non-binding plan to combat illicit activity, including that by flag-of-convenience vessels. Developed under the auspices of the UN Food and Agriculture Organization, the plan calls for improved oversight of vessels and coastal waters by flag states, better coordination and sharing of information

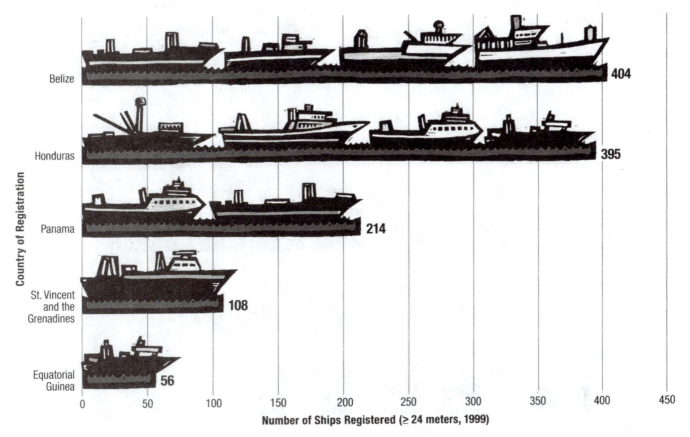

Country of Registration

Belize **404**

Honduras **395**

Panama **214**

St. Vincent and the Grenadines **108**

Equatorial Guinea **56**

0 50 100 150 200 250 300 350 400 450

Number of Ships Registered (≥ 24 meters, 1999)

False Colors
Treaties to control overfishing and other marine crimes generally require the nation where a ship is registered to do the policing. Commercial fishers often register their ships with countries unable, or deliberately reluctant, to enforce the treaties. Such flag-of-convenience (FOC) vessels can operate with little risk of prosecution, which leads to overfishing and depletion of fish stocks. As much as one-tenth of the world's fishing vessels fly flags of convenience. The top five FOC registration countries are shown here, and the opposite page shows the top five countries where firms owning the vessels are based.

among countries, and stronger efforts to ratify, implement, and enforce existing fisheries accords. The plan is expected to work in conjunction with other international efforts to protect global fisheries, including a 1993 pact known as the Compliance Agreement and the 1995 UN Agreement Relating to the Conservation and Management of Straddling Fish Stocks and Highly Migratory Fish Stocks.

Dumping on Land

A Japanese district court last March sentenced Hiromi Ito, the president of a waste disposal company, to four years in prison and slapped him with a 5 million yen ($40,000) fine. Ito and his accomplices had masterminded an elaborate scheme to illegally dump some 2,700 tons of industrial and medical waste in the Philippines, in containers marked 'paper for recycling.' When the Filipino importer opened the 122 shipping containers, each 40 feet long, he found not just paper but also hazardous materials, including contaminated hypodermic needles and bandages, used plastic sheeting, and old electronics equipment.

With Ito's conviction, the Japanese government closed the books on an embarrassing international incident. But the episode points to a much larger global challenge: effectively regulating the 300 to 500 million tons of hazardous waste generated

worldwide each year. This vast waste mountain includes everything from used batteries, electronic wastes, old ships, and toxic incinerator ash to industrial sludge and contaminated medical and military equipment.

Roughly 10 percent of this waste is shipped legally across international borders, under the terms of the 1989 Basel Convention on the Control of Transboundary Movements of Hazardous Wastes and Their Disposal. The agreement requires member countries to obtain prior consent from the importing country for waste shipments and uses a system of permits to track the pathway to disposal. Most of the waste originates in and moves among industrial countries, but some also travels to and within the developing world.

As waste disposal problems mount in countries like Japan, so does the likelihood of a lucrative illegal trade in hazardous wastes. Asia is thought to be one of the biggest destinations for the illicit waste. Greenpeace India reports that more than 100,000 tons of unauthorized wastes entered India in 1998 and 1999, including toxic zinc ash and residues, lead waste, used batteries, and scraps of chromium, cadmium, thallium, and other heavy metals. The illicit imports, originating in places like Australia, Belgium, Germany, Norway, and the United States, violated both the Basel treaty's notification rules and a 1997 Indian government ban on waste imports.

31

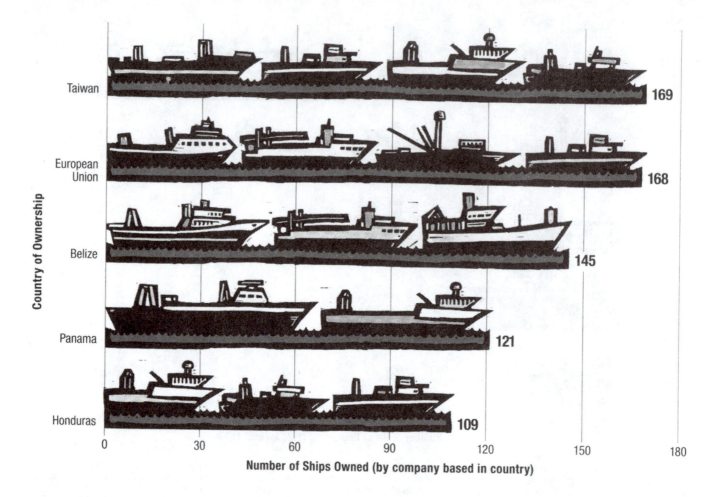

Taiwan **169**

European Union **168**

Belize **145**

Panama **121**

Honduras **109**

Country of Ownership

0   30   60   90   120   150   180

Number of Ships Owned (by company based in country)

China receives a massive flow of illegally imported hazardous electronic waste each year. A recent study by the Basel Action Network (BAN), a watchdog group that monitors implementation of the Basel treaty, and the California-based Silicon Valley Toxics Coalition reported that workers in Chinese recycling factories risk serious exposure to heavy metals and other poisonous chemicals as they salvage components from old computer circuitboards, monitors, batteries, and other equipment. This toxic trade continues despite a Chinese import ban on the material, and hence violates Basel rules that forbid the export of wastes to countries that have banned their import.

Roughly half of this "e-waste" originates in the United States, where an estimated 20 million computers become obsolete each year. Yet the U.S. government doesn't consider the high-tech shipments to be illegal because the waste isn't technically classified as hazardous. The United States is also the only industrial country that hasn't ratified the Basel Convention.

Developing countries are particularly vulnerable to the health and environmental effects associated with the illegal waste trade. Many governments lack the infrastructure or equipment to dispose of or recycle waste safely, prevent exposure to workers and communities, clean up dumpsites, or monitor waste movements. For this reason, a bloc of developing countries secured passage in 1995 of a far-reaching amendment to the Basel Convention, known as the Basel Ban, which would outlaw all transfers of hazardous wastes from industrial countries to the developing world. Though many countries already observe its terms voluntarily and the European Union has already implemented it, the Ban is not yet in strict legal force and still faces serious opposition from a few industrialized countries, including the United States.

Even so, the Basel Ban has helped slow the flow of hazardous wastes from industrial countries to the developing world, says Jim Puckett, coordinator of BAN. And once the amendment enters into legal force (it needs 32 more ratifications), it should be even harder for waste traders to operate, as violators would face strict criminal penalties. If the ban fails to enter into force, however, or if it is weakened through continued opposition, the waste flood could resume.

But the ban alone would likely fail to wipe out the illegal waste trade. Smugglers rely on a wide range of tactics, including false permits, bribes, and mislabeling of wastes as raw materials, less dangerous substances, or other products, to evade the laws. Moreover, no port in the world can check all sea-going containers, let alone developing-country ports.

One growing trend is the export of hazardous wastes under the pretext of "recycling." Like Hiromi Ito, illicit waste dealers increasingly pass off waste as recyclable material, which in many cases frees them from strict government oversight. Greenpeace estimates that as much as 90 percent of waste shipped to the developing world—particularly plastics and

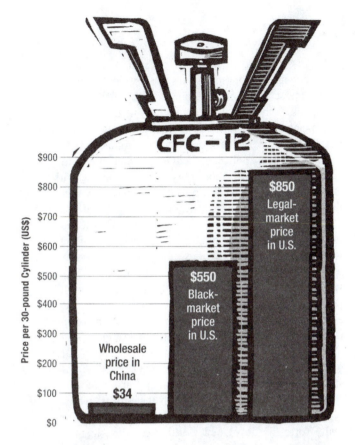

The Price of Cool
The Montreal Protocol ended CFC manufacture in the industrialized world, but developing nations have until 2010 to phase it out. Some of their output feeds illicit demand in the United States and Europe.

countries by agreeing to set up a centralized system for reporting suspect activity. The treaty also requires members to pass laws to curb and punish illegal waste traffic and outlines how to handle illegally traded waste once it is discovered.

A new liability protocol to the convention, negotiated in December 1999, could further discourage illegal activity—if it ever comes into force. It makes exporters and disposers of hazardous waste liable for any harm that might occur during transport, both legal and illegal. It also requires dealers to be insured against the damage and to provide financial compensation to those affected. But the protocol will not be legally binding until 20 ratifications are received (none have been registered so far). Moreover, it still only covers harm that occurs in transit, not after disposal, and only applies to damage suffered in the jurisdiction of a treaty member.

CFCs on the Loose

The landmark Montreal Protocol on Substances That Deplete the Ozone Layer, adopted in 1987, mandated far-reaching restrictions in the use of certain chemicals that damage the thin, vital veil of stratospheric ozone that protects the earth and its inhabitants from excessive ultraviolet radiation. The Protocol and its later amendments set target dates for the phaseout of 96 different ozone-depleting substances, most notably CFCs and halons, chemicals once widely used in a range of industrial applications. Industrialized countries were required to halt production and import of CFCs in 1996, while developing countries have until 2010 to complete the phaseout.

Considered one of the world's most successful environmental treaties, the Protocol has resulted in a dramatic decline in the overall use of ozone-depleting substances. But starting in the mid-1990, the different phaseout schedules for industrial and developing countries helped stoke a flourishing illegal trade in the banned chemicals. CFCs that were still legally produced in the developing world began to make their way to lucrative black markets in the United States and Europe, where demand for substances like Freon (used in older-model auto air conditioners) remained high.

Government and industry reports suggest that in the mid-1990s, as much as 15 percent of global annual production of the chemicals, or 38,000 tons, was smuggled into industrial countries. Early shipments originated in Russia, but today the bulk of the smuggled CFC supply is thought to originate in China and India, which together account for more than half of the world's remaining CFC production.

CFC importers resort to fraud and other evasive tactics to smuggle the banned chemicals. For instance, traders abuse existing loopholes in the Montreal Protocol and domestic laws to pass off shipments of new CFCs as recycled material or as CFC replacements, neither of which are restricted under the treaty. The chemicals are typically colorless and odorless, making them easy to disguise and virtually impossible to differentiate without chemical analysis.

In the United States, the illegal CFC trade is believed to have peaked in the mid-1990s, just after the initial phaseout. At that

heavy metals—is now labeled as destined for recycling. Much of this waste, however, is never recycled—or, as with e-waste in China, is recycled in highly polluting operations that are little better than dumping. Typically, the end result is the same: the export of a serious pollution problem from a rich country to a poor one.

In many cases, organized crime is thought to be behind large-scale waste trading, which can be closely linked with money laundering, the illegal arms trade, and other criminal activities. The Italian mafia is reportedly a key player in the robust trade in radioactive metal waste from Eastern Europe and the former Soviet Union, which it re-sells to smelters as "safe" scrap metal.

Efforts to combat the illegal waste trade face serious challenges. A recent survey by the Basel secretariat revealed that many countries lack adequate—or any—legislation for preventing and punishing illegal waste traffic. On a global scale, the absence of uniform definitions of hazardous waste and of coordinated enforcement efforts among customs officers and port authorities has contributed to the spread of illegal waste trading. Even for countries that do have the resources, the lack of hard data on the extent or geographic flow of this trade makes it hard for officials to know how to allocate the resources properly.

The member countries of the Basel Convention have taken some steps to give the treaty sharper teeth. In 1994, the Basel secretariat strengthened information sharing among member

LOGGING ILLOGIC

Money doesn't grow on trees, but some trees might as well be pure gold. The world's voracious (and growing) appetite for wood, paper, and other forest products is driving a stampede to mow down forests.

Much of this logging is illegal. Illegally cut wood accounted for up to 65 percent of world supply in 2000, according to the World Resources Institute. Estimates of illegal logging as a share of the total range from 35 percent in Malaysia, to 50 percent in Cameroon, 50 percent in eastern Russia, 70 percent in Gabon, 73 percent in Indonesia, and up to 80 percent in parts of Brazil. About 40 percent of the wood processed in the pulp and paper industry in Indonesia is of questionable origin, and up to 46 percent of the domestic demand in the Philippines is supplied from illegal sources. Precise global data are unavailable, but in terms of commodity value, illegal logging may be the most serious transnational environmental crime.

Illegal logging activities include logging in national parks and protected areas; harvesting protected timber species; overcutting and underreporting of timber volume, grade, and species; logging for illegal commercial charcoal and fuelwood production; logging without permits; smuggling; and violating forest laws and restrictions. Like money, timber can be "laundered;" in May the BBC reported on illegally logged Indonesian timber that was imported openly into Malaysia, processed into garden furniture, and exported as of Malaysian origin.

National governments have taken a range of measures against these crimes—overhaul of forest legislation, reforming permit processes, adjusting tax codes and royalty systems, various bans—with mixed success. Indonesia, for instance, banned log exports last year and in May declared a temporary moratorium on logging concessions. Logging persists, however, to supply pulp and paper mills and (via bulldozing and burning) to clear forests for palm oil plantations. Logging bans often increase illegal logging in neighboring countries; a recent Chinese logging ban has boosted demand for timber from Cambodia, Vietnam and far eastern Russia. Thailand's logging ban in the 1990s encouraged the extraction of timber in Laos and Myanmar, especially teak.

On the plus side, in northern Tanzania, pilot village forest management programs were so successful in controlling illegal forest activities, including logging, that Tanzania's forest policy now promotes community involvement in forest management nationwide. Ecuador recently launched Vigilancia Verde, a coalition of the Environment Ministry, the armed forces, police, and environmental NGOs, to collaborate in monitoring the transport of timber to markets and mills. In its first year it seized five times the volume of timber confiscated by the government the year before.

Controlling illegal logging is dangerous. After Brazil banned mahogany logging, one of the promoters of the ban was gunned down at his house. In Guerrero, Mexico, the army shot one farmer and jailed and tortured two others who protested logging abuses. A campaign by the Philippine Department of Environment and Natural Resources against illegal logging and its own corruption has made progress—in 2001, over 12 thousand cubic meters of illegal timber confiscated, 76 criminal cases filed, and 14 DENR personnel suspended—but at the cost of nearly 80 DENR staff killed in the line of duty over the years.

The risk is commensurate with the profits, which are high because of strong demand for timber products, especially form the developed world. According to the nonprofit Environmental Investigation Agency, the European Union imported about $1.5 billion worth of illegal timber in 1999; in 2000 the United States imported an estimated $330 million worth of illegal timber from Indonesia alone. The G8 nations (Britain, Canada, France, Germany, Italy, Japan, Russia, and the United States) account for about three-quarters of global timber and wood products imports, yet to date no G8 nation has laws requiring routine seizure of illegal timber imports.

—Christine Haugen

Ms. Haugen (chaugen976@aol.com) is a consultant on resource and energy issues. She is the author of the Directory of Tropical Forestry Journals and Newsletters *(FAO-Bangkok, 1997) and co-author of* Keeping it Green; Tropical Forestry Opportunities for Mitigating Climate Change *(World Resources Institute, 1995).*

time, as much as 10,000 tons of the chemicals entered the country each year. By 1995, CFCs were considered the most valuable contraband entering Miami, after cocaine. Following a crackdown on large consignments through East Coast ports, much of the illegal trade shifted to the Mexican border.

The U.S. Department of Justice estimates that in total, some 10,000–20,000 tons of CFCs have been smuggled into the United States since 1992. Despite stronger enforcement efforts, officials have recovered only a fraction of this contraband. So far, activities under the North American CFC-Anti-Smuggling Initiative, an interagency task force established in 1994, have

led to 114 convictions and the seizure of some 1,125 tons of smuggled CFCs.

Today, a second, smaller, spike in black market activity is occurring as the remaining U.S. stockpiles of legal CFC-12 are depleted. Traders can once again earn a high profit on the contraband supplies: a 30-pound cylinder of CFC-12, bought in China for as little as $30 to $60, can be resold on the U.S. black market for as much as $600. For auto repair shops and other end users, obtaining this illegal product can still be cheaper than buying the legal supplies, which now cost as much as $1,000 per cylinder.

Europe has been another significant market for illegal CFCs. In the mid 1990s, researchers with the London-based Environmental Investigation Agency (EIA) uncovered a thriving regional trade amounting to between 6,000 and 20,000 tons of the chemicals annually. Well after the phaseout deadline, supplies were still abundant and prices disproportionately low, suggesting that the market was being swamped with illegal imports. Meanwhile, regional sales of CFC replacements were slower than expected.

The European black market has thrived in part because regional refrigerant management programs have been poorly organized, and because consumers perceive alternatives to CFCs to be too costly and less efficient. In Central and Eastern Europe, where the illegal CFC trade is thought to be increasing, a major problem is that border officials are typically untrained in identifying the chemicals and have difficulty deciphering their often vague customs codes.

European efforts to control CFC smuggling have generally lagged behind those of the United States, but there are signs that European enforcement is improving. In 1997, authorities in Belgium, Germany, the Netherlands, and the United Kingdom jointly nabbed a multimillion-dollar crime ring that had illegally imported more than 1,000 tons of Chinese-made CFCs for redistribution in Europe and the United States. And in an unprecedented move, in September 2000 the European Union adopted a regional ban on CFC sales and use.

Demand for contraband CFCs has also been high in Japan, especially as retail prices for the chemicals have skyrocketed. Although the Japanese government banned the use of Freon in new cars in 1994, 15 to 20 million vehicles in the country still use the refrigerant. The *Japan Times* reports that auto repair shops in the country circulated more than 100,000 canisters of illicit CFC-12 in 2001, most of them originating in China and other developing countries.

As the CFC phaseout begins to take hold in the developing world (countries were required to freeze consumption in July 1999), black markets are beginning to emerge in places like Asia, where there are still millions of users of CFC-based equipment, including old cars and refrigerators that have been exported from the industrial world. In October 2001, EIA reported on a growing multi-million dollar market for illegal CFCs in India, Pakistan, Bangladesh, Malaysia, the Philippines, and Vietnam. Between early 1999 and March 2000, smugglers slipped some 880 tons of ozone-depleting substances into India, representing an estimated 12 percent of national consumption.

To improve monitoring of the CFC trade and head off future black markets, parties to the Montreal Protocol recently adopted a new licensing amendment that entered into force in 1999. It requires member countries to issue licenses or permits for the import and export of all new, used, and recycled ozone-depleting substances and to exchange information regularly about these activities. By identifying who is and is not licensed to trade, the system should make it easier for police and customs officials to track the movement of the chemicals worldwide.

From Words to Action

Four years ago, the environment ministers of the leading economic powers expressed "grave concern about the ever-growing evidence of violations of international environmental agreements," and called for a range of cooperative actions aimed at stepped-up enforcement. This initiative was followed by the adoption last February of international guidelines to promote compliance with multilateral environmental agreements and prevent cross-border environmental crime. This August's World Summit on Sustainable Development will focus renewed international attention on the importance of adequately implementing and enforcing international environmental treaties and other agreements.

In other promising developments, the World Customs Organization is working with governments to harmonize classification systems for waste and other environmental contraband. The international police organization INTERPOL is training national enforcement officers and customs agents to identify illicitly traded goods more easily, and is also working with national police forces to bring international environmental criminals to justice. Both institutions have established close working relations with UNEP and with the CITES and Basel Convention secretariats.

NGOs are playing a strong role as well. Brazil's RENCTAS is cooperating with the police and the federal environment ministry to train officers in wildlife inspection and is investigating anonymous tips about wildlife smuggling left on its Web site. At the international level, TRAFFIC is tracking national customs enforcement efforts, documenting areas of unsustainable wildlife trade, identifying trade routes for wildlife commodities, and investigating smuggling allegations.

New technologies are also being deployed against international environmental crime. Remote sensing, for instance, was used by the U.S. Coast Guard in the late 1990s to gather the evidence of illegal dumping of oil in international waters that helped to bring Royal Carribean to justice. Satellite-linked Vessel Monitoring Systems are also increasingly being used to monitor the movement of fishing boats in order to detect illegal harvesting. DNA tracing is being used to monitor both fishing and wildlife trade by enabling researchers to link seafood items and wildlife parts back to the species or even the animal of origin.

Targeted industries and countries have bowed to the combined might of NGOs and consumers in several cases. In 1999, pressure from the World Wide Fund for Nature and increased public awareness of the threats that traditional medicine usage poses to wildlife caused leading practitioners and retailers in China to pledge not to prescribe or promote medicines containing parts from tigers, rhinos, bears, and other endangered species. Late last year, Belize, which Greenpeace calls "the world's most fish pirate-friendly country," struck five notorious pirate vessels from its shipping register after Greenpeace showcased the plight of the Patagonian toothfish.

FOR FURTHER INFORMATION

Wildlife Trade
CITES: www.cites.org
TRAFFIC: www.traffic.org
RENCTAS: www.renctas.org.br

Illegal Fishing
Take a Pass on Chilean Sea Bass campaign: http://environet.policy.net/marine/csb

International Transport Workers' Federation FOC campaign: www.itf.org.uk/seafarers/foc/foc.htm
FAO Fisheries Department: www.fao.org/fi

Hazardous Waste Trade
Basel Convention: www.basel.int
Basel Action Network: www.ban.org

CFC Smuggling
Montreal Protocol: www.unep.org/ozone

Miscellany
Environmental Investigation Agency: www.eia-international.org
UN Environment Programme: www.unep.org
World Summit on Sustainable Development: www.johannesburgsummit.org
INTERPOL: www.interpol.int
World Customs Organization: www.wcoomd.org

Although so far they are exceptions rather than the rule, these examples demonstrate that international environmental crime can be controlled through the combined efforts of governments, international institutions, businesses, NGOs, and ordinary citizens. That's good news for the integrity of the Earth's protective ozone layer, numerous threatened species, and the health of communities poisoned by hazardous wastes.

Lisa Mastny is a research associate at Worldwatch Institute. Hilary French is director of the Institute's Global Governance Project.

From *World Watch*, September/October 2002, pp. 12-23. © 2002 by Worldwatch Institute, www.worldwatch.org.

Toward a SUSTAINABILITY Transition

The International Consensus

by Thomas M. Parris

Sustainable development has broad appeal and little specificity, but some combination of development, equity, and environment is found in most attempts to describe it. Defining sustainability requires a clearly articulated consensus on what to develop, what to sustain, and for how long. It also requires thought about how to make a transition from behaviors that trend toward the unsustainable to ones that are more likely to be sustainable.

In recent years, the scientific community has begun to focus on the concept of a twenty-first-century sustainability transition. The U.S. National Research Council (NRC) defines a sustainability transition as one in which a stabilizing world population meets its needs, reduces hunger and poverty, and maintains the planet's life-support systems and living resources.[1] However, for this concept to prove useful in both policy and science, it is essential to define it in terms that can be measured. Without such a mechanism, it is impossible to know whether we are making genuine progress toward or away from a sustainability transition.

This is not a new idea. Much effort has been devoted to defining sustainability indicators at local, regional, and global scales.[2] Despite this large body of work, NRC's Board on Sustainable Development found that "there is no consensus on the appropriateness of the current set of indicators or the scientific basis for choosing them. Their effectiveness is limited by the lack of agreement on what to develop, what to sustain, and for how long."[3]

How can the notion of a sustainability transition be more precisely characterized? To answer this question, this article provides a thorough review of the existing body of international agreements and plans of action to identify a core set of consensus goals and targets. Where such consensus exists, quantitative measures are identified for monitoring progress toward meeting these goals and targets. This article also addresses whether and by whom these measures are being applied and assessed.

In conducting such a review, it is important to distinguish between goals, indicators, targets, and trends. Goals are broad and nonquantitative statements about objectives, such as the Habitat Agenda's goal of "adequate shelter for all" and the United Nations Framework Convention on Climate Change's (FCCC) goal of "stabilization of greenhouse gas concentrations in the atmosphere at a level that would prevent dangerous anthropogenic interference with the climate system."[4] Indicators are quantitative measures for assessing progress toward or away from a stated goal. Agenda 21 (from the 1992 Rio Earth Summit) sets forth two indicators of adequate shelter: access to improved drinking water and access to improved sanitation services.[5] Targets use indicators to quantify goals with specific thresholds and timetables. Examples include reducing the childhood mortality rate to one-third of the 1990 rate by 2015, from

the International Development Goals, and the Kyoto Protocol's target of reducing overall emissions of greenhouse gases by at least 5 percent below 1990 levels by 2008–12.[6] Trends are the values of indicators as a function of time.

Varying Degrees of International Consensus

The structure of this article follows the basic concept of a sustainability transition. It starts by reviewing the state of consensus on meeting human needs and reducing hunger and poverty, and then it covers the consensus on maintaining Earth's life-support systems and living resources. The most striking finding from this review is that consensus goals and targets for meeting human needs and reducing hunger and poverty are more clearly articulated and better institutionalized than those for maintaining life-support systems and living resources.

The consensus

REGARDING WHAT TO SUSTAIN IS HIGHLY DIFFUSE AND POORLY INSTITUTIONALIZED.

A relatively small body of agreements defines a concrete set of goals and targets

Table 1. Priority human needs indicators

Goal	Indicator	Target	Quality of reporting and assessment
Improve health	Childhood mortality	Reduce to one-third of 1990 rate by 2015.[a]	A
Provide education	Literacy	Reduce illiteracy to half of the 2000 rate by 2015.[b]	A
	Male-female secondary school enrollment rates	Eliminate gender disparities in primary and secondary education by 2005.[c]	A
Reduce hunger	Prevalence of undernourishment	Reduce to half of 2000 levels by 2015.[d]	C
	Prevalence of vitamin A deficiency	Virtually eliminate vitamin A deficiency and its consequences (including blindness) by 2000.[e]	D
Reduce poverty	Poverty rate	Reduce the proportion of the world's people whose income is less than one dollar per day to half of the 2000 rate by 2015.[f]	C
Provide housing	Access to improved sanitation services	Ensure that 75 percent of the urban population is provided with on-site or community facilities for sanitation by 2000.[g]	B

[a] International Monetary Fund, Organisation for Economic Co-operation and Development, United Nations, and World Bank Group, *2000—A Better World For All: Progress towards the International Development Goals*, June 2000 (see "Millennium Development Goals" at http://www.developmentgoals.org).

[b] Ibid.

[c] Ibid.

[d] UN General Assembly, *United Nations Millennium Declaration*, A/RES/55/2 (18 September 2000), accessible via http://www.un.org/documents/ga/res/55/a55r002.pdf.

[e] UN Children's Fund, "Goals for Children and Development in the 1990s," World Summit for Children (New York, 30 September 1990), accessible via http://www.unicef.org/wsc/goals.htm.

[f] UN General Assembly, note *d* above; and UN General Assembly, *Further Initiatives for Social Development*, A/RES/S-24/2 (15 December 2000), accessible via http://www.un.org/esa/socdev/docs/english.pdf.

[g] Ibid.

NOTE: The last column grades the quality of reporting and assessment for each indicator. A letter grade of "A" indicates that the indicator has been and continues to be routinely measured, reported, and assessed on a global basis (so it is possible to establish long-term trends). "B" indicates that the indicator is currently being measured, reported, and assessed on a global basis and is likely to be in the future. "C" indicates that global and regional estimates are possible only through extensive modeling and extrapolation. "D" indicates that only rough contemporary estimates through proxies are available.

SOURCE: T. M. Parris.

for improving health, providing education, and reducing hunger and poverty. Somewhat weaker goals and targets have been established for providing housing and employment. International efforts to monitor trends, assess progress, and implement programs associated with each of these goals to meet human needs are institutionalized in specialized agencies and programs of the UN such as the World Health Organization (WHO), the UN Educational, Scientific and Cultural Organization (UNESCO), the Food and Agriculture Organization (FAO), the World Bank, the UN Development Programme, the UN Human Settlements Programme, and the International Labour Organization. Although these institutions vary in effectiveness, they do provide international focal points for such efforts.

In contrast, hundreds of different agreements cover key components of the Earth's life-support systems and living resources.[7] The consensus regarding what to sustain is highly diffuse and poorly institutionalized. Global treaties with clearly articulated targets, such as the Montreal Protocol on Substances That Deplete the Ozone Layer and the Kyoto Protocol to FCCC, are rare exceptions to a more general rule of broadly conceived but vaguely defined objectives. Most environmental agreements are regional and apply to only a small number of countries. Few of them are institutionalized to the same degree as agreements to meet human needs. As one example of the disparities in overall institutional capacity, the annual operating budget of the World Conservation Union

(IUCN) is less than one-tenth that of FAO.[8] This relative lack of focal institutions diminishes the ability to maintain Earth's systems and resources.

There are at least three complementary explanations for this fragmentary approach toward the life-support systems components of a sustainability transition. First is the relative immaturity of international efforts to address environmental issues. Most international environmental diplomacy has occurred since the 1972 UN Conference on the Human Environment—a period during which countries have been reluctant to establish major new intergovernmental organizations.[9] In comparison, efforts to address human needs are firmly embedded in the Charter of the United Nations signed in 1945 and were rapidly institutionalized in the specialized agencies mentioned above. Second, many environmental issues were identified initially as regional problems that were best addressed by regional institutions. The need to address environmental issues on a global scale has emerged slowly in response to the pervasive nature and improved scientific understanding of the reach of these problems. Third, international diplomacy remains dominated by anthropocentric concerns that are typically one level removed from the underlying health of the life-support system in question. Agreements to maintain the system tend to focus on just those aspects that most immediately serve human needs. For example, the vast majority of fisheries treaties are structured to produce maximum sustainable yields of specific economically valuable species rather than to nurture the marine ecosystems in which these species live.

Meeting Human Needs

By focusing on indicators for which there are well-established goals and targets and removing indicators that are highly related to one another (except where needed to reiterate an important consensual statement), one can identify a core set of seven indicators that capture the essence of the consensus targets for meeting human needs and reducing hunger and poverty. Table 1 on this page lists these indicators, their associated goals and targets, and the quality of reporting and assessment for each of them. Despite the fact that measurable progress has been made toward many of these targets, few have been attained fully. As a result, there is a clear tendency to reaffirm existing goals and targets—but with longer timetables for completion. For example,

many targets originally established for 2000 recently were pushed forward to 2015.

Survival

The most fundamental human need is simple survival, including the basic health and development opportunities that accompany it. This notion is embodied in several international accords that establish quantitative goals and targets for infant, childhood, and maternal mortalities as well as life expectancy.[10] These indicators are all highly correlated with one another.[11] Of these four indicators, childhood mortality was selected for Table 1 due to the preference several sources give to the wellbeing of children.[12] There also are consensus goals and targets for numerous disease-specific measures, including poliomyelitis, neonatal tetanus, and acute respiratory infections.[13] However, all of these can be considered components of the overall childhood mortality rate.

Globally, the childhood mortality rate has steadily declined from approximately 190 deaths per 1,000 live births in 1960 to 78 per 1,000 in 1999.[14] Nevertheless, this rate of decline is well short of what would be required to meet the consensus target. This is primarily attributable to slower-than-necessary rates of decline in childhood mortality in South Asia, sub-Saharan Africa, and the former Soviet Union. In sub-Saharan Africa, childhood mortality rates actually increased from 155 per 1,000 in 1990 to 161 per 1,000 in 1999.[15] The quality of reporting and assessment for childhood mortality rates is good. Data are routinely collected as a vital statistic, and long-term trends are available for most countries.

Education

The international community has spoken with clarity on education, repeatedly calling for universal access to primary education, decreasing illiteracy to half of 2000 levels by 2015, and reducing educational disparities between genders.[16] Indicators for literacy and primary school enrollment are correlated with one another.[17] Therefore, the main output of the educational process—literacy—is highlighted in the table instead of input measures on school enrollment and completion. As with childhood mortality, the trends are promising but do not meet the agreed-upon targets. Globally, adult illiteracy rates fell from roughly 45 percent in 1970 to 21 percent in 2000.[18] Most countries measure and assess literacy as part of their decennial censuses. Figures

on school enrollment are reported and assessed routinely by UNESCO.[19]

The international consensus also emphasizes the important role of women in development and establishes targets for eliminating male-female educational disparities. The most recent targets associated with this goal have been expressed in terms of eliminating the gender gap in primary and secondary school enrollment ratios.[20] Secondary school enrollment is used here as the more encompassing of the two targets. The gap between male and female secondary school enrollment declined from approximately 11 percent in 1970 to just less than 7 percent in 1995.[21]

Hunger

Reducing hunger continues to be a major emphasis of the international development community. Consensus statements about hunger call for eliminating episodic or "crisis" hunger due to famine, war, and financial crisis; reducing chronic household hunger by half; reducing iron deficiencies; and eliminating iodine and vitamin A deficiencies.[22] Chronic household hunger was selected as a core indicator because it is more pervasive than episodic hunger and better represents the capacity of society to feed its people on a continuing basis. The number of undernourished people in the developing world declined from roughly 950 million people (29 percent of the developing-world population) in 1965 to roughly 800 million people (14 percent) in 1995. Current trends will reduce this number to between 550 million and 600 million people (10 percent) by 2015—which is well short of the target of 400 million to 440 million.[23]

*R*educing hunger
CONTINUES TO BE A
MAJOR EMPHASIS OF
THE INTERNATIONAL
DEVELOPMENT COMMUNITY.

Vitamin A deficiency was chosen as another core indicator—primarily because, considering other micronutrient deficiencies, iron deficiency is a complex issue that arises from multiple sources and is difficult to address, and iodine deficiency is not as widespread and has a well-known technical solution: iodizing salt.[24] Between 1985 and 1995, the prevalence of clinical vitamin A deficiency in children 5 years old and younger in developing countries

Table 2. Priority life-support system indicators

Goal	Indicator	Target	Quality of reporting and assessment
Reduce emissions of atmospheric pollutants	Greenhouse gas emissions	Reduce overall emissions of greenhouse gases by at least 5 percent below 1990 levels by 2008–12.[a]	A/B
	Sulfur oxides emissions	Targets vary by agreement.[b]	C
Stabilize ocean productivity	Condition of marine ecosystems	Not stated.	D
Maintain freshwater availability	Consumptive freshwater withdrawals	Not stated.	D
Reduce land use and land cover change	Land use and land cover change	Not stated.	C
Maintain biodiversity	Land use and land cover change in biodiversity hotspots	Not stated.	D
Reduce emissions of toxic substances	Dioxin and furan emissions	Reduce or eliminate releases from unintentional production as measured by toxic equivalency units.[c]	D

[a] United Nations Framework Convention on Climate Change, *Kyoto Protocol to the United Nations Framework Convention on Climate Change*, 1997, accessible via http://unfccc.int/resource/docs/convkp/kpeng.html.

[b] See the Helsinki Protocol on the Reduction of Sulphur Emissions or Their Transboundary Fluxes by at Least 30 Per Cent (1985); the Oslo Protocol on Further Reduction of Sulphur Emissions (1994); the Gothenburg Protocol to Abate Acidification, Eutrophication and Ground-Level Ozone (1999); and the Canada–United States Air Quality Agreement (1991).

[c] UN Environment Programme, *Stockholm Convention on Persistent Organic Pollutants*, 2001.

SOURCE: T. M. Parris.

declined from 1.06 percent to 0.63 percent—a reduction of just more than 40 percent.[25] Estimates of chronic household hunger and vitamin A deficiency rely extensively on models to extrapolate very sparse data. The quality of data for other indicators of hunger is even poorer.

Poverty

The UN Millennium Summit and the World Summit for Social Development+5 each have articulated a concrete set of goals and targets for reducing poverty—defined as the proportion of people living on less than one dollar per day—to half of 2000 levels by 2015.[26] The proportion of people in developing countries living in poverty declined from 28.3 percent in 1987 to 23.4 percent in 1998—a decline of approximately 1.44 percent per year.[27] However, to meet the stated goal, the annualized rate of decline would have to nearly double to 2.7 percent per year. The quality of data on poverty rates is poor. In addition, it is not certain that an international poverty standard measures the same degree of need or deprivation across countries.[28] Measures of employment are not explicitly included here because they are closely related to poverty and no universal measures seem likely.

Housing

Compared with the fairly precise goals and targets described above, the international consensus on goals and targets for providing housing is relatively weak. Although the UN Conference on Human Settlements (dubbed Habitat II), held in Istanbul in 1996, established an ambitious agenda to provide adequate shelter for all, it did not set forth any quantified goals or targets.[29] Of the international consensus documents that address housing, only Agenda 21 provides a concrete goal to ensure that all urban residents have access to at least 40 liters per capita per day of safe drinking water and that 75 percent of the urban population is provided with on-site or community sanitation facilities.[30] Of these two indicators, access to improved sanitation services is used here as the core indicator of housing quality. This measure is favored over access to improved water services because it is more closely related to many other forms of important household services such as electric-ity and communications. In addition, efforts to improve sanitation services tend to lag behind efforts to improve access to water.

Maintaining Earth's Life-Support Systems

At first glance, it may appear hopeless to attempt to extract a coherent sense of international consensus on priorities for maintaining Earth's life-support systems and living resources from the plethora of environmental agreements. However, by using common themes that appear in regional agreements as likely forerunners of more comprehensive global agreements, the many pieces of this puzzle fit loosely together to form a more or less complete whole from which one can infer an international consensus. Progress on this inferred consensus can be monitored effectively with a core set of seven indicators, shown in Table 2 on this page. The goals of this consensus are primarily structured by environmental media: atmosphere and climate, oceans, freshwater, land use and land cover, terrestrial biodiversity, and toxics.

Atmosphere and Climate

The climate system is a strong regulator of life on Earth, and perturbations to the climate system therefore are of great concern. The major international agreements regarding the atmosphere and climate address emissions of greenhouse gases; substances that deplete stratospheric ozone; sulfur oxides (SO_x); tropospheric ozone precursors such as nitrogen oxides (NO_x), volatile organic compounds (VOCs), and ammonia; and toxics (mostly metals).[31]

Greenhouse gas emissions, as measured by global warming potential, provide the best overall indicator of anthropogenic stress on the global climate system. Because substances that deplete the ozone layer also are potent greenhouse gases, progress in reducing their emissions is tracked as a component of overall greenhouse gas emissions. Progress in reducing emissions has been slow, even if one considers the FCCC target of capping carbon dioxide (CO_2) emissions at 1990 levels (rather than the more ambitious Kyoto Protocol target of reducing overall emissions of greenhouse gases by at least 5 percent below 1990 levels by 2008–12). Aggregate worldwide CO_2 emissions from the consumption and flaring of fossil fuels were 8.3 percent greater in 2000 than in 1990.[32] While trends for greenhouse gases that are well-mixed by the atmosphere—such as CO_2, nitrous oxide, and methane—are relatively well understood, much less is known about trends for other substances that increase warming, such as soot.[33] Furthermore, there is significant disagreement about how best to aggregate the effects of the many different greenhouse substances into a single overall measure such as global warming potential.[34] Although this is not necessary to monitor past and current climate forcing, it is required to estimate future forcing given projections for emissions.

Tropospheric air pollution has a significant impact on human health, ecosystem health, and agricultural productivity. There are many components of this type of air pollution, but the focus here is on SO_x emissions because they were the first to be addressed by regional consensus agreements.[35] There is some overlap in the forces driving SO_x and NO_x emissions (such as electric power production) as well as in the forces driving emissions of ozone precursors and greenhouse gases (such as automobile use). By 1990, SO_x emissions had increased to more than 5.5 times their levels in 1900.[36] However, they appear to

have peaked in 1989 and declined by 2.6 percent from 1990 to 2000.[37] The quality of reporting for SO_x emissions is relatively weak. FCCC encourages but does not require Annex 1 parties (mostly industrialized countries) to report current national emissions.[38] Historical trends for SO_x emissions have been reconstructed using proxy variables. International reporting of NO_x and other ozone precursor emissions is even weaker.

Oceans

There are three principal threads of international agreements regarding oceans: protection of marine fisheries; land-based sources of marine pollution; and ocean dumping from ships, aircraft, and exploration and mining.[39] Considering these threads as a group, one can infer an underlying consensus to stabilize ocean productivity, as represented by the overall condition of ocean ecosystems.

FAO produces annual fisheries assessments, but there are inherent difficulties in monitoring the health of fish stocks, and these assessments focus almost exclusively on selected species that have economic value rather than on the health of the ecosystems that support them. More systematic attention to ecosystems instead of selected species can be found in agreements such as the Convention on the Conservation of Antarctic Marine Living Resources, which includes specific goals to maintain the ecological relationships between harvested, dependent, and related populations of marine living resources and the restoration of depleted populations—as well as goals to prevent or minimize the risk of changes in the marine ecosystem that are not potentially reversible over two or three decades.[40] However, reporting of marine ecosystem conditions is idiosyncratic, and there currently are no accepted measures to compare assessments from habitat to habitat.

Freshwater

Freshwater is addressed primarily by national and subnational legislative and regulatory regimes. Of course, when rivers, lakes, and aquifers cross national boundaries, the issue becomes international. While the number of countries participating in any given agreement is small, the number of transboundary water bodies is large. There are more than 260 international river basins and about 150 multilateral treaties on transboundary water bodies.[41] These treaties typically address three major sets of issues: freshwater fish-

eries, pollution, and water withdrawals. From these many agreements, one can infer an emerging international consensus to stabilize consumptive freshwater withdrawals, defined as "water withdrawn from a source and made unusable for reuse in the same basin, such as through irrecoverable losses like evaporation, seepage to a saline sink, or contamination."[42] Global consumptive freshwater withdrawals have increased about 5.5-fold since 1900.[43] International reporting for freshwater is poor. Past trends have been reconstructed from proxy variables and there has been little coordinated effort to monitor current trends systematically. The Global International Waters Assessment was designed to address these shortcomings but is currently conceived as a onetime endeavor to be completed in 2003.[44]

Land Use and Land Cover Change

As is the case with freshwater, the vast majority of legislation for and regulation of land use and land cover change is addressed within national and subnational contexts. The international community becomes involved only in cases of large-scale land use and land cover change phenomena such as desertification, decline of large wetlands, deforestation, and the transformation of internationally recognized landmarks such as the Chapada dos Veadeiros and Emas National Parks in Brazil. Where international agreements do exist, they typically seek to slow the rate of harmful anthropogenic land use and land cover change.[45] There is little agreement on how to measure change on a global scale. Most efforts focus on specific types of change such as tropical deforestation, and few efforts attempt to balance negative changes with positive changes such as the reemergence of forests in New England.

One approach is to monitor the net annual CO_2 flux into the atmosphere due to anthropogenic land use and land cover change—a crude measure of the change in biomass that accounts for tropical, temperate-zone, and boreal deforestation; cultivation of mid-latitude grassland soils; accumulation of carbon in wood products and woody debris; losses of carbon from oxidation of wood products, woody debris, and soil organic matter; and the accumulation of carbon in forests recovering from harvest and in the fallows of shifting cultivation. This figure has increased by a factor of 3.6 since 1900.[46]

Another approach is to look at long-term changes in net primary productivity (NPP)—a measure of the amount of new biomass produced by photosynthetic organisms and other producers during a specified time period.[47] One such effort found that the greatest losses in NPP have occurred in Africa, the Indian subcontinent, and the Middle East—but that NPP in temperate nonforested areas converted to cropland has increased.[48]

Terrestrial Biodiversity

Numerous international agreements address the protection, management, and restoration of living resources. The major efforts to protect marine and freshwater living resources are discussed above; the focus here is on land-based living resources. These efforts range from general attempts to preserve biodiversity, including the Convention on International Trade in Endangered Species of Wild Fauna and Flora and the Convention on Biological Diversity, to a variety of more focused efforts, including the establishment of reserves, the management of certain highly valued species, and the control of invasive or pest species.[49] Most agreements focus on land use and land cover change, with a special emphasis on protecting habitats in biologically diverse regions of the world. This suggests a consensus to maintain biodiversity by slowing the rate of land use and land cover change in biodiversity hotspots.[50] As with land cover and land use change in general, there is little agreement about how best to consistently measure such changes within a variety of very different biodiversity hotspots. While it would be possible to produce measures such as net annual CO_2 flux into the atmosphere due to anthropogenic land use and land cover change or long-term changes in NPP for biodiversity hotspots, such measures are not currently produced.

Unfortunately, THERE IS NO WAY OF INTEGRATING THE WIDE RANGE OF TOXICS INTO A SINGLE OVERALL MEASURE OF EMISSIONS AND IMPACTS.

Toxics

The international community has agreed upon the need to live in an environment free of toxic substances. The international consensus has focused on heavy metals (such as lead, mercury, and cadmium), persistent organic pollutants, and

nuclear waste and radiation. In addition, several of the agreements that cover climate and atmosphere, oceans, and freshwater address toxics.[51] Unfortunately, there is no way of integrating the wide range of toxics into a single overall measure of emissions and impacts, and global reporting and assessment are uniformly poor for all toxic pollutants.

Emissions of dioxins and furans are used here because they are explicitly subject to an international agreement (the Stockholm Convention on Persistent Organic Pollutants), they are synthetic and highly toxic, and they bioaccumulate in fish and meat. Dioxins and furans are environmental contaminants detectable in almost all compartments of the global ecosystem in trace amounts. They have never been produced intentionally and have never served any useful purpose. Rather, they are formed as unwanted byproducts in many industrial and combustion processes. Global emissions of dioxins and furans are estimated to be 50 ± 10 kilograms international toxic equivalency units (a measure that compares the relative toxicity of various chemical compounds). These emissions constitute a multiple of roughly 100 trillion times the tolerable daily intake of a 150-pound adult. These estimates rely extensively on models and do not assess trends over time.[52]

The Quality of Life-Support System Indicators

In striking contrast to the progress made on priority human needs indicators, only one of the indicators for Earth's life-support systems—overall emissions of greenhouse gases—is operationally produced on a global basis. Some indicators such as the condition of ocean ecosystems require additional basic research before they can be generated economically on a global scale. Others such as consumptive freshwater withdrawals are produced sporadically by individual scholars and are subject to the whims of scientific funding agencies. In principle, these indicators could be produced economically on an operational basis with modest investments in the capacity of intergovernmental organizations to routinely collect and process the relevant data. Other indicators, such as emissions of dioxins and furans, require significant new institutional arrangements to establish procedures for reporting and aggregating data at the relevant jurisdictional scale.

The technical corollary to the need for focal institutions that correspond to the major valued life-support subsystems and ecosystem services is the need to develop indicators that aggregate multiple conceptually related trends to a common measure. The current practice for producing life-support system indicators mirrors the fragmentation of existing institutions. For example, indicators for land use and land cover change focus on specific types of change such as tropical deforestation, desertification, and urbanization. Although each of these types of change is important, efforts to monitor progress toward a sustainability transition require some way of aggregating these individual indicators to a common measure. Aggregate measures allow regional and global trends and projections to be produced for the life-support system as a whole.

In deciding WHAT TO DEVELOP, WHAT TO SUSTAIN, AND FOR HOW LONG, SCIENCE CAN GUIDE THE VALUE JUDGMENT, BUT IT IS ULTIMATELY A SOCIAL CHOICE.

In this regard, experience with the Montreal and Kyoto Protocols is instructive. In both cases, the environmental impacts associated with the emissions of many different trace gases were analyzed and assessed using common measures—ozone-depleting potential and global warming potential, respectively. These aggregate measures played an important role in enabling the negotiation and implementation of each agreement. Virtually all of the indicators discussed above for Earth's life-support systems and living resources are aggregate indicators—the only exception is SO_x emissions. Even in this case, a method for aggregating emissions of tropospheric air pollutants would be welcome.

It is important to note that aggregating conceptually related indicators to a common measure is substantially different from developing composite indices that arithmetically combine disparate measures into a single overall score, such as the Environmental Sustainability Index and the Ecosystem Well Being Index.[53] The difference is that aggregating related indicators uses scientific methods to establish equivalencies to a common unit of measure, whereas composite indices use subjective methods to define an overall grade with no associated units of measure.

In deciding what to develop, what to sustain, and for how long, science can guide the value judgment, but it is ultimately a social choice. While basic features of this choice—meeting human needs, reducing hunger and poverty, and maintaining life-support systems—are largely consistent across scale, the relative importance, specific indicators, and policy options with which to implement change vary widely.[54] For example, in small regions it is not necessary to aggregate multiple types of land use and land cover change into a single overall measure, as global-scale analysis requires. Instead, one type of change is the dominant concern in each region. In Mexico's Yucatan peninsula, for example, the dominant land use and land cover issue is deforestation (followed by invasive species); in Mexico's Yaqui valley it is coastal-zone aquaculture; and in the Arctic Circle the dominant issue is the melting of permafrost due to climate change.[55]

This review has focused primarily on the global scale and on international policy agreements. At this scale, it often is counterproductive to use countries as the unit of analysis, because individual countries fear that consensus on indicators will be applied selectively and punitively. Indeed, this was a major impediment in the UN Commission on Sustainable Development's effort to develop a consensus set of sustainability indicators. From this perspective, it is more productive to analyze sustainability transitions regionally, so that each region consists of multiple countries and is defined by a "peer" relationship such as geography or income. Regionalization prevents analysts from ignoring important geographic variation, and it minimizes the likelihood that specific countries will be singled out as "bad actors."

The aggregate indicators discussed in this article should be considered first approximations that serve as useful placeholders. Additional work is required to improve the scientific basis for these indicators, develop methods for producing them economically and at the necessary spatial and temporal scales, and establish the necessary institutional arrangements to produce them regularly. Although meeting consensus goals and targets is a necessary condition to a sustainability transition, it is most likely not sufficient. There are key aspects of the nature-society system for which there has yet to evolve a strong political or scientific consensus or for which

the current consensus may be unsound. (For example, the vast majority of the international agreements regarding fisheries are based on the concept of maximum sustainable yield, yet emerging scientific literature challenges the use of single-species assessments of maximum sustainable yield as the sole metric with which to manage fishery stocks.[56] There always will be a need to extend the existing international consensus by drawing on emerging political and scientific literature. As a result, the definition of a sustainability transition will almost certainly change over time as new insights are incorporated into the international consensus.

NOTES

1. National Research Council (NRC), Board on Sustainable Development, *Our Common Journey: A Transition toward Sustainability* (Washington, D.C.: National Academy Press, 1999), chapter 1, accessible via http://www.nap.edu/catalog/9690.html.

2. Major examples include United Nations Division for Sustainable Development, "Indicators of Sustainable Development: Framework and Methodologies," Background Paper No. 3 for the Ninth Session of the Commission on Sustainable Development, DESA/DSD/2001/3 (New York, 16–27 April 2001), accessible via http://www.un.org/esa/sustdev/csd9/csd9_indi_bp3.pdf; Consultative Group on Sustainable Development Indicators, "The Dashboard of Sustainability," Version 2.0 (Winnipeg, Canada: International Institute for Sustainable Development, 2001), accessible via http://iisd1.iisd.ca/cgsdi/dashboard.htm; R. PrescottAllen, *The Wellbeing of Nations: A Country-by-Country Index of Quality of Life and the Environment* (Washington, D.C.: Island Press, 2001); World Economic Forum, *2002 Environmental Sustainability Index* (Davos, Switzerland, 2002), which can be downloaded from the Center for International Earth Science Information Network (CIESIN) web site at http://www.ciesin.org/indicators/ESI/downloads.html; and World Wide Fund for Nature (WWF), *Living Planet Report 2002* (Gland, Switzerland: WWF International, 2002), accessible via http://www.panda.org/downloads/general/LPR_2002.pdf. Several more indicator efforts have been characterized in UN Division for Sustainable Development, "Report on the Aggregation of Indicators

of Sustainable Development," Background Paper for the Ninth Session of the Commission on Sustainable Development (New York, 16–27 April 2001), accessible via http://www.un.org/esa/sustdev/csd9/csd9-aisd-bp.pdf. For descriptions of other efforts, see International Institute for Sustainable Development, "Compendium of SD Indicator Initiatives," accessible via http://iisd.ca/measure/compindex.asp.

3. NRC, note I above, page 243.

4. UN Conference on Human Settlements, *Report of the United Nations conference an Human Settlements (Habitat II)*, A/CONF.165/14 (7 August 1996), accessible via http://www.unchs.org/unchs/english/hagenda/index.htm; and UN, *United Nations Framework Convention on Climate Change* (Rio de Janeiro, 1992), accessible via http://unfecc.int/resource/docs/convkp/conveng.pdf.

5. UN Conference on Environment and Development, *Report of the United Nations Conference on Environment and Development*, A/CONF.151/26, Rio de Janeiro, 3–14 June 1992, vols. 1–3 (12 August 1992), accessible via http://www.un.org/esa/sustdevagenda2ltext.htm.

6. International Monetary Fund (IMF), Organisation for Economic Co-operation and Development (OECD), UN, and World Bank Group, *2000—A Better World For All: Progress towards the International Development Goals*, June 2000 (see "Millennium Development Goals" at http://www.developmentgoals.org); and UN Framework Convention on Climate Change (FCCC), *Kyoto Protocol to the United Nations Framework Convention on Climate Change*, 1997, accessible via http://www.unfccc.int/resource/docs/convkp/kpeng.html.

7. The Register of International Treaties and Other Agreements in the Field of Environment lists 170 documents, most of which address small fragments of the overall problem: and the World Conservation Union's (IUCN) Environmental Law Information System contains descriptions of more than 470 agreements. See UN Environment Programme (UNEP) and IUCN, *ECOLEX: A Gateway to Environmental Law* (Jarrow, U.K.: DJL Software Consultancy, Ltd., 1997), accessible via http://www.ecolex.org/.

8. IUCN, "Finances," 2002, accessible via http://www.iucn.org/about/finances.htm; and Food and Agriculture Organization (FAO), "FAO: What It Is—

What It Does," 2001, accessible via http://www.fao.org/UNFAO/e/wstruc-e.htm.

9. See, for example, M. Hamer, "The Filthy Rich: How Developed Nations Plotted to Undermine Global Pollution Controls," *New Scientist*, 5 January 2002, 7.

10. This consensus can be seen in the negotiated outcomes of the Convention on the Rights of the Child (1989), the World Summit for Children (1990), the International Conference on Population and Development (1994), the World Summit for Social Development (1995), and A Better World for All (2000).

11. Correlations were constructed using country statistics for 2000. See World Bank, *World Development Indicators 2002 on CD-ROM* (Washington, D.C.: The International Bank for Reconstruction and Development (IBRD), 2002).

12. For example, see UN Children's Fund (UNICEF), "Goals for Children and Development in the 1990s," World Summit for Children (New York, 30 September 1990), accessible via http://www.unicef.org/wsc/goals.htm; and UN General Assembly, *A World Fit for Children* (New York, May 2002), accessible via http://www.unicef.org/specialsession/documentation/documents/A-S27-l9-RevlE-annex.pdf.

13. This consensus can be seen in the negotiated outcomes of the World Summit for Children (1990) and Agenda 21 (1992).

14. World Bank, *World Development Indicators 2001 on CD-ROM* (Washington, D.C.: IBRD, 2001).

15. Ibid.

16. Examples include the World Conference on Education for All (1990), the World Summit for Children (1990), the World Summit for Social Development (1995), the International Conference on Population and Development+5 (1999), and the World Summit for Social Development+5 (2000).

17. Correlations were constructed using country statistics for 1998 (see World Bank, note 14 above).

18. World Bank, *World Development Indicators 2000 on CD-ROM* (Washington, D.C.: IBRD, 2000); UN Educational, Scientific and Cultural Organization (UNESCO), *Statistical Document: Education for All—2000 Assessment* (Paris, 2000), accessible via http://unesdoc.unesco.org/images/0012/001204/120472e.pdf; and UNICEF, *The State of the World's Children 2002* (New York, 2002), accessible via http://www.unicef.

org/pubsgen/sowc02/sowc2002-eng-full.pdf.

19. See, for example, UNESCO, ibid.

20. IMF, OECD, UN, and World Bank Group, note 6 above.

21. World Bank, note 18 above.

22. For statements on crisis hunger, see the Fourth UN Development Decade (1990) and the World Food Summit (1996). For statements on chronic household hunger, see the Convention on the Rights of the Child (1989), the World Summit for Children (1990), the World Conference on Human Rights (1993), the World Summit for Social Development (1995), the Millennium Summit (2000), and the World Food Summit. For statements on micronutrient deficiencies, see the World Summit for Children and the Fourth UN Development Decade.

23. FAO, *The State of Food Insecurity in the World 2000* (Rome, 2000), accessible via http://www.fao.org/FOCUS/E/SOF100/img/sofirep-e.pdf.

24. This choice is not intended to minimize the importance of efforts to implement salt iodization programs in many parts of the world. For more information, see the International Council for the Control of Iodine Deficiency Disorders web site at http://www.people.virginia.edu/~jtd/iccidd/.

25. UN Administrative Committee on Coordination Sub-Committee on Nutrition, *Third Report on the World Nutrition Situation* (Geneva, 1997), 27–31.

26. UN General Assembly, *United Nations Millennium Declaration*, A/RES/55/2 (18 September 2000), accessible via http://www.un.org/documents/ga/res/55/a55r002.pdf; and UN General Assembly, *Further Initiatives for Social Development*, A/RES/S-24/2 (15 December 2000), accessible via http://www.un.org/esa/socdev/docs/english.pdf.

27. World Bank, *Global Economic Prospects and the Developing Countries 2001* (Washington, D.C.: IBRD, 2001), accessible via http://www.worldbank.org/prospects/gep2001/full.htm; and S. Chen and M. Ravaillion, "How Did the World's Poorest Fare in the 1990s?" World Bank Development Research Group, Policy Research Working Paper WPS 2409 (Washington, D.C.: IBRD, 2001), accessible via http://www-wds.worldbank.org/servlet/WDS_IBank_Servlet?pcont=details&eid=000094946_00081406502730.

28. World Bank, note 11 above (see http://www.worldbank.org/poverty/data/2_6wdi2002.pdf).

29. UN Conference on Human Settlements, note 4 above.

30. See UN Conference on Environment and Development, note 5 above.

31. For greenhouse gases, see FCCC (1992) and the Kyoto Protocol to FCCC (1997). For sulfur oxides, see the Helsinki Protocol on the Reduction of Sulphur Emissions or Their Transboundary Fluxes by at Least 30 Per Cent (1985): the Oslo Protocol on Further Reduction of Sulphur Emissions (1994); the Gothenburg Protocol to Abate Acidification, Eutrophication and Ground-Level Ozone (1999); and the Canada-United States Air Quality Agreement (1991). For nitrogen oxides and other ozone precursors, see the Sofia Protocol concerning the Control of Emissions of Nitrogen Oxides or Their Transboundary Fluxes (1988), the Geneva Protocol concerning the Control of Emissions of Volatile Organic Compounds or Their Transboundary Fluxes (1991), the Gothenburg Protocol, the Canada-United States Air Quality Agreement, and the Ozone Annex to the 1991 Canada-United States Air Quality Agreement (2000).

32. Energy Information Administration, *International Energy Annual 2000* (Washington, D.C.: U.S. Department of Energy, 2002), appendix H, accessible via http://www.eia.doe.gov/emeu/iea/tableh1.html.

33. J. E. Hansen and M. Sato, "Trends in Measured Climate Forcing Agents," *Proceedings of the National Academy of Sciences* 98, no. 26 (2001): 14,778–83.

34. See, for example, S. J. Smith and T. M. L. Wigley, "Global Warming Potentials: 1. Climatic Implications of Emissions Reductions," *Climatic Change* 44 (2000): 445–57; and S. J. Smith and T. M. L. Wigley, "Global Warming Potentials: 2. Accuracy," *Climatic Change* 44 (2000): 459–69.

35. See note 31 above.

36. A. S. Lefohn, J. D. Husar, and R. Husar, "Estimating Historical Anthropogenic Global Sulfur Emission Patterns for the Period 1850–1990," *Atmospheric Environment* 33, no. 21 (1999): 3,435–44; J. Dignon and S. Hameed, "Global Emissions of Nitrogen and Sulfur Oxides from 1860 to 1980," *Journal of the Air & Waste Management Association* 39, no. 2 (1989): 180–86; and S. Hameed and J. Dignon, "Global Emissions of Nitrogen and Sulfur Oxides in Fossil Fuel Combustion 1970–1986." *Journal of the Air & Waste Management Association* 42, no.2 (1992): 159–63.

37. Intergovernmental Panel on Climate Change, *Special Report on Emissions Scenarios* (Cambridge, U.K.: Cambridge University Press, 2000), Version 1.1 data downloaded from the CIESIN web site at http://sres.ciesin.org/ on 19 July 2001.

38. FCCC, *Review of the Implementation of Commitments and of Other Provisions of the Convention, UNFCCC Guidelines on Reporting and Review*, FCCC/CP/1997/7 (16 February 2000), accessible via http://www.unfcce.int/resource/docs/cop5/07.pdf.

39. For protection of marine fisheries, see the Convention for the Regulation of Whaling (1931); the International Convention for the Regulation of Whaling (1946); the Protocol to the International Convention for the Regulation of Whaling (1956); the Convention on Fishing and Conservation of the Living Resources of the High Seas (1958); the Convention concerning Fishing in the Black Sea (1959); the Agreement for the Establishment of the Asia-Pacific Fishery Commission (1948, 1976, 1993, and 1996); the Agreement for the Establishment of a General Fisheries Commission for the Mediterranean (1949, 1963, and 1976); the Convention for the Establishment of an Inter-American Tropical Tuna Commission (1949); the Agreement concerning Measures for the Protection of the Stocks of Deep Sea Prawns, European Lobsters, Norway Lobsters and Crabs (1952); the International Convention for the Conservation of Atlantic Tunas (1966); the Agreement on the Protection of the Salmon in the Baltic Sea (1966 and 1972); the Convention on Fishing and Conservation of the Living Resources in the Baltic Sea and the Belts (1973); the Convention on Future Multilateral Cooperation in the Northwest Atlantic Fisheries (1978); the South Pacific Forum Fisheries Agency Convention (1979); the Convention on the Conservation of Antarctic Marine Living Resources (1980); the Convention on Multilateral Cooperation in North-East Atlantic Fisheries (1980); the Convention for the Conservation of Salmon in the North Atlantic Ocean (1982); the UN Convention on the Law of the Sea (1982); the Convention for the Conservation of Southern Bluefin Tuna (1993); the Eastern Pacific Ocean Tuna Fishing Agreement (1983); the Agreement for the Establishment of the Indian Ocean Tuna Commission (1993); and the Agreement on the Conservation of Cetaceans of the Black Sea, Mediterranean Sea and Contiguous Atlantic Area

(1996). For land-based sources of marine pollution, see the Convention for the Prevention of Marine Pollution from Land-Based Sources (1974 and 1986), the Protocol for the Protection of the Mediterranean Sea against Pollution from Land-Based Sources (1980), the Protocol concerning Mediterranean Specially Protected Areas (1982), the Protocol concerning Specially Protected Areas and Biological Diversity in the Mediterranean (1995), the UN Convention on the Law of the Sea (1982), the Protocol for the Protection of the South East Pacific against Pollution from Land-Based Sources (1983), and the Protocol for the Protection of the Marine Environment against Pollution from Land-Based Sources (1990), which covers the Persian Gulf. For ocean dumping, see the Convention on the High Seas (1958); the International Convention for the Prevention of Pollution of the Sea by Oil (1954, 1962, 1969, superseded in 1983); the International Convention on Civil Liability for Oil Pollution Damage (1969); the International Convention Relating to Intervention on the High Seas in Cases of Oil Pollution Casualties (1969); the Protocol Relating to Intervention on the High Seas in Cases of Pollution by Substances Other Than Oil (1973); the Protocol for the Prevention of Pollution of the Mediterranean Sea by Dumping from Ships and Aircraft (1976); the Protocol concerning Co-operation in Combating Pollution of the Mediterranean Sea by Oil and Other Harmful Substances in Cases of Emergency (1976); the Convention on Civil Liability for Oil Pollution Damage Resulting from Exploration for and Exploitation of Seabed Mineral Resources (1977); the International Convention for the Prevention of Pollution from Ships (1978); the Protocol concerning Regional Co-operation in Combating Pollution by Oil and Other Harmful Substances in Cases of Emergency (1982); the Protocol concerning Co-operation in Combating Oil Spills in the Wider Caribbean Region (1983); the Agreement for Cooperation in Dealing with Pollution of the North Sea by Oil and Other Harmful Substances (1983); and the International Convention on Oil Pollution Preparedness, Response and Co-operation (1990).

40. Commission for the Conservation of Antarctic Marine Living Resources, *Convention on the Conservation of Antarctic Marine Living Resources* (Canberra, Australia, 1980), accessible via http://www.ccamir.org/pu/E/pubs/bd/pt1.pdf.

41. P. H. Gleick, *The World's Water: The Biennial Report on Freshwater Resources 1998–1999* (Washington, D.C.: Island Press, 1998), 219–38; A. T. Wolf, J. A. Natharius, J. J. Danielson. B. S. Ward, and J. K. Pender, "International River Basins of the World," *International Journal of Water Resources Development* 15, no. 4 (1999), accessible via http://www.transboundarywaters.orst.edu/publications/register; and S. Yoffe, B. Ward, and A. Wolf, "The Transboundary Freshwater Dispute Database Project: Tools and Data for Evaluating International Water Conflict" (forthcoming), abstract accessible via http://www.transboundarywaters.orst.edu/publications/tools_data_.

42. P. H. Gleick, *The World's Water: The Biennial Report on Freshwater Resources 2000–2001* (Washington, D.C.: Island Press, 2000), 41.

43. I. A. Shiklomanov, "Assessment of Water Resources and Water Availability in the World," Report for the Comprehensive Assessment of the Freshwater Resources of the World, UN, data archive on CD-ROM from the State Hydrological Institute, St. Petersburg, Russia, 1998 (as reproduced in Gleick, ibid.).

44. UNEP, *Global International Waters Assessment*, GEF Project Brief Cover Page as Approved by the GEF Project Council Meeting November 1997, accessible via http://www.giwa.net/giwa_doc/projdoc1.pdf; and J. C. Pernetta and L. D. Mee, "Global International Waters Assessment, GIWA" (Kalmar, Sweden: Global International Waters Assessment, 2001), accessible via http://www.giwa.net/giwa_doc/documents_article.phtml.

45. See the UNESCO Conference on the Conservation and Rational Use of the Biosphere (1968), the UN Conference on Desertification (1977), the UN Convention to Combat Desertification (1994), the Convention on Wetlands of International Importance Especially as Waterfowl Habitat (1971, 1982, and 1994), the International Tropical Timber Agreement (1983 and 1994), and the Convention concerning the Protection of the World Cultural and Natural Heritage (1972).

46. R. A. Houghton, "The Annual Net Flux of Carbon to the Atmosphere from Changes in Land Use 1950–1990," *Tellus Series B—Chemical and Physical Meteorology* 51, no. 2 (1999): 298–313; and R. A. Houghton and J. L. Hackler "Carbon Flux to the Atmosphere from Land-Use Changes: 1850 to 1990," ORNL/CDIAC-131, NDP-050/R1, Carbon Dioxide Infor-

mation Analysis Center, U.S. Department of Energy (Oak Ridge, Tenn.: Oak Ridge National Laboratory, 2001), accessible via http://cdiac.esd.orn.gov/trends/landuse/houghton/houghton.html.

47. M. Cain, H. Damman, R. Lue, and C. Kaesuk-Yoon, *Discover Biology Online* (New York: W. W. Norton, 2000), accessible via http://www.wwnorton.com/cdly/.

48. R. DeFries, "Past and Future Sensitivity of Primary Production to Human Modification of the Landscape," *Geophysical Research Letters* 29. no. 7 (2002): 10.1029/2001GL13620.

49. For agreements on the establishment of reserves, see the Convention Relative to the Preservation of Fauna and Flora in Their Natural State (1933), the Convention on Nature Protection and Wildlife Preservation in the Western Hemisphere (1940), the Protocol on Environmental Protection to the Antarctic Treaty (1991), the African Convention on the Conservation of Nature and Natural Resources (1968), the Convention on Conservation of Nature in the South Pacific (1976), the ASEAN Agreement on the Conservation of Nature and Natural Resources (1985), and the Convention on the Conservation of European Wildlife and Natural Habitats (1979). For species-specific agreements, see the International Convention for the Protection of Birds (1950), the Agreement on the Conservation of African-Eurasian Migratory Waterbirds (1995), and the Convention on Conservation of North Pacific Fur Seals (1976). For agreements on the control of invasive species, see the Convention for the Establishment of the European and Mediterranean Plant Protection Organisation (1951), the International Plant Protection Convention (1951), and the Agreement for the Establishment of the Near East Plant Protection Organization (1993). For control of pest species, see the Conven-

tion on the African Migratory Locust (1962), the Agreement for the Establishment of a Commission for Controlling the Desert Locust in the Eastern Region of Its Distribution Area in South-West Asia (1963), the Agreement for the Establishment of a Commission for Controlling the Desert Locust in the Near East (1965), and the Agreement for the Establishment of a Commission for Controlling the Desert Locust in North-West Africa (1970).

50. N. Meyers, R. A. Mittermeier, C. G. Mittermeier, G. A. B. da Fonseca, and J. Kent, "Biodiversity Hotspots for Conservation Priorities," *Nature*, 24 February 2000, 853–58; and R. A. Mittermeier, N. Meyers, and C. G. Mittermeier, *Hotspots* (Mexico City: CEMEX and Conservation International, 1999).

51. See the Convention concerning the Use of White Lead in Painting (1921); the Convention concerning the Protection of Workers against Ionizing Radiations (1960); the Convention on Third Party Liability in the Field of Nuclear Energy (1960); the Convention on Civil Liability for Nuclear Damage (1963); the Treaty Banning Nuclear Weapon Tests in the Atmosphere, in Outer Space and under Water (1963); the Aarhus Protocol on Persistent Organic Pollutants (1998); and the Stockholm Convention on Persistent Organic Pollutants (2001).

52. UNEP, *Dioxin and Furan Inventories; National and Regional Emissions of PCDD/PCDF* (Geneva: Inter-Organizations Programme for the Sound Management of Chemicals, 1999), accessible via http://www.irptc.unep.ch/pops/pdf/dioxinfuran/difurpt.pdf.

53. Prescott-Allen, note 2 above; and World Economic Forum, note 2 above.

54. For example, indigenous people often are underrepresented in the national and international forums that establish

goals, indicators, and targets. It has been suggested that these groups place a much greater emphasis on cultural issues (such as language preservation) than on issues such an poverty reduction.

55. See B. L. Turner II et al., "Deforestation in the Southern Yucatán Peninsular Region: An Integrative Approach." *Forest Ecology and Management* 5521 (2001): 1–18; P. A. Matson, R. Naylor, and I. Ortiz-Monasterio, "Integration of Environmental. Agronomic, and Economic Aspects of Fertilizer Management, *Science*, 3 April 1998, 112–15; and Arctic Council and the International Arctic Science Committee, *Arctic Climate Impact Assessment* (forthcoming).

56. See, for example, Reykjavik Conference on Responsible Fisheries in the Marine Environment, "Abstracts of Papers to Be Presented to the Scientific Symposium" (Reykjavik; FAO, August 2001), accessible via ftp://ftp.fao.org/fi/document/reykjavik/Y1498E.doc.

Thomas M. Parris is executive director of and research scientist at the Boston office of ISciences, LLC, in Jamaica Plain, Massachusetts, and a contributing editor of *Environment*. The author thanks his colleagues in the Research and Assessment Systems for Sustainability Project (http://sust.harvard.edu/), particularly Robert W. Kates, for their encouragement and constructive criticism of earlier drafts. This paper is based on research supported in part by a grant from the National Science Foundation (award BCS-0004236) with contributions from the National Oceanic and Atmospheric Administration's office of Global Programs for the Research and Assessment Systems for Sustainability Program. Parris can be reached at (617) 524-8041 or by e-mail at parris@isciences.com.

MAKING THE GLOBAL LOCAL

RESPONDING TO CLIMATE CHANGE CONCERNS FROM THE GROUND UP

BY ROBERT W. KATES AND THOMAS J. WILBANKS

We live on Earth, but we reside, work, and play in local places. Increasingly we speak of—and often worry about—global change, yet we perceive few phenomena in which local actions are directly connected to what happens globally. This is especially true of positive local actions such as recycling wastes or assuring energy-efficient homes and vehicles, as opposed to such negative, headline-grabbing actions as terrorism.

Perhaps the most widely recognized connection between global environmental processes and local actions is the chain of causality that drives climate change. In this chain of cause and consequences, societal forces such as population, affluence, or technology drive the varied human activities that produce greenhouse gas (GHG) emissions. All three of the above societal forces, for example, drive energy, manufacturing, and transportation needs that ultimately produce the emissions from electric power plants, industries, and vehicles. GHG emissions then enhance solar radiative forcing of the climate (in which more solar radiation is kept within the Earth's atmosphere than is being reradiated back to space), inducing climate change, which in turn impacts nature and society through such effects as warming, changes in precipitation and storm behavior, and sea-level rise. Finally, the anticipation and experience of effects of climate change encourage a range of human responses to prevent climate change, mitigate it, or adapt to it.

While climate change is truly a global phenomenon, most of the specific actions that lead to climate change and its impacts on nature and society take place at smaller scales. These scales vary geographically more than a billion-fold, from as small an area as a household, farm, or factory to the Earth as a whole. Figure 1 on the following page illustrates this range, indicating where major actions take place for each element of the chain of causes and consequences. In this simplified representation, four spatial scales divide the range: global, regional (continental, subcontinental, economic and political unions, and large nations), large area (small nations, states, provinces, large river basins, and areas that constitute 5 to 10 equatorial latitude and longitude degree grids),[1] and local (1 degree grid squares, small river basins, cities, households, farms, firms, and factories).

Action scales vary widely across the causal chain. Driving forces occur at all scales; for example, population serves as a driving force across all four scales. Atmospheric processes tend to fall into large-area or regional categories, while emissions, impacts, and responses are primarily local. It is important to note that although the entire assemblage of processes is commonly referred to as global climate change, only atmospheric concentrations of greenhouse gases that mix rapidly and the resulting radiative forcing are truly global in scale.

Global Change and Local Places Research Group

In *Global Change and Local Places: Estimating, Understanding and Reducing Greenhouse Gases*, the Association of American Geographers (AAG) Global Change and Local Places (GCLP) Research Group reports on a five-year study of four local places in the United States.[2] The research group had three main goals: to identify the bundle, or total package, of different greenhouse gas emissions (carbon dioxide, methane, and nitrous oxide) generated by each place over time, the forces that seem to have driven these emissions, and the potentials and limitations for those who live and work in each place to reduce or alter emissions.[3] This article summarizes the Global Change and Local Places Research Group's findings on making a global issue—in this case climate change—local.

Four Local Places

Each of the local places in the AAG study covers approximately one degree of equatorial latitude and longitude, an area of about 110 kilometers (69 miles) on a side—about the size of the state of Connecticut. An area this size provides a degree of land use diversity at each study site and enables the studies to nest conveniently into the 5 to 10 degree grid often used in climate modeling. The four areas offer varied climatic and socioeconomic conditions, ranging in climate from arid to humid, in livelihoods from industry to agriculture and forestry, and in economic well-being from vigor to stagnation. In choosing sites, it was also important that a study team could be found that was

Figure 1. Scale domains of climate change and consequences

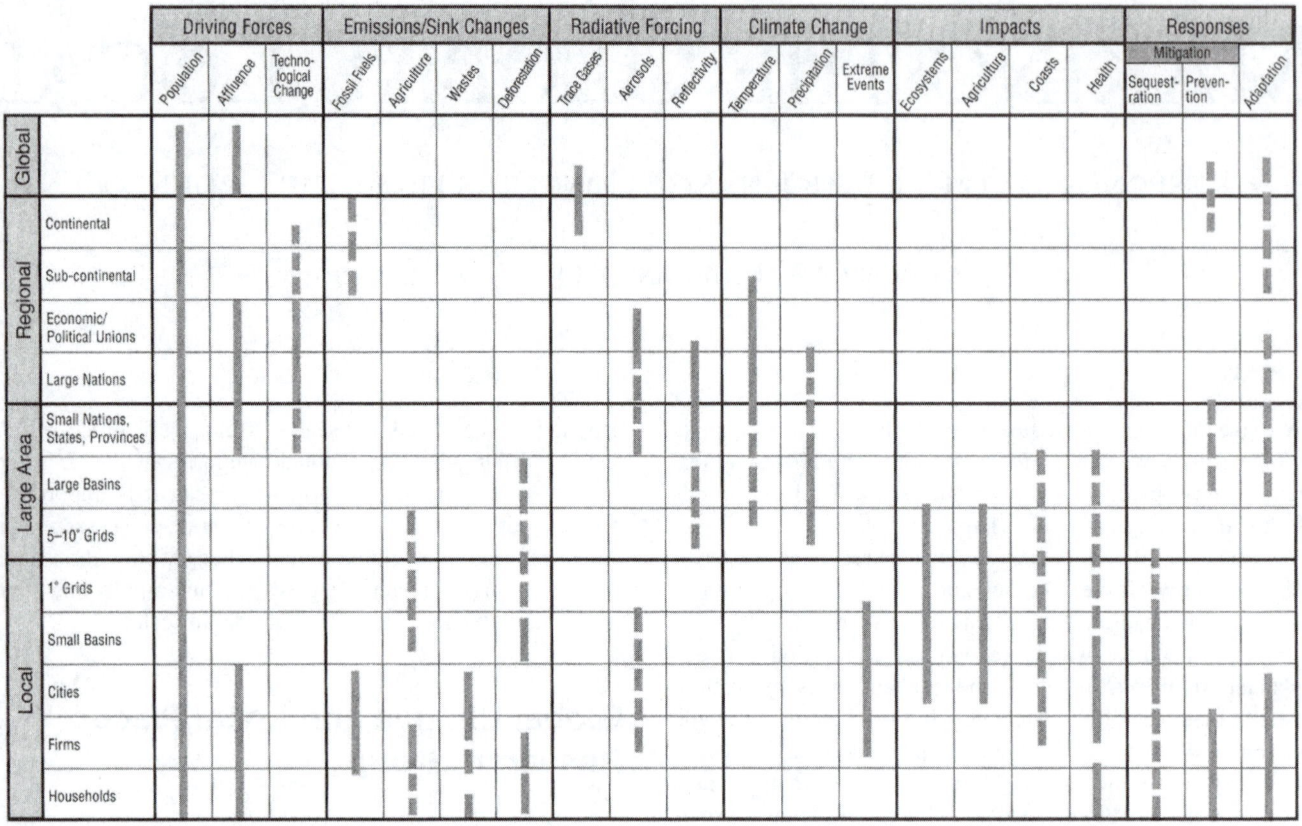

| | | Driving Forces | | | Emissions/Sink Changes | | | | Radiative Forcing | | | Climate Change | | | Impacts | | | | Responses | | |
| | | | | | | | | | | | | | | | | | | | Mitigation | | |
		Population	Affluence	Techno-logical Change	Fossil Fuels	Agriculture	Wastes	Deforestation	Trace Gases	Aerosols	Reflectivity	Temperature	Precipitation	Extreme Events	Ecosystems	Agriculture	Coasts	Health	Sequest-ration	Preven-tion	Adaptation
Global																					
Regional	Continental																				
	Sub-continental																				
	Economic/Political Unions																				
	Large Nations																				
Large Area	Small Nations, States, Provinces																				
	Large Basins																				
	5–10° Grids																				
Local	1° Grids																				
	Small Basins																				
	Cities																				
	Firms																				
	Households																				

NOTE: This chart depicts the scale of actions, not necessarily the focus of decision making. The dashed lines indicate occasional consequences or a lower level of confidence.

SOURCE: Association of American Geographers Global Change and Local Places Research Group, *Global Change in Local Places: Estimating, Understanding, and Reducing Greenhouse Gases.* (Cambridge, U.K.: Cambridge University Press, 2003)

based in or near the study site, with extensive local knowledge and contact networks. The lead researchers at each site were for the most part faculty and students in geography, planning, and environmental science departments.[4] The four places were named after the region in which they lay: southwestern Kansas, northwestern North Carolina, northwestern Ohio, and central Pennsylvania.

Southwestern Kansas

The southwestern Kansas study area lies in the center of the High Plains of the United States and includes six counties with the major towns of Garden City, Dodge City, and Liberal. The region, in the heart of the 1930s dust bowl, is flat, arid, and sparsely populated—in 2000, there were 110,333 people and more than 900,000 cows. Most of these cows are housed in large feedlots, fed by forage crops irrigated from the underground Ogallala Aquifer, and processed in five large meat-packing plants. This industry, plus the major natural gas field underlying the region, constitutes its economic base. The area is distant from major population centers, and its residents tend to view themselves as independent and skeptical of external interference. In recent years, however, Latin American and Asian immigrants seeking jobs in the area's agricultural processing facilities have added some diversity to what was a relatively homogeneous non-immigrant population.

Northwestern North Carolina

The northwestern North Carolina study area covers 12 counties with 870,246 people in 2000. The Blue Ridge Mountains and the rolling Piedmont—a region of foothills that lie between the mountains and the coastal plain—stretch through the area. Winston-Salem is the major urban center, although there are other small urban areas as well. In the mountains, forestry and tourism form the economic base. The Piedmont's economy includes manufacturing of small-scale furniture, electronics, and textiles as well as tobacco, beef cattle, and poultry production. Using a well-developed transportation network, many rural dwellers commute to work in urban industries. Population is growing with an influx of new residents, attracted both by job opportunities in a growing regional economy and the pleasant setting, with abundant recreational and scenic attractions.

Northwestern Ohio

The northwestern Ohio study area sits in the nation's industrial heartland. It contains 23 counties, a major city (Toledo), and in 2000, a population of 1,675,468. Although it contains significant agriculture, the area is dominated economically by a manufacturing economy that has undergone a major restructuring. The expansion of auto parts production and the replacement of older iron, steel, and chemical industrial plants by newer, cleaner plants are the main contributors to this economic change. Socially, the site has a high degree of ethnic and racial diversity, as is the case in most other North American urban-industrial centers. Politically, given its rust-belt history, there is a focus on issues of industrial economic viability and job creation, and these concerns give the area's major employers considerable political influence.

Central Pennsylvania

The central Pennsylvania study area's five counties lie among the hills and mountains of Appalachia, with a historic dependence on coal mining, logging, and small-scale agriculture. While still a major coal producer, the area also includes State College, the area's largest settlement and home to the main campus of Penn State University. The population of 336,224 (in 2000) now depends largely on the university and associated high-tech manufacturing and service industries for employment. Socially and politically, there are sharp contrasts between the conservative, traditional cultures of most counties (some hosting Amish and Mennonite enclaves) and the diversity and liberal attitudes of the large university community.

Three Questions

The study sought answers to three questions: What was the bundle of greenhouse gas emissions generated from each site from 1970 to 1990? What were the driving forces for these greenhouse gases and how are they likely to change between 1990 and 2020? What are the potentials and limitations for occupants of each site to reduce or alter present and future emissions?

To identify the bundle of greenhouse gas emissions, changes in the three major GHG emissions were estimated for 1970, 1980, and 1990. The estimations were for emissions for each county of each site, by major sources and sinks, including power plants, transportation, industry, households, agriculture, and forestry. Study teams converted these data to a common measure of greenhouse gas potential and compared the local data to state and national data. This was accomplished by adapting the U.S. Environmental Protection Agency (EPA)

statewide methodology to county-scale analysis by finding credible proxies for missing county-scale data such as a county's consumption of fossil fuels. Projections of emissions were made for 2000, 2010, and 2020 based on widely used socioeconomic forecasts adjusted by the use of local knowledge and insight.

> ## While climate change is truly a global phenomenon, most of the specific actions that lead to climate change and its impacts on nature and society take place at smaller scales.

To better understand the driving forces producing GHG emissions, the portion of these emissions used within a study site was allocated to three different user-groups, each responsible for different greenhouse gas-generating processes. The groups were defined as residential, industrial-commercial, and agricultural. Study teams considered variables that were specific to each user-group's consumption patterns to explain the growth or decline in emissions associated with each group. Generally, they followed the E=PAT Kaya identity in identifying variables, where the growth and decline in emissions are generally a function of population, affluence or economic growth, and production and consumption technologies.[5]

Study teams created two inventories of local emissions to identify how much control the local community had over different portions of the site's emissions bundle—and therefore how much control local people would have in reducing emissions. These inventories consisted of a source or production inventory, and a user or consumption inventory. To estimate local knowledge of global warming issues and the willingness to act to reduce emissions, site teams undertook hands-on investigative work. They surveyed a sample of householders through the mail and by telephone and directly interviewed managers or representatives of major commercial and industrial emitters. In addition, team members studied historic local analogs to greenhouse gas mitigation for insights about local responses to environmental challenges in the past. For example, the study group in Kansas found significant local responses to a problem of groundwater depletion, and in Ohio, they recorded responses to pollution in Lake Erie. Finally, site teams combined local knowledge with information from generic national and international assessments of greenhouse gas reduction potential. With this information, they assessed the potential for future mitigation in each site for conceivable emissions reduction targets.

Changing Emissions

Emissions at the four sites, using the international baseline year of 1990, were substantially but not dramatically different from global, national, or state emissions. When compared with state, national, and global averages for that year, local emissions differed moderately in the mix of greenhouse gases, somewhat more so in the mix of sources, and considerably in the amount of emissions per person or unit of area. Land use and land cover changes, which in principle reflect economic activities associated with emissions, as well as affecting an area's capacity for absorbing emissions and heat if the changes are significant, accounted for only minor differences in emissions and reflectivity. A major finding, however, was that emissions at the four sites differed greatly both between sites and compared to national trends in emissions between 1970 and 1990. Between 1970 and 1990, GHG emissions in the United States rose by 12 percent; 7 percent in the first decade and 5 percent more in the second. In contrast, emissions at the Kansas site declined in the first decade and rose in the second, declined steadily in both decades in Ohio, and more than doubled (164 percent increase) in the North Carolina site. Only emissions in central Pennsylvania mirrored the U.S. trajectory, albeit at a rate of increase twice that of the nation as a whole.

Emissions at the four sites differed greatly both between sites and compared to national trends in emissions between 1970 and 1990.

Five factors largely determine these differences in sources and rates of emission: whether there is a major source of electricity generation within the site; whether natural resource extraction or production is a significant part of the site's economy; whether the local economy grew or declined; whether such change was reflected in the number of households; and what technologies were in use over time. In a sense, the first two of these factors are random elements, a kind of spatial lottery that places a utility generating plant within the boundaries of a local area or situates primary production (agriculture, forestry, or mining) as a major part of the local economy. The final three, however, are a local expression of the same set of trends found at national and international scales: changes in population, affluence-consumption, and technology.

Given these modest differences, are local inventories necessary, or are national inventories and the increasing availability of state inventories—available for 40 U.S. states—sufficient for policy purposes? Site comparisons by category of emission tell us that national and state gas estimates for carbon dioxide in 1990 might suffice to inform local efforts to reduce emissions, although they would overlook local details that could be important. They would not be sufficient for methane or nitrous oxide, and using a national inventory guide to estimate local sources would clearly miss the importance of agriculture in Kansas, biomass burning in North Carolina, particular industrial processes in Ohio, and the preponderance of coal burning in Pennsylvania. But the major contribution of local inventories is in identifying trends in emissions through time. The study's estimates

of greenhouse gas emissions from 1970, 1980, and 1990 constitute a unique data set. This is true not only because they complement national and state estimates (most of which begin with 1990 rather than providing a longer time line), but because local emission trends through time cannot be determined from readily available sources, even though they reveal local causes and consequences that can be important for policymaking.

Who Is in Charge?

In the best of analytical worlds, it would be possible to take the bundle of greenhouse gas emissions emanating from a study site and simply ask, Who is in charge of the bundle? What portion of total emissions is beyond the control or decisions of the local communities? What portion is under the control of local residents? What portion is a joint decision? A federal regulation that required constructing a coal-fired plant atop a natural gas field in southwestern Kansas, after all, would seem to be outside local control. On the other hand, residents would appear to be responsible for the ten-degree range of heating and cooling they have as they set their home thermostats. And the purchase of an automobile seems to be a good example of a joint decision: local residents choose what kind of car or truck to buy, but their choices are constrained by styles and fuel efficiencies offered that are determined in Japan, Germany, or Detroit and influenced by fuel economy regulations set in Washington.

While the methodology used by the GCLP group reflects international and national practices and in fact sets a new standard for assessments at a local level, it may actually over- or underestimate the bundle of emissions produced directly by the local communities. Some activities and products that contribute to local emissions, while providing income to local people, might also benefit non-locals. Examples of this include natural gas, automobile, and furniture production. Much local consumption of goods, services, and electricity generated elsewhere is associated with emissions that often originate far from the study site. Moreover, a small but perhaps significant portion of the bundle may come from outside sources, as in the case of emissions from vehicles simply passing through the area.

In reality, most of the emissions bundle is almost certainly jointly controlled. History is a major participant. For instance, residents can alter the emissions emanating from their homes but only within a small range governed by who built the structure and how and when they did so. Also, renters in multifamily dwellings have little influence over such issues as insulation. On the other hand, even such external decisions as a distant utility office's choice of location for an electricity generating plant may have been influenced by local desires for economic development or by local demand. Mindful of these variables, the GCLP study focused on that portion of local emissions that might be reduced either through local actions or by local participation in joint multi-scale actions.

Emissions-Reducing Opportunities

Two standards for emissions reduction served as illustrative goals—the standard agreed to by the United States in Kyoto in

Table 1. Opportunities for GHG emissions reduction

Sector	IPPCC (Global)	DOE (National)	ICLEI (Local)	GCLP (Local)
Buildings	L	M-S	L	S
Industry	M-L	M	M-L	S-L
Transportation	M	M	M-L	M
Agriculture	M	N/A	N/A	L-N
Waste	S	N/A	S-L	N/A
Energy	M	L-M	N/A	L

NOTE: Opportunities were designated as large (L), moderate (M), small (S), negligible (N), not estimated (N/A), or are recorded as a range (For example, M-L indicates that opportunities ranged from medium to large). Target date for reductions set as 2020.

SOURCES: IPCC: Intergovernmental Panel on Climate Change, *Climate Change 2001: Mitigation* (Cambridge, U.K.: Cambridge University Press, 2001); DOE: M. A. Brown and M. Levine 1997, *Scenarios of U.S. Carbon Reductions: Potential Impacts of Energy-Efficient and Low-Carbon Technologies by 2010 and Beyond,* ORNL/CON-444; National Laboratory Directors 2001, *Technology Opportunities to Reduce U.S. Greenhouse Gas Emissions,* ORNL/FPO-01/2, accessible via http:// www.ornl.gov/climate_change; and Interlaboratory Working Group 2000, *Scenarios for a Clean Energy Future,* ORNL/CON-476; ICLEI: Regarding the international program, see http://www.iclei.org/co2/; for the U.S. program, see http://www.iclei.org/us/ccp/; and GCLP: Association of American Geographers Global Change and Local Places Research Group, *Global Change in Local Places: Estimating, Understanding, and Reducing Greenhouse Gases* (Cambridge, U.K.: Cambridge University Press, 2003).

1997 and a more challenging target undertaken by some of the world's cities. The Kyoto agreement specified a 7 percent reduction in 1990 emissions by 2010, while the more stringent target, sometimes referred to as the Toronto goal after the city that adopted it, suggested a 20 percent reduction over a similar period. Actual emission reductions for the four sites in the AAG study would need to be much larger than these percentages, of course, because they would also have to address growth in emissions associated with economic and demographic change that is currently projected for the next two decades. Based on the most plausible of the three emissions projections, overall reductions in emissions levels of 13 to 37 percent in 2020, depending on the site, would be required to reach a 7 percent reduction target, and 25 to 46 percent in 2020 to reach a 20 percent target. How difficult would it be to achieve such reductions?

Four different analyses identified opportunities to make reductions of this magnitude: an international study conducted by the Intergovernmental Panel for Climate Change (IPCC),[6] U.S. nationwide studies conducted by national laboratories for the Department of Energy,[7] a site-specific analysis using the methodology of the Cities for Climate Protection Program of the International Council of Local Environmental Initiatives (ICLEI),[8] and a local knowledge analysis by study site teams.[9] These studies differed in their technological and economic optimism and in their assumptions about political will, which led to different estimates of potentials from particular technology applications. Table 1 on this page summarizes these different perspectives. For instance, emitters in the energy production sector are among the major emitters in the GCLP and Department of Energy studies but were not considered in the ICLEI

analyses; buildings-related emissions are a major issue in the ICLET studies but not in the GCLP studies. Delving deeper into the four studies further illustrates the different perspectives. For example, the ICLEI analysis found that if two of the GCLP study areas applied known emissions reduction strategies, emphasizing energy efficiency improvement, emissions resulting from building and transportation energy use would drop significantly by 2020—20 to 50 percent for buildings and 50 percent for transportation. In contrast, the local area study teams were less optimistic about such prospects. The Department of Energy studies were prepared by scientists and engineers who are optimistic about technology, as were the IPCC studies. The Department of Energy conclusions, however, were seemingly pessimistic about prospects for strong governmental initiatives to mandate or encourage the adoption of such technology. Subsequent developments have borne out their caution; in recent years the United States has refused to accede to even the mild commitments of the Kyoto Protocol.

For the immediate future, then, the potentials for significant reductions in GHG emissions in the United States will likely depend on the decentralized actions of state, local, and corporate entities. Because the study sites offer a glimpse of what might happen on the local level, the next question that begs answering is whether local communities have the capacity—or the will—to take on such a challenge.

Can Local Places Reduce GHGs?

There are many technological possibilities to address major sources of greenhouse gases in the four study areas. Solutions include substituting natural gas for coal in electricity generating plants in North Carolina and Kansas, energy efficiency improvements in industrial plants in Pennsylvania and Ohio, improvements in transportation efficiencies in North Carolina, Ohio, and Kansas, and reducing methane emissions from cattle in Kansas. More specifically, opportunities include benefits from bio-mass burning (in place of fossil fuels consumption for industrial production) in North Carolina[10] and improving the efficiency of natural gas pipeline compressors in Kansas. In Ohio, shifting the industrial mix toward "cleaner" production activities and changing locational patterns to reduce distances to work were found as significant opportunities. Other options that address larger challenges for sustainable development in the localities can also lead to emissions reductions, such as growth management in North Carolina, irrigation efficiency in Kansas, sprawl limitation in Ohio, and diversification of the economy in Pennsylvania.

Willingness to Change

Local people at all four sites have significant opportunities to reduce greenhouse gas emissions. But how willing are localities to act upon them? How willing and able are local people to reduce emissions? Clues may be found in surveys of major emitters and households, an analysis of case studies of analogous response to a major environmental concern, and an examination of current participation in local efforts to reduce emissions.

At each of the four sites, local managers or representatives of major emitters were surveyed through face-to-face or telephone interviews. The emitters included local energy providers (electric power generation, natural gas, and the university), industrial producers (manufacturing and oil companies), transporters (trucking companies), and agriculture processors (feedlots and meat packers). These people were asked to describe their industry or activity and were asked about their knowledge and concern for global warming. Other questions involved their understanding of specific opportunities for emissions reduction, and where emissions generating and reduction decisions were made.

Local people at all four sites have significant opportunities to reduce greenhouse gas emissions. But how willing are localities to act upon them?

Those who were best informed about climate change issues and potential regulations associated with compliance with the Kyoto Protocol were the industrial emitters. Concerns were generally greatest in Ohio, because of the larger numbers of industrial producers there, and least in southwestern Kansas. But even for the most informed, there was little connection between global warming and the specifics of their local activities. At all of the four sites, local industries tended to focus on air pollution concerns rather than on climate change as affecting their activities. Perceived emission reduction opportunities differed considerably between the four sites, but for the most part they involved some form of energy efficiency improvement, and all those involved in electricity generation recognized the benefits of switching fuels and upgrading technologies. There was a widespread preference for state and local regulatory oversight rather than by the federal government, because state and local regulators were seen as being more trustworthy and understanding of local problems and capabilities.

In most cases, however, the decisions about applications of many of the technological opportunities could not be made by local managers. Although emissions abatement is primarily a local activity, it takes place within the context of larger-scale government and corporate policies. Few of the needed larger-scale government and corporate policies that might enable local action are currently in place in the four study areas. Exceptions included Toledo's membership in the Cities for Climate Protection program and the important local roles of BP-Amoco and Sun Oil, corporations committed to emissions reduction.

Types of incentives or mandates that would make a difference include tax credits for emissions-reducing investments, fees for carbon emissions, mandatory emissions limitations, and more stringent vehicle emissions standards. One combination of mandates and incentives would be an emissions cap defined in terms of permits to emit, allowing permits to be traded. In other words, companies that exceed their level of allowable emissions could purchase permits from companies that produce fewer emissions. In such a scenario, another incentive would be to allow U.S. emitters to receive credit for emissions reductions either through their own operations or by reducing emissions (possibly including biomass sequestration) elsewhere in the world.

A sample of householders was surveyed at each site to gauge local levels of public concern. The first survey was carried out in 1997, when global warming was beginning to emerge as a subject of public interest. Global warming at that time ranked low among householders' concerns. This finding seems to contradict national opinion polls of the time that showed higher levels of concern and willingness to reduce emissions, even at some cost. Opportunities for households to reduce emissions were generally perceived by householders as being mainly in transportation, space heating, and electricity use. But given the low level of concern, local residents were often unaware of opportunities for emission reduction, and there was little evidence of an inclination to act upon them once they were pointed out. An abbreviated survey was later undertaken in 2000, finding little change in the four local areas.

If, however, global warming and climate change become a greater public concern at the local scale in the United States, what might be the local response? Looking at the case studies that the site teams compiled is instructive here. These case studies, of an historic analog, show an effective local response to a major environmental problem in each area. In Kansas, researchers noted the local community's response to declining irrigation water levels in the Ogallala Aquifer. The other cases involved watershed protection in North Carolina, an effort to clean up pollution in Lake Erie, and air pollution abatement in Pennsylvania. These studies looked at the elements of success in each case and estimated the capacity and propensity of the local place to respond to global warming. Given the success of local participation in the issues, the analogs suggested that if localities become seriously concerned about global warming and aware of links to their own activities, they will be able to undertake major actions. For example, in Kansas the farming community widely adopted water conservation practices. In North Carolina, improvements were made in water quantity and quality, and the cleanup of Lake Erie in Ohio is an ongoing effort.

If global warming and climate change become a greater public concern at the local scale in the United States, what might be the local response?

Finally, the GCLP study examined real-time efforts in support of emissions reduction in the study sites and the relationship of study site teams to them. In northwestern Ohio, local branches of industrial firms that have recognized global warming as a real issue undertook major emissions reducing actions and encouraged the city of Toledo to join the International Council on Local Environmental Initiatives's Cities for Climate

Recent State, Corporate, and Local Actions to Reduce GHG Emissions

The United States is unique among the signatories of the Framework Convention on Climate Change in not seeking ratification of the Kyoto Protocol to reduce greenhouse gas (GHG) emissions. However, U.S. states, cities, corporations, and congregations, enabled or encouraged by their global and national counterparts, are moving to reduce GHGs, often by a greater degree than specified by the Kyoto quotas.[1]

In August 2001, the New England governors and eastern Canadian premiers adopted the Regional Climate Change Action Plan to jointly reduce regional GHG emissions to 1990 levels by 2010 and 10 percent below that by 2020. To do so, New Hampshire declared carbon dioxide a pollutant and will seek to regulate emissions for its power plants. Massachusetts capped carbon dioxide emissions of its six major emitting power plants. Other states have applied similar measures. New Jersey adopted a 2005 goal to reduce GHG emissions to 3.5 percent below 1990 levels, and Oregon has pioneered mandatory offsets for 17 percent of electric utility emissions. Especially noteworthy, California will now regulate carbon dioxide emissions from automobile exhaust, and New York state is considering a similar policy.[2]

Looking more at the local scale, it is impressive to note that 138 U.S. cities and counties, accounting for 16 percent of the U.S. population and about 16 percent of GHG emissions, participate in International Council for Local Environmental Initiatives's Cities for Climate Protection Campaign. Together they have already reduced their emissions by an estimated 10.4 million tons.[3] To date, 38 cities have completed the second of five milestones—setting an emissions target—with the most common target being a 20 percent reduction by 2010 below their 1990 emissions.

At least 38 major corporations have adopted targets for energy and/or emissions reduction, and a number of companies, such as BP-Amoco Oil (which has a target of reducing corporate emissions by 10 percent by 2010), are currently conducting intra-corporation emissions "cap and trade" programs.[4] In addition, 18 State Councils of Churches are participating in an Interfaith Global Climate Change Campaign.[5] In Maine, for example, individuals and families create carbon dioxide savings accounts equivalent to at least a 7 percent (Kyoto) reduction in 1990 levels of personal and household emissions.[6] These "bottom-up" initiatives not only act locally and globally but also create the interest and involvement that presage needed political change.

1. Regional, state, and corporate examples cited here are taken from Pew Center on Global Climate Change, "Climate Change Activities in the United States" (Arlington, VA. 2002), accessible via http://www.pewclimate.org/projects/us_activities.pdf (last accessed 21 February 2003).
2. Pew Center on Global Climate Change, note 1 above.
3. Susan Ode, Director of Outreach for the U.S. Office of Cities for Climate Protection, International Council for Local Environmental Initiatives, personal communication with authors, Berkeley, CA. 21 October 2002; International Council for Local Environmental Initiatives. U.S. Office, *U.S. CCP Milestone Progress*, 16 October 2002.
4. See the Interfaith Global Climate Change Campaigns web site, accessible via http://www.webofcreation.org/climate.html.
5. Pew Center on Global Climate Change, note 1 above.
6. See the Maine Interfaith Climate Change Initiative web site, accessible via http://www.mainecouncilofchurches.org/climate2.html.

Protection Program. In North Carolina, the GCLP study area team, with EPA funding, prepared an *Action Plan for Reducing North Carolina's Greenhouse Gas Emissions*.[11] At the southwestern Kansas site, research findings encouraged members of the state's congressional delegation to explore carbon sequestration in the soils of the area's agricultural fields.

Local "thinking" is insufficient for action because, for the most part, decisions about major emissions-reducing actions are made far from the local community.

Encouraging Local Initiatives

Given the assessment of both the capacity and willingness to undertake emissions reduction, it seems that significant emissions reductions, on the order of the changes required to meet either of the hypothetical targets, will be difficult for these four local places on their own. Between now and 2020, emissions will continue to grow at all of the sites, as well as for the country as a whole. Meeting even the modest illustrative target of Kyoto appears to be a major challenge for local places. To tackle this challenge would require at least three things:

• greater control for local communities over a significant portion of their emissions;
• local understanding and will derived from a shared belief that emissions reduction is in the interest of the area; and
• access for the community to technological and institutional means that are not currently available.

At the time of the study, most of the political or economic decisionmaking authority was beyond the scope of local residents and managers. The community's awareness of global warming was low in 1997 and only slightly higher in 2000. Even well-informed major emitters were, for the most part, reluctant or unable to undertake reduction actions on their own. At several sites, needed technologies for reduction were not available. At all sites the needed institutional framework of incentives and mandates is still absent. But it is also apparent, looking at the study of analogs and the community experience of Toledo, that a combination of global and local actions can be very effective under certain conditions.

Next Steps

Looking to the future, at least four actions would help to encourage local action to reduce greenhouse emissions. The growing evidence of impacts from global climate change, negative and positive, needs to be presented to the local community in a way that vulnerabilities are made clear. One starting point was the recent U.S. assessment *Climate Change Impacts on the*

United States: The Potential Consequences of Climate Variability and Change,[12] which identifies impacts relevant to the GCLP local sites such as increased drying of agricultural and ranch lands in southwestern Kansas and lowered Great Lakes levels for northwestern Ohio. But more can be done to involve local stakeholders in assessing their own vulnerabilities to climate change. More can be done as well to facilitate local community understanding of local expressions of climatic variability and of physical and biological changes due to climate change that are already under way.

GHG emissions reduction also needs to be linked to economic and environmental benefits to the local area, directly or indirectly. Most of the cities that have joined the Cities for Climate Protection campaign (see the box on previous page) have done so because of such local concerns as air pollution, energy costs, traffic, and urban sprawl. Cities have also signed on to the campaign to be more competitive for business and industrial growth—members of Cities for Climate Protection can claim that they are cleaner and more progressive than non-members.

Local innovations need to be encouraged by combining broader incentives and technical assistance with a high degree of local participation in designing and implementing strategies. Biomass burning in North Carolina or feedlot diet improvement in southwestern Kansas, as examples, are the result of local innovation.

Finally, technology improvements that are appropriate to local conditions need to be developed without significantly increasing costs or inconvenience. Ideally, these improvements would provide local opportunities—such as the potential for sequestering carbon in the prairie and plains soils of southwestern Kansas—to reduce net emissions from the area.

Thinking and Acting Globally and Locally

Simply stated, the Global Change in Local Places project shows that the beguiling slogan "Think globally and act locally" is insufficient to deal with climate change and its causes and consequences. Climate change is a global phenomenon, but global or even national "thinking" averages together too many distinctive local trajectories of greenhouse gas emissions and their driving forces, missing opportunities to reduce emissions and making local actions less specific. But local "thinking" is also insufficient for action because, for the most part, decisions about major emissions-reducing actions are made far from the local community. The GCLP experience suggests that thinking and acting both globally and locally involves at least three imperatives to succeed:

• *Make the global local.* To be able to think globally in a local place, the global must first be made local. To make such global knowledge local, locally based and trusted sources such as the collaborating GCLP regional institutions—with support from external information sources—can communicate current scientific understanding, identify local sources and responsibilities for greenhouse gas emissions, and suggest local sectors and places vulnerable to climate change. Local studies therefore become important as a way to transmit information and understanding. They then create support for national and state actions, and enable informed local action.[13]

• *Look beyond the local.* Local places must also think globally. Local studies, knowledge, and "thinking" are important, but understanding how and why a place creates its bundle of greenhouse gases requires an appreciation of driving forces beyond the local. The driving forces of local economic development, the associated increases in the number of households, and the technologies in use vary widely between sites. These driving forces are proximate, however, and are motivated by larger regional, national, and global processes related to industrial and economic organization. In addition, they are affected by social change in the workplace, household structure, and associated norms and values. The specifics of energy supply and price, environmental and land use regulation, and consumer and market demand for products that encapsulate emissions in their production and consumption are also factors. Leaders in local areas need to be fully aware of processes that operate at a larger scale. In this way, local communities can realize potentials that such processes support and avoid disillusionment from efforts that they undermine.

• *Act globally to act locally.* Reducing emissions is primarily rooted in local activities, but these actions take place within a context of corporate policy and regional, national, and global government. For local places to act to reduce their bundle of greenhouse gas emissions, they must have a desire to act, and they must have some control over a significant portion of their emissions. Most importantly, they must have access to the technological and institutional means of doing so. This requires enabling actions at larger scales—from market forces and corporate policies to actions of states, nations, and international agreements—that encourage localities to take action. The current framework of incentives and mandates does not encourage such local action, and the portfolio of available technological opportunities is often a poor match with the needs of specific places. Especially in this time when progress with greenhouse gas emission reduction through international agreements has been so disappointing,[14] localized action may be the most promising path toward addressing a global challenge if its potential can be realized. Recent actions by U.S. states, localities, corporations, and congregations offer some reasons to be encouraged (see the box on previous page). But the GCLP project strongly suggests that this prospect will remain largely a tantalizing dream unless government and business leaders at national and global scales are willing to give local communities more control over their activities, to develop more persuasive rewards for emission reduction initiatives, and to give communities technology options and other tools suited for local conditions.

NOTES

1. 5 to 10 equatorial latitude and longitude degree grids are commonly used in global climate models.

2. Association of American Geographers Global Change and Local Places Research Group, *Global Change in Local*

Places: Estimating, Understanding, and Reducing Greenhouse Gases (Cambridge, U.K.: Cambridge University Press, forthcoming). We are truly indebted to the NASA's Office of Earth Science for providing research support, to the book's lead editor, Ronald Abler, and to our fellow authors: David P. Angel, Clark University; Samuel A. Aryeetey-Attoh, University of Toledo; Susan L. Cutter, University of South Carolina; Jennifer DeHart, Allegheny College; Andrea S. Denny, Environmental Protection Agency; William E. Easterling, Pennsylvania State University; Douglas G. Goodin, Kansas State University; John Harrington Jr., Kansas State University; Lisa M. B. Harrington, Kansas State University; Arleen A. Hill, University of South Carolina; David G. Howard, University of Toledo; Sylvia-Linda Kaltins, Kansas State University; Robert W. Kates, independent scholar; C. Gregory Knight, Pennsylvania State University; David E. Kromm, Kansas State University; Peter S. Linquist, University of Toledo; Neal G. Lineback, Appalachian State University; Michael W. Mayfield, Appalachian State University; Jerry T. Mitchell, Bloomsburg University of Pennsylvania; William A. Muraco, University of Toledo; Colin Polsky, Harvard University; Neil Reid, University of Toledo; Audrey Reynolds, University of Texas, Austin; Robin Shudak, U.S. Environmental Protection Agency; Stephen E. White, Kansas State University; Thomas J. Wilbanks, Oak Ridge National Laboratory; and Brent Yarnal, Pennsylvania State University.

3. Excluded from the GCLP schema were local impacts of global climate change. At the time the project began, capabilities to produce reliable impact estimates for small areas did not exist. Were a similar project being designed today, of course, impact estimates would most likely be included as a fourth module.

4. These were: Kansas: the Department of Geography at Kansas State University; North Carolina: the Department of Geography and Planning at Appalachian State College; Ohio: the Department of Geography and Planning at the University of Toledo; and Pennsylvania: the Department of Geography and the Center for Integrated Regional Assessment at Pennsylvania State University.

5. Y. Kaya, "Impact of Carbon Dioxide Emission Control on GNP Growth: Interpretation of Proposed Scenarios" (paper presented to the IPCC Energy and Industry subgroup, Response Strategies Subgroup, Paris (mimeo), 1990).

6. Intergovernmental Panel on Climate Change, *Climate Change 2001: Mitigation* (Cambridge, U.K.: Cambridge University Press, 2001).

7. M. A. Brown and M. Levine 1997, *Scenarios of U. S. Carbon Reductions: Potential Impacts of Energy-Efficient and Low-Carbon Technologies by 2010 and Beyond*, ORNL/CON-444; National Laboratory Directors 2001, *Technology Opportunities to Reduce U.S. Greenhouse Gas Emissions*, ORNL/FPO-0l/2, accessible via http://www.ornl.gov/climate_change; and Interlaboratory Working Group 2000, *Scenarios for a Clean Energy Future*, ORNL/CON-476.

8. Regarding the international program, see http://www.iclei.org/ co2/; for the U.S. program, see http://www.iclei.org/us/ccp/.

9. Association of American Geographers, note 2 above.

10. N. G. Lineback et al., "Industrial Greenhouse Gas Emissions: Does CO_2 from Burning Biomass Really Matter?" *Climate Research*, 13, no. 3 (1999): 221–9.

11. Appalachian State University Department of Geography and Planning, *North Carolina's Sensible Greenhouse Gas Reduction Strategies* (Boone. N.C., 1999).

12. National Assessment Synthesis Team, *Climate Change Impacts on the United States: The Potential Consequences of Climate Variability and Change*, U.S. Global Change Research Program (Washington, D.C., 2000).

13. The GCLP study groups are listed in note 4, above.

14. J. G. Speth, "A New Green Regime: Attacking the Root Causes of Global Environmental Deterioration," *Environment*, September 2002, 16–25.

Robert W. Kates, an independent scholar, is currently co-convenor of the Initiative for Science and Technology for Sustainability and co-principal investigator of the Association of American Geographers–Global Change in Local Places (AAG-GCLP) research group. He is a university professor emeritus at Brown University, where he served as director of the Feinstein World Hunger Program. Kates was awarded the National Medal of Science in 1991 for his work on hunger, environment, and natural hazards. He is an executive editor of *Environment* and can be contacted at rkates@acadia.net. Thomas J. Wilbanks is a corporate research fellow and leader of Global Change and Developing Country Programs at Oak Ridge National Laboratory in Oak Ridge, Tennessee. His research focuses on sustainable development, responses to concerns about impacts of global change, issues of geographic scale and scale interactions. He is currently co-principal investigator of the AAG-GCLP research group. He has previously served as chair of geography committees at the National Academy of Sciences and has authored or co-authored numerous books and journal articles on sustainable development and global change. Wilbanks is a contributing editor at *Environment* and can be contacted at (865) 574-5515 or via e-mail at twz@ornl.gov. The authors retain copyright.

From *Environment*, April 2003. Published by Heldref Publications, 1319 Eighteenth St., NW, Washington, DC 20036-1802. © 2003.

UNIT 2
Population, Policy, and Economy

Unit Selections

6. **Population and Consumption: What We Know, What We Need to Know**, Robert W. Kates
7. **An Economy for the Earth**, Lester R. Brown
8. **Factory Farming in the Developing World**, Danielle Nierenberg
9. **Common Ground for Farmers and Forests**, Joyce Gregory Wyels
10. **Where the Sidewalks End**, Molly O'Meara Sheehan

Key Points to Consider

- Why should policy makers in the more developed countries of the world become more aware of the true dimensions of the world's food problem? How can increased awareness of food scarcity and misallocation lead to solutions for both food production and environmental protection?

- How are present economic theories insufficient in addressing environmental issues? How might an "eco-economy" take on different characteristics than the traditional economies that exist in the world's developing countries?

- How does "factory farming," which may increase food production, also increase problems of environmental contamination? What is the relationship between regulations developed to control factory farming and the migration of this agrobusiness system to new areas?

- How can the development of community-based sustainable agriculture slow the rate of deforestation in tropical areas of the world? Are some of the oldest forms of tropical agriculture really the best in terms of sound ecological principles?

 **Links: www.dushkin.com/online/**
These sites are annotated in the World Wide Web pages.

The Hunger Project
http://www.thp.org
Poverty Mapping
http://www.povertymap.net
World Health Organization
http://www.who.int
World Population and Demographic Data
http://geography.about.com/cs/worldpopulation/
WWW Virtual Library: Demography & Population Studies
http://demography.anu.edu.au/VirtualLibrary/

One of the greatest setbacks on the road to the development of more stable and sensible population policies came about as a result of inaccurate population growth projections made in the late 1960s and early 1970s. The world was in for a population explosion, the experts told us back then. But shortly after the publication of the heralded works *The Population Bomb* (Paul Ehrlich, 1975) and *Limits to Growth* (D. H. Meadows et al., 1974), the growth rate of the world's population began to decline slightly. There was no cause and effect relationship at work here. The decline in growth was simply demographic transition at work, a process in which declining population growth tends to accompany increasing levels of economic development. Since the alarming predictions did not come to pass, the world began to relax a little. However, two facts still remain: population growth in biological systems must be limited by available resources, and the availability of Earth's resources is finite.

That population growth cannot continue indefinitely is a mathematical certainty. But it is also a certainty that contemporary notions of a continually expanding economy must give way before the realities of a finite resource base. Consider the following: In developing countries, high and growing rural population densities have forced the use of increasingly marginal farmland once considered to be too steep, too dry, too wet, too sterile, or too far from markets for efficient agricultural use. Farming this land damages soil and watershed systems, creates deforestation problems, and adds relatively little to total food production. In the more developed world, farmers also have been driven—usually by market forces—to farm more marginal lands and to rely more on environmentally harmful farming methods utilizing high levels of agricultural chemicals (such as pesticides and artificial fertilizers). These chemicals create hazards for all life and rob the soil of its natural ability to renew itself.

The increased demand for economic expansion has also created an increase in the use of precious groundwater reserves for irrigation purposes, depleting those reserves beyond their natural capacity to recharge and creating the potential for once-fertile farmland and grazing land to be transformed into desert. The continued demand for higher production levels also contributes to a soil erosion problem that has reached alarming proportions in all agricultural areas of the world, whether high or low on the scale of economic development. The need to increase the food supply and its consequent effects on the agricultural environment are not the only results of continued population growth. For industrialists, the larger market creates an almost irresistible temptation to accelerate production, requiring the use of more marginal resources and resulting in the destruction of more fragile ecological systems, particularly in the tropics. For consumers, the increased demand for products means increased competition for scarce resources, driving up the cost of those resources until only the wealthiest can afford what our grandfathers would have viewed as an adequate standard of living.

The articles selected for this second unit all relate, in one way or another, to the theory and reality of population growth and its relationship to public policy and economic growth. In the first se-

lection, "Population and Consumption: What We Know and What We Need to Know," geographer and MacArthur Fellow Robert Kates argues that the present set of environmental problems is tied to both the expanding human population in strict numerical terms and the tendency of that growing population to demand more per capita shares of the world's dwindling resources. In the following selection, "An Economy for Earth," Lester Brown takes a theoretical approach in contending that economic theory does not explain why so many environmental problems exist, whereas ecological theory does. What is required is a merger of the two sets of theories.

The next two articles in this section move from the theoretical to the practical and from global to local scales in addressing issues of agriculture and environment. In "Factory Farming in the Developing World," Danielle Nierenberg of the World Watch Institute notes that, while factory farming has allowed meat to become more common in the diet of developing countries, the farming system itself produces significant damage at both local and global scales. She suggests that part of the problem is in thinking about factory farms—or the presence of meat in the diet—as symbols of wealth. Certain types of farming as symbols of wealth also occupy travel writer Joyce Gregory Wyels, who notes, in "Common Ground for Farmers and Forests," that tropical deforestation in Central America in general and Costa Rica in particular is the consequence of commercial farming systems that are viewed as economically "good." Wyels suggests that the tourism value of Costa Rica's tropical forests exceeds the value of farming and recommends the development of agricultural policies that would maintain small-scale forest farming and environmental integrity while still achieving economic stability.

The last article in the unit addresses the question of poverty, particularly that existing in the world's urban slums—another manifestation of the globalization of economic systems. Worldwatch Institute researcher Molly O'Meara Sheehan, in "Where the Sidewalk Ends," describes the plight of the urban poor. As more agricultural, mineral, and forest production is demanded by economic growth, the traditional patterns of subsistence and village life are broken down and enormous numbers of unskilled and destitute rural people flee the countryside for the cities. Here they exist without benefit of the urban infrastructure of sewers, utilities, and transportation that makes urban life livable for millions of the world's people. These urban poor and the slums they inhabit pose an enormous economic, social, and political challenge for developing regions.

All of the authors of selections in this unit make it clear that the global environment is being stressed by population growth as well as environmental and economic policies that result in more environmental pressure and degradation. While it should be evident that we can no longer afford to permit the unplanned and unchecked growth of the planet's dominant species, it should also be apparent that the unchecked growth of economic systems without some kind of environmental accounting systems is just as dangerous.

Population and Consumption

What We Know, What We Need to Know

by Robert W. Kates

T hirty years ago, as Earth Day dawned, three wise men recognized three proximate causes of environmental degradation yet spent half a decade or more arguing their relative importance. In this classic environmentalist feud between Barry Commoner on one side and Paul Ehrlich and John Holdren on the other, all three recognized that growth in population, affluence, and technology were jointly responsible for environmental problems, but they strongly differed about their relative importance. Commoner asserted that technology and the economic system that produced it were primarily responsible.[1] Ehrlich and Holdren asserted the importance of all three drivers: population, affluence, and technology. But given Ehrlich's writings on population,[2] the differences were often, albeit incorrectly, described as an argument over whether population or technology was responsible for the environmental crisis.

Now, 30 years later, a general consensus among scientists posits that growth in population, affluence, and technology are jointly responsible for environmental problems. This has become enshrined in a useful, albeit overly simplified, identity known as IPAT, first published by Ehrlich and Holdren in *Environment* in 1972[3] in response to the more limited version by Commoner that had appeared earlier in *Environment* and in his famous book *The Closing Circle*.[4] In this identity, various forms of environmental or resource impacts (I) equals population (P) times affluence (A) (usually income per capita) times the impacts per unit of income as determined by technology (T) and the institutions that use it. Academic debate has now shifted from the greater or lesser importance of each of these driving forces of environmental degradation or resource depletion to debate about their interaction and the ultimate forces that drive them.

However, in the wider global realm, the debate about who or what is responsible for environmental degradation lives on. Today, many Earth Days later, international debates over such major concerns as biodiversity, climate change, or sustainable development address the population and the affluence terms of Holdrens' and Ehrlich's identity, specifically focusing on the character of consumption that affluence permits. The concern with technology is more complicated because it is now widely recognized that while technology can be a problem, it can be a

solution as well. The development and use of more environmentally benign and friendly technologies in industrialized countries have slowed the growth of many of the most pernicious forms of pollution that originally drew Commoner's attention and still dominate Earth Day concerns.

A recent report from the National Research Council captures one view of the current public debate, and it begins as follows:

For over two decades, the same frustrating exchange has been repeated countless times in international policy circles. A government official or scientist from a wealthy country would make the following argument: The world is threatened with environmental disaster because of the depletion of natural resources (or climate change or the loss of biodiversity), and it cannot continue for long to support its rapidly growing population. To preserve the environment for future generations, we need to move quickly to control global population growth, and we must concentrate the effort on the world's poorer countries, where the vast majority of population growth is occurring.

Government officials and scientists from low-income countries would typically respond:

If the world is facing environmental disaster, it is not the fault of the poor, who use few resources. The fault must lie with the world's wealthy countries, where people consume the great bulk of the world's natural resources and energy and cause the great bulk of its environmental degradation. We need to curtail overconsumption in the rich countries which use far more than their fair share, both to preserve the environment and to allow the poorest people on earth to achieve an acceptable standard of living.[5]

It would be helpful, as in all such classic disputes, to begin by laying out what is known about the relative responsibilities of both population and consumption for the environmental crisis, and what might need to be known to address them. However, there is a profound asymmetry that must fuel the frustration of the developing countries' politicians and scientists: namely, how much people know about population and how little they know about consumption. Thus, this article begins by examining these differences in knowledge and action and concludes with the alternative actions needed to go from more to enough in both population and consumption.[6]

Population

What population is and how it grows is well understood even if all the forces driving it are not. Population begins with people and their key events of birth, death, and location. At the margins, there is some debate over when life begins and ends or whether residence is temporary or permanent, but little debate in between. Thus, change in the world's population or any place is the simple arithmetic of adding births, subtracting deaths, adding immigrants, and subtracting outmigrants. While whole subfields of demography are devoted to the arcane details of these additions and subtractions, the error in estimates of population for almost all places is probably within 20 percent and for countries with modern statistical services, under 3 percent—better estimates than for any other living things and for most other environmental concerns.

Current world population is more than six billion people, growing at a rate of 1.3 percent per year. The peak annual growth rate in all history—about 2.1 percent—occurred in the early 1960s, and the peak population increase of around 87 million per year occurred in the late 1980s. About 80 percent or 4.8 billion people live in the less developed areas of the world, with 1.2 billion living in industrialized countries. Population is now projected by the United Nations (UN) to be 8.9 billion in 2050, according to its medium fertility assumption, the one usually considered most likely, or as high as 10.6 billion or as low as 7.3 billion.[7]

A general description of how birth rates and death rates are changing over time is a process called the demographic transition.[8] It was first studied in the context of Europe, where in the space of two centuries, societies went from a condition of high births and high deaths to the current situation of low births and low deaths. In such a transition, deaths decline more rapidly than births, and in that gap, population grows rapidly but eventually stabilizes as the birth decline matches or even exceeds the death decline. Although the general description of the transition is widely accepted, much is debated about its cause and details.

The world is now in the midst of a global transition that, unlike the European transition, is much more rapid. Both births and deaths have dropped faster than experts expected and history foreshadowed. It took 100 years for deaths to drop in Europe compared to the drop in 30 years in the developing world. Three is the current global average births per woman of reproductive age. This number is more than halfway between the average of five children born to each woman at the post World War II peak of population growth and the average of 2.1 births required to achieve eventual zero population growth.[9] The death transition is more advanced, with life expectancy currently at 64 years. This represents three-quarters of the transition between a life expectancy of 40 years to one of 75 years. The current rates of decline in births outpace the estimates of the demographers, the UN having reduced its latest medium ex-

pectation of global population in 2050 to 8.9 billion, a reduction of almost 10 percent from its projection in 1994.

Demographers debate the causes of this rapid birth decline. But even with such differences, it is possible to break down the projected growth of the next century and to identify policies that would reduce projected populations even further. John Bongaarts of the Population Council has decomposed the projected developing country growth into three parts and, with his colleague Judith Bruce, has envisioned policies that would encourage further and more rapid decline.[10] The first part is unwanted fertility, making available the methods and materials for contraception to the 120 million married women (and the many more unmarried women) in developing countries who in survey research say they either want fewer children or want to space them better. A basic strategy for doing so links voluntary family planning with other reproductive and child health services.

Yet in many parts of the world, the desired number of children is too high for a stabilized population. Bongaarts would reduce this desire for large families by changing the costs and benefits of childrearing so that more parents would recognize the value of smaller families while simultaneously increasing their investment in children. A basic strategy for doing so accelerates three trends that have been shown to lead to lower desired family size: the survival of children, their education, and improvement in the economic, social, and legal status for girls and women.

However, even if fertility could immediately be brought down to the replacement level of two surviving children per woman, population growth would continue for many years in most developing countries because so many more young people of reproductive age exist. So Bongaarts would slow this momentum of population growth by increasing the age of childbearing, primarily by improving secondary education opportunity for girls and by addressing such neglected issues as adolescent sexuality and reproductive behavior.

How much further could population be reduced? Bongaarts provides the outer limits. The population of the developing world (using older projections) was expected to reach 10.2 billion by 2100. In theory, Bongaarts found that meeting the unmet need for contraception could reduce this total by about 2 billion. Bringing down desired family size to replacement fertility would reduce the population a billion more, with the remaining growth—from 4.5 billion today to 7.3 billion in 2100—due to population momentum. In practice, however, a recent U.S. National Academy of Sciences report concluded that a 10 percent reduction is both realistic and attainable and could lead to a lessening in projected population numbers by 2050 of upwards of a billion fewer people.[11]

Consumption

In contrast to population, where people and their births and deaths are relatively well-defined biological events, there is no consensus as to what consumption includes. Paul Stern of the National Research Council has described the different ways physics, economics, ecology, and sociology view consumption.[12] For physicists, matter and energy cannot be consumed, so consumption is conceived as transformations of matter and energy with increased entropy. For economists, consumption is spending on consumer goods and services and thus distinguished from their production and distribution. For ecologists, consumption is obtaining energy and nutrients by eating something else, mostly green plants or other consumers of green plants. And for some sociologists, consumption is a status symbol—keeping up with the Joneses—when individuals and households use their incomes to increase their social status through certain kinds of purchases. These differences are summarized in the box below.

In 1977, the councils of the Royal Society of London and the U.S. National Academy of Sciences issued a joint statement on consumption, having previously done so on population. They chose a variant of the physicist's definition:

> *Consumption is the human transformation of materials and energy. Consumption is of concern to the extent that it makes the transformed materials or energy less available for future use, or negatively impacts biophysical systems in such a way as to threaten human health, welfare, or other things people value.*[13]

On the one hand, this society/academy view is more holistic and fundamental than the other definitions; on the other hand, it is more focused, turning attention to the environmentally damaging. This article uses it as a working definition with one modification, the addition of information to energy and matter, thus completing the triad of the biophysical and ecological basics that support life.

In contrast to population, only limited data and concepts on the transformation of energy, materials, and information exist.[14] There is relatively good global knowledge of energy transformations due in part to the common units of conversion between different technologies. Between 1950 and today, global energy production and use increased more than fourfold.[15] For material transformations, there are no aggregate data in common units on a global basis, only for some specific classes of materials including materials for energy production, construction, industrial minerals and metals, agricultural crops, and water.[16] Calculations of material use by volume, mass, or value lead to different trends.

Trend data for per capita use of physical structure materials (construction and industrial minerals, metals, and forestry products) in the United States are relatively complete. They show an inverted S shaped (logistic) growth pattern: modest doubling between 1900 and the depression of the 1930s (from two to four metric tons), followed by a steep quintupling with economic recovery until the early 1970s (from two to eleven tons), followed by a leveling off since then with fluctuations related to economic downturns (see Figure 1).[17] An aggregate analysis of all current material production and consumption in the United States averages more than 60 kilos per person per day (excluding water). Most of this material flow is split between energy and related products (38 percent) and minerals for construction (37 percent), with the remainder as industrial minerals (5 percent), metals (2 percent), products of fields (12 percent), and forest (5 percent).[18]

A massive effort is under way to catalog biological (genetic) information and to sequence the genomes of microbes, worms, plants, mice, and people. In contrast to the molecular detail, the number and diversity of organisms is unknown, but a conservative estimate places the number of species on the order of 10 million, of which only one-tenth have been described.[19] Although there is much interest and many anecdotes, neither concepts nor data are available on most cultural information. For example, the number of languages in the world continues to decline while the number of messages expands exponentially.

What Is Consumption?

Physicist: "What happens when you transform matter/energy"

Ecologist: "What big fish do to little fish"

Economist: "What consumers do with their money"

Sociologist: "What you do to keep up with the Joneses"

Trends and projections in agriculture, energy, and economy can serve as surrogates for more detailed data on energy and material transformation.[20] From 1950 to the early 1990s, world population more than doubled (2.2 times), food as measured by grain production almost tripled (2.7 times), energy more than quadrupled (4.4 times), and the economy quintupled (5.1 times). This 43-year record is similar to a current 55-year projection (1995–2050) that assumes the continuation of current trends or, as some would note, "business as usual." In this 55-year projection, growth in half again of population (1.6 times) finds almost a doubling of agriculture (1.8 times), more than twice as much energy used (2.4 times), and a quadrupling of the economy (4.3 times).[21]

Thus, both history and future scenarios predict growth rates of consumption well beyond population. An attractive similarity exists between a demographic transition that moves over time from high births and high deaths to low births and low deaths with an energy, materials, and information transition. In this transition, societies will use increasing amounts of energy and materials as consumption increases, but over time the energy and materials input per unit of consumption decrease and information substitutes for more material and energy inputs.

Some encouraging signs surface for such a transition in both energy and materials, and these have been variously labeled as decarbonization and dematerialization.[22] For more than a century, the amount of carbon per unit of energy produced has been decreasing. Over a shorter period, the amount of energy used to produce a unit of production has also steadily declined. There is also evidence for dematerialization, using fewer materials for a unit of production, but only for industrialized countries and for some specific materials. Overall, improvements in technology

Figure 1. Consumption of physical structure materials in the United States, 1900-1991

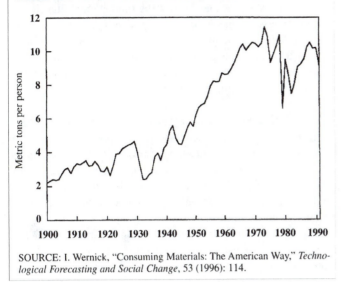

SOURCE: I. Wernick, "Consuming Materials: The American Way," *Technological Forecasting and Social Change*, 53 (1996): 114.

and substitution of information for energy and materials will continue to increase energy efficiency (including decarbonization) and dematerialization per unit of product or service. Thus, over time, less energy and materials will be needed to make specific things. At the same time, the demand for products and services continues to increase, and the overall consumption of energy and most materials more than offsets these efficiency and productivity gains.

What to Do about Consumption

While quantitative analysis of consumption is just beginning, three questions suggest a direction for reducing environmentally damaging and resource-depleting consumption. The first asks: *When is more too much for the life-support systems of the natural world and the social infrastructure of human society?* Not all the projected growth in consumption may be resource-depleting— "less available for future use"—or environmentally damaging in a way that "negatively impacts biophysical systems to threaten human health, welfare, or other things people value."[23] Yet almost any human-induced transformations turn out to be either or both resource-depleting or damaging to some valued environmental component. For example, a few years ago, a series of eight energy controversies in Maine were related to coal, nuclear, natural gas, hydroelectric, biomass, and wind generating sources, as well as to various energy policies. In all the controversies, competing sides, often more than two, emphasized environmental benefits to support their choice and attributed environmental damage to the other alternatives.

Despite this complexity, it is possible to rank energy sources by the varied and multiple risks they pose and, for those concerned, to choose which risks they wish to minimize and which they are more willing to accept. There is now almost 30 years of experience with the theory and methods of risk assessment and 10 years of experience with the identification and setting of en-

vironmental priorities. While there is still no readily accepted methodology for separating resource-depleting or environmentally damaging consumption from general consumption or for identifying harmful transformations from those that are benign, one can separate consumption into more or less damaging and depleting classes and *shift* consumption to the less harmful class. It is possible to *substitute* less damaging and depleting energy and materials for more damaging ones. There is growing experience with encouraging substitution and its difficulties: renewables for nonrenewables, toxics with fewer toxics, ozone-depleting chemicals for more benign substitutes, natural gas for coal, and so forth.

The second question, *Can we do more with less?*, addresses the supply side of consumption. Beyond substitution, shrinking the energy and material transformations required per unit of consumption is probably the most effective current means for reducing environmentally damaging consumption. In the 1997 book, *Stuff: The Secret Lives of Everyday Things*, John Ryan and Alan Durning of Northwest Environment Watch trace the complex origins, materials, production, and transport of such everyday things as coffee, newspapers, cars, and computers and highlight the complexity of reengineering such products and reorganizing their production and distribution.[24]

Yet there is growing experience with the three Rs of consumption shrinkage: reduce, recycle, reuse. These have now been strengthened by a growing science, technology, and practice of industrial ecology that seeks to learn from nature's ecology to reuse everything. These efforts will only increase the existing favorable trends in the efficiency of energy and material usage. Such a potential led the Intergovernmental Panel on Climate Change to conclude that it was possible, using current best practice technology, to reduce energy use by 30 percent in the short run and 50–60 percent in the long run.[25] Perhaps most important in the long run, but possibly least studied, is the potential for and value of substituting information for energy and materials. Energy and materials per unit of consumption are going down, in part because more and more consumption consists of information.

The third question addresses the demand side of consumption—*When is more enough?*[26] Is it possible to reduce consumption by more satisfaction with what people already have, by *satiation*, no more needing more because there is enough, and by *sublimation*, having more satisfaction with less to achieve some greater good? This is the least explored area of consumption and the most difficult. There are, of course, many signs of *satiation* for some goods. For example, people in the industrialized world no longer buy additional refrigerators (except in newly formed households) but only replace them. Moreover, the quality of refrigerators has so improved that a 20-year or more life span is commonplace. The financial pages include frequent stories of the plight of this industry or corporation whose markets are saturated and whose products no longer show the annual growth equated with profits and progress. Such enterprises are frequently viewed as failures of marketing or entrepreneurship rather than successes in meeting human needs sufficiently and efficiently. Is it possible to reverse such views, to create a standard of satiation, a satisfaction in a need well met?

Can people have more satisfaction with what they already have by using it more intensely and having the time to do so? Economist Juliet Schor tells of some overworked Americans who would willingly exchange time for money, time to spend with family and using what they already have, but who are constrained by an uncooperative employment structure.[27] Proposed U.S. legislation would permit the trading of overtime for such compensatory time off, a step in this direction. *Sublimation*, according to the dictionary, is the diversion of energy from an immediate goal to a higher social, moral, or aesthetic purpose. Can people be more satisfied with less satisfaction derived from the diversion of immediate consumption for the satisfaction of a smaller ecological footprint?[28] An emergent research field grapples with how to encourage consumer behavior that will lead to change in environmentally damaging consumption.[29]

A small but growing "simplicity" movement tries to fashion new images of "living the good life."[30] Such movements may never much reduce the burdens of consumption, but they facilitate by example and experiment other less-demanding alternatives. Peter Menzel's remarkable photo essay of the material goods of some 30 households from around the world is powerful testimony to the great variety and inequality of possessions amidst the existence of alternative life styles.[31] Can a standard of "more is enough" be linked to an ethic of "enough for all"? One of the great discoveries of childhood is that eating lunch does not feed the starving children of some far-off place. But increasingly, in sharing the global commons, people flirt with mechanisms that hint at such—a rationing system for the remaining chlorofluorocarbons, trading systems for reducing emissions, rewards for preserving species, or allowances for using available resources.

A recent compilation of essays, *Consuming Desires: Consumption, Culture, and the Pursuit of Happiness*,[32] explores many of these essential issues. These elegant essays by 14 well-known writers and academics ask the fundamental question of why more never seems to be enough and why satiation and sublimation are so difficult in a culture of consumption. Indeed, how is the culture of consumption different for mainstream America, women, inner-city children, South Asian immigrants, or newly industrializing countries?

Why We Know and Don't Know

In an imagined dialog between rich and poor countries, with each side listening carefully to the other, they might ask themselves just what they actually know about population and consumption. Struck with the asymmetry described above, they might then ask: "Why do we know so much more about population than consumption?"

The answer would be that population is simpler, easier to study, and a consensus exists about terms, trends, even policies. Consumption is harder, with no consensus as to what it is, and with few studies except in the fields of marketing and advertising. But the consensus that exists about population comes from substantial research and study, much of it funded by governments and groups in rich countries, whose asymmetric concern readily identifies the troubling fertility behavior of others and only reluctantly considers their own consumption behavior. So while consumption is harder, it is surely studied less (see Table 1).

The asymmetry of concern is not very flattering to people in developing countries. Anglo-Saxon tradition has a long history of dominant thought holding the poor responsible for their con-

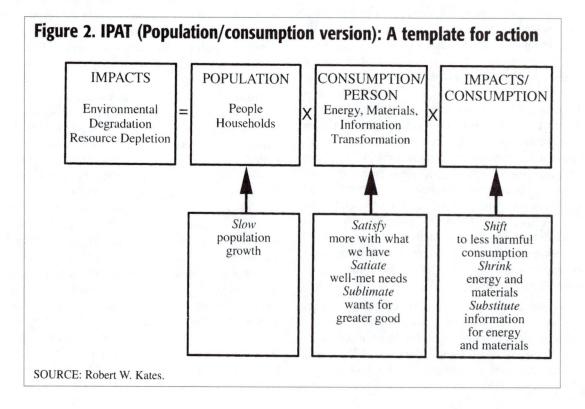

Figure 2. IPAT (Population/consumption version): A template for action

IMPACTS		POPULATION		CONSUMPTION/ PERSON		IMPACTS/ CONSUMPTION
Environmental Degradation Resource Depletion	=	People Households	X	Energy, Materials, Information Transformation	X	

| | | *Slow* population growth | | *Satisfy* more with what we have *Satiate* well-met needs *Sublimate* wants for greater good | | *Shift* to less harmful consumption *Shrink* energy and materials *Substitute* information for energy and materials |

SOURCE: Robert W. Kates.

dition—they have too many children—and an even longer tradition of urban civilization feeling besieged by the barbarians at their gates. But whatever the origins of the asymmetry, its persistence does no one a service. Indeed, the stylized debate of population versus consumption reflects neither popular understanding nor scientific insight. Yet lurking somewhere beneath the surface concerns lies a deeper fear.

Table 1. A comparison of population and consumption

Population	Consumption
Simpler, easier to study	More complex
Well-funded research	Unfunded, except marketing
Consensus terms, trends	Uncertain terms, trends
Consensus policies	Threatening policies

SOURCE: Robert W. Kates.

Consumption is more threatening, and despite the North–South rhetoric, it is threatening to all. In both rich and poor countries alike, making and selling things to each other, including unnecessary things, is the essence of the economic system. No longer challenged by socialism, global capitalism seems inherently based on growth—growth of both consumers and their consumption. To study consumption in this light is to risk concluding that a transition to sustainability might require profound changes in the making and selling of things and in the opportunities that this provides. To draw such conclusions, in the absence of convincing alternative visions, is fearful and to be avoided.

What We Need to Know and Do

In conclusion, returning to the 30-year-old IPAT identity—a variant of which might be called the Population/Consumption (PC) version—and restating that identity in terms of population and consumption, it would be: $I = P*C/P*I/C$, where I equals environmental degradation and/or resource depletion; P equals the number of people or households; and C equals the transformation of energy, materials, and information (see Figure 2).

With such an identity as a template, and with the goal of reducing environmentally degrading and resource-depleting influences, there are at least seven major directions for research and policy. To reduce the level of impacts per unit of consumption, it is necessary to separate out more damaging consumption and shift to less harmful forms, *shrink* the amounts of environmentally damaging energy and materials per unit of consumption, and *substitute* information for energy and materials. To reduce consumption per person or household, it is necessary to *satisfy* more with what is already had, *satiate* well-met consumption needs, and *sublimate* wants for a greater good. Finally, it is possible to *slow* population growth and then to *stabilize* population numbers as indicated above.

However, as with all versions of the IPAT identity, population and consumption in the PC version are only proximate

driving forces, and the ultimate forces that drive consumption, the consuming desires, are poorly understood, as are many of the major interventions needed to reduce these proximate driving forces. People know most about slowing population growth, more about shrinking and substituting environmentally damaging consumption, much about shifting to less damaging consumption, and least about satisfaction, satiation, and sublimation. Thus the determinants of consumption and its alternative patterns have been identified as a key understudied topic for an emerging sustainability science by the recent U.S. National Academy of Science study.[33]

But people and society do not need to know more in order to act. They can readily begin to separate out the most serious problems of consumption, shrink its energy and material throughputs, substitute information for energy and materials, create a standard for satiation, sublimate the possession of things for that of the global commons, as well as slow and stabilize population. To go from more to enough is more than enough to do for 30 more Earth Days.

Robert W. Kates is an independent scholar in Trenton, Maine; a geographer; university professor emeritus at Brown University; and an executive editor of *Environment*. The research for "Population and Consumption: What We Know, What We Need to Know" was undertaken as a contribution to the recent National Academies/National Research Council report, *Our Common Journey: A Transition Toward Sustainability*. The author retains the copyright to this article. Kates can be reached at RR1, Box 169B, Trenton, ME 04605.

NOTES

1. B. Commoner, M. Corr, and P. Stamler, "The Causes of Pollution," *Environment*, April 1971, 2–19.

2. P. Ehrlich, *The Population Bomb* (New York: Ballantine, 1966).

3. P. Ehrlich and J. Holdren, "Review of The Closing Circle," *Environment*, April 1972, 24–39.

4. B. Commoner, *The Closing Circle* (New York: Knopf, 1971).

5. P. Stern, T. Dietz, V. Ruttan, R. H. Socolow, and J. L. Sweeney, eds., *Environmentally Significant Consumption: Research Direction* (Washington, D.C.: National Academy Press, 1997), 1.

6. This article draws in part upon a presentation for the 1997 De Lange-Woodlands Conference, an expanded version of which will appear as: R. W. Kates, "Population and Consumption: From More to Enough," in *In Sustainable Development: The Challenge of Transition*, J. Schmandt and C. H. Wards, eds. (Cambridge, U.K.: Cambridge University Press, forthcoming), 79–99.

7. United Nations, Population Division, *World Population Prospects: The 1998 Revision* (New York: United Nations, 1999).

8. K. Davis, "Population and Resources: Fact and Interpretation," K. Davis and M. S. Bernstam, eds., in *Resources, Environment and Population: Present Knowledge, Future Options*, supplement to *Population and Development Review*, 1990: 1–21.

9. Population Reference Bureau, *1997 World Population Data Sheet of the Population Reference Bureau* (Washington, D.C.: Population Reference Bureau, 1997).

10. J. Bongaarts, "Population Policy Options in the Developing World," *Science*, 263: (1994), 771–776; and J. Bongaarts and J. Bruce, "What Can Be Done to Address Population Growth?" (unpublished background paper for The Rockefeller Foundation, 1997).

11. National Research Council, Board on Sustainable Development, *Our Common Journey: A Transition Toward Sustainability* (Washington, D.C.: National Academy Press, 1999).

12. See Stern, et al., note 5 above.

13. Royal Society of London and the U.S. National Academy of Sciences, "Towards Sustainable Consumption," reprinted in *Population and Development Review*, 1977, 23 (3): 683–686.

14. For the available data and concepts, I have drawn heavily from J. H. Ausubel and H. D. Langford, eds., *Technological Trajectories and the Human Environment*. (Washington, D.C.: National Academy Press, 1997).

15. L. R. Brown, H. Kane, and D. M. Roodman, *Vital Signs 1994: The Trends That Are Shaping Our Future* (New York: W. W. Norton and Co., 1994).

16. World Resources Institute, United Nations Environment Programme, United Nations Development Programme, World Bank, *World Resources*, 1996–97 (New York: Oxford University Press, 1996); and A. Gruebler, *Technology and Global Change* (Cambridge, Mass.: Cambridge University Press, 1998).

17. I. Wernick, "Consuming Materials: The American Way," *Technological Forecasting and Social Change*, 53 (1996): 111–122.

18. I. Wernick and J. H. Ausubel, "National Materials Flow and the Environment," *Annual Review of Energy and Environment*, 20 (1995): 463–492.

19. S. Pimm, G. Russell, J. Gittelman, and T. Brooks, "The Future of Biodiversity," *Science*, 269 (1995): 347–350.

20. Historic data from L. R. Brown, H. Kane, and D. M. Roodman, note 15 above.

21. One of several projections from P. Raskin, G. Gallopin, P. Gutman, A. Hammond, and R. Swart, *Bending the Curve: Toward Global Sustainability*, a report of the Global Scenario Group, Polestar Series, report no. 8 (Boston: Stockholm Environmental Institute, 1995).

22. N. Nakicénovíc, "Freeing Energy from Carbon," in *Technological Trajectories and the Human Environment*, eds., J. H. Ausubel and H. D. Langford. (Washington, D.C.: National Academy Press, 1997); I. Wernick, R. Herman, S. Govind, and J. H. Ausubel, "Materialization and Dematerialization: Measures and Trends," in J. H. Ausubel and H. D. Langford, eds., *Technological Trajectories and the Human Environment* (Washington, D.C.: National Academy Press, 1997), 135–156; and see A. Gruebler, note 16 above.

23. Royal Society of London and the U.S. National Academy of Science, note 13 above.

24. J. Ryan and A. Durning, *Stuff: The Secret Lives of Everyday Things* (Seattle, Wash.: Northwest Environment Watch, 1997).

25. R. T. Watson, M. C. Zinyowera, and R. H. Moss, eds., *Climate Change 1995: Impacts, Adaptations, and Mitigation of Climate Change—Scientific-Technical Analyses* (Cambridge, U.K.: Cambridge University Press, 1996).

26. A sampling of similar queries includes: A. Durning, *How Much Is Enough?* (New York: W. W. Norton and Co., 1992); Center for a New American Dream, *Enough!: A Quarterly Report on Consumption, Quality of Life and the Environment* (Burlington, Vt.: The Center for a New American Dream, 1997); and N. Myers, "Consumption in Relation to Population, Environment, and Development," *The Environmentalist*, 17 (1997): 33–44.

27. J. Schor, *The Overworked American* (New York: Basic Books, 1991).

28. A. Durning, *How Much Is Enough?: The Consumer Society and the Future of the Earth* (New York: W. W. Norton and Co., 1992); Center for a New American Dream, note 26 above; and M. Wackernagel and W. Ress, *Our Ecological Footprint: Reducing Human Impact on the Earth* (Philadelphia. Pa.: New Society Publishers, 1996).

29. W. Jager, M. van Asselt, J. Rotmans, C. Vlek, and P. Costerman Boodt, *Consumer Behavior: A Modeling Perspective in the Contest of Integrated Assessment of Global Change*, RIVM report no. 461502017 (Bilthoven, the Netherlands: National Institute for Public Health and the Environment, 1997); and P. Vellinga, S. de Bryn, R. Heintz, and P. Molder, eds., *Industrial Transformation: An Inventory of Research*. IHDP-IT no. 8 (Amsterdam, the Netherlands: Institute for Environmental Studies, 1997).

30. H. Nearing and S. Nearing. *The Good Life: Helen and Scott Nearing's Sixty Years of Self-Sufficient Living* (New York: Schocken, 1990); and D. Elgin, *Voluntary Simplicity: Toward a Way of Life That Is Outwardly Simple, Inwardly Rich* (New York: William Morrow, 1993).

31. P. Menzel, *Material World: A Global Family Portrait* (San Francisco: Sierra Club Books, 1994).

32. R. Rosenblatt, ed., *Consuming Desires: Consumption, Culture, and the Pursuit of Happiness* (Washington, D.C.: Island Press, 1999).

33. National Research Council, Board on Sustainable Development, *Our Common Journey: A Transition Toward Sustainability* (Washington, D.C.: National Academy Press, 1999).

AN
Economy
FOR THE
Earth

by LESTER R. BROWN

In 1543, Polish astronomer Nicolaus Copernicus published "On the Revolutions of the Celestial Spheres," in which he challenged the view that the sun revolves around the Earth, arguing instead that the Earth revolves around the sun. With his new model of the solar system, he began a wide-ranging debate among scientists, theologians, and others. His alternative to the earlier Ptolemaic model, which had the Earth at the center of the universe, led to a revolution in thinking, to a new worldview.

Today we need a similar shift in our worldview in how we think about the relationship between the Earth and the economy. The issue now is not which celestial sphere revolves around the other but whether the environment is part of the economy or the economy is part of the environment. Economists see the environment as a subset of the economy. Ecologists, on the other hand, see the economy as a subset of the environment.

Like Ptolemy's view of the solar system, the economists' view is confusing efforts to understand our modern world. This has resulted in an economy that is out of sync with the ecosystem on which it depends.

Economic theory and economic indicators don't explain how the economy is disrupting and destroying the Earth's natural systems. Economic theory doesn't explain why Arctic Sea ice is melting, why grasslands are turning into desert in northwestern China, why coral reefs are dying in the South Pacific, or why the Newfoundland cod fishery collapsed. Nor does it explain why we are in the early stages of the greatest extinction of plants and animals since the dinosaurs disappeared 65 million years ago. Yet economics is essential to measuring the cost to society of these excesses.

Evidence that the economy is in conflict with the Earth's natural systems can be seen in the daily news reports of shrinking forests, eroding soils, deteriorating rangelands, expanding deserts, rising carbon dioxide levels, falling water tables, rising temperatures, more destructive storms, melting glaciers, rising sea level, dying coral reefs, collapsing fisheries, and disappearing species. These trends, which mark an increasingly stressed relationship between the economy and the ecosystem, are taking a growing economic toll. At some point this could overwhelm the worldwide forces of progress, leading to economic decline. The challenge for our generation is to reverse these trends before environmental deterioration leads to long-term economic decline—as it did for so many earlier civilizations.

These increasingly visible trends indicate that, if the operation of the subsystem—the economy—is not compatible with the behavior of the larger system—the ecosystem—both will eventually suffer. The larger the economy becomes relative to the ecosystem, and the more it presses against the Earth's natural limits, the more destructive this incompatibility will be.

An environmentally sustainable economy—an *eco-economy*—requires that the principles of ecology establish the framework for the formulation of economic policy and that economists and ecologists work together to fashion the new economy. Ecologists understand that all economic activity, indeed all life, depends on the Earth's ecosystem—the complex of individual species living together, interacting with each other and their physical habitat. These millions of species exist in an intricate balance, woven together by food chains, nutrient cycles, the hydrological cycle, and the climate system. Economists know how to translate goals into policy. Ecologists and economists working together can design and build an eco-economy that can sustain progress.

Just as recognition that the Earth is not the center of the solar system set the stage for advances in astronomy, physics, and related sciences, so will recognition that the economy isn't the center of our world create the conditions to sustain economic progress and improve the human condition.

Converting the world economy into an eco-economy, however, is a monumental undertaking, as the current gap between economists and ecologists in their perception of the world could not be wider. There is no precedent for transforming an economy shaped largely by market forces into one shaped by the principles of ecology.

The scale of projected economic growth outlines the dimensions of the challenge. The growth in world output of goods and services from $6 trillion in 1950 to $43 trillion in 2000 has caused environmental devastation on a scale that we could not easily have imagined a half-century ago. If the world economy continues to expand at 3 percent annually, the output of goods and services will increase fourfold over the next half-century, reaching $172 trillion.

Building an eco-economy in the time available requires rapid systemic change. We won't succeed with a project here and a project there. We are winning occasional battles, but we are losing the war because we don't have a strategy for the systemic economic change that will put the world on a development path that is environmentally sustainable.

Although the concept of environmentally sustainable development evolved a quarter-century ago, not one country has a strategy to build an eco-economy—to restore balances, to stabilize population and water tables, and to conserve forests, soils, and diversity of plant and animal life. We can find individual countries that are succeeding with one or more elements of restructuring but not one that is progressing satisfactorily on all fronts.

Nevertheless, glimpses of the eco-economy are clearly visible in some countries. For example, thirty-one nations in Europe, as well as Japan, have stabilized their population size, satisfying one of the most basic conditions of an eco-economy. Europe has stabilized its population within its food-producing capacity, leaving it with an exportable surplus of grain to help meet the deficits in developing countries. Furthermore, China—the world's most populous country—now has lower fertility than the United States and is moving toward population stability.

Currently Denmark is the eco-economy leader. It has stabilized its population, banned the construction of coal-fired power plants, banned the use of nonrefillable beverage containers, and is now getting 15 percent of its electricity from wind. In addition, it has restructured its urban transport network; now 32 percent of all trips in Copenhagen are on bicycle. Denmark is still not close to balancing carbon emissions and fixation, but it is moving in that direction.

Other countries have also achieved specific goals. A reforestation program in South Korea, begun more than a generation ago, has blanketed that country's hills and mountains with trees. Costa Rica has a plan to shift entirely to renewable energy by 2025. Iceland plans to be the world's first hydrogen-powered economy.

Building an eco-economy will affect every facet of our lives. It will alter how we light our homes, what we eat, where we live, how we use our leisure time, and how many children we have. It will give us a world in which we are a part of nature instead of estranged from it.

According to Seth Dunn in the November/December 2000 issue of *World Watch* magazine, a consortium of corporations led by Shell Hydrogen and DaimlerChrysler reached an agreement in 1999 with the government of Iceland to establish this new economy. Shell is interested because it wants to begin developing its hydrogen production and distribution capacity, and DaimlerChrysler expects to have the first fuel cell-powered automobile on the market. Shell plans to open its first chain of hydrogen stations in Iceland.

So we can see pieces of the eco-economy emerging, but systemic change requires a fundamental shift in market signals—signals that respect the principles of ecological sustainability. Unless we are prepared to shift taxes from income to environmentally destructive activities—such as carbon emissions and the wasteful use of water—we won't succeed in building an eco-economy.

It is a huge undertaking to restore the balances of nature. For energy, this will depend on shifting from a carbon-based economy to a hydrogen-based one. Even the most progressive oil companies, such as British Petroleum and Royal Dutch Shell, which are all talking extensively about building a solar/hydrogen energy economy, are still investing overwhelmingly in oil, with funds going into climate-benign sources accounting for a minute share of their investment.

Reducing soil erosion to the level of new soil formation will require changes in farming practices. In some situations, it will mean shifting from intense tillage to minimum tillage or no tillage. Agroforestry will loom large in an eco-economy.

Restoring forests that recycle rainfall inland and control flooding is itself a huge undertaking. It means reversing decades of tree cutting and land clearing with forest restoration—an activity that will require millions of people planting billions of trees.

Building an eco-economy will affect every facet of our lives. It will alter how we light our homes, what we eat, where we live, how we use our leisure time, and how many children we have. It will give us a world in which we are *a part of* nature instead of estranged from it.

In May 2001, the Bush White House released with great fanfare a twenty-year plan for the United States' energy economy. It disappointed many people because it largely overlooked the enormous potential for raising energy efficiency. It also overlooked the huge potential of wind power, which is likely to add more to U.S. generating capacity over the next twenty years than coal does. The plan was indicative of the problems some governments are having in fashioning an energy economy that is compatible with the Earth's ecosystem.

Prepared under the direction of Vice-President Dick Cheney, the administration's plan centers on expanding production of fossil fuels—something more appropriate for the early twentieth century than the early twenty-first. It emphasizes the role of coal, but its architects were apparently unaware that world coal use peaked in 1996 and has declined some 7 percent since then as other countries have turned away from this fuel. Even China, which rivals the United States as a coal-burning country, has reduced its coal use by an estimated 14 percent since 1996.

Bush's energy plan notes that the 2 percent of U.S. electricity generation that today comes from renewable sources, excluding hydropower, would increase to 2.8 percent in 2020. But months before the Bush plan was released, the American Wind Energy Association was projecting a staggering 60 percent growth in U.S. wind-generating capacity in 2001. Worldwide, use of wind power alone has multiplied nearly fourfold over the last five years—a growth rate matched only by the computer industry.

The solar cell is a relatively new source of alternative energy and, after wind power, is already the second fastest growing source. In 1952, three scientists at Bell Labs in Princeton, New Jersey, discovered that sunlight striking a silicon-based material produces electricity. The discovery of this photovoltaic, or solar cell, opened up a vast new potential for power generation. Initially very costly, solar cells were used mostly for high-value purposes such as providing the electricity to operate satellites. As the solar cell became economical, it opened the potential for providing electricity to remote sites not yet linked to an electrical grid. It is already more economical in remote areas to install solar cells than to build a power plant and connect villages by grid. By the end of 2000, about a million homes worldwide were getting their electricity from solar cell installations. An estimated 700,000 of these were in villages in developing countries.

Today, as the cost of solar cells continues to decline, this technology is becoming competitive with large, centralized power sources. For many of the two billion people in the world who don't have access to conventional electricity sources, small solar cell arrays provide an affordable shortcut. In the developing world, in some communities not serviced by a centralized power system, local entrepreneurs are investing in solar cell generating facilities and selling the energy to village families.

Perhaps the most exciting technological advance has been the development of a photovoltaic roofing material in Japan. A joint effort involving the construction industry, the solar cell manufacturing industry, and the Japanese government plans to have 4,600 megawatts of electrical generating capacity in place by 2010—enough to satisfy all of the electricity needs of a country like Estonia. With photovoltaic roofing material, the roof of a building becomes the power plant. In some countries, including Germany and Japan, buildings now have a two-way meter—selling electricity to the local utility when they have an excess and buying it when they don't have enough.

In contrast to these other sources of renewable energy, geothermal energy comes from within the Earth itself. Produced radioactively and by the pressures of gravity, it is a vast resource, most of which is deep within the planet. Geothermal energy can be economically tapped when it is relatively close to the surface, as evidenced by hot springs, geysers, and volcanic activity. It is used directly both to supply heat and generate electricity.

Geothermal energy is much more abundant in some parts of the world than in others. The richest region is the vast Pacific Rim: along the western coastal regions of Latin America, Central America, and North America; and widely distributed in eastern Russia, Japan, the Korean Peninsula, China, and island nations like the Philippines, Indonesia, New Guinea, Australia, and New Zealand.

This energy source is essentially inexhaustible. Hot baths, for example, have been used for millennia. It is possible to extract heat faster than it is generated at any local site, but this is a matter of adjusting the extraction of heat to the amount generated. In contrast to oil fields, which are eventually depleted, properly managed geothermal fields keep producing indefinitely. In a time of mounting concern about climate change, many governments are beginning to exploit the geothermal potential—as in Iceland, where it heats some 85 percent of buildings; as in Japan, for hot baths when springs bring geothermal energy to the surface; as in the United States for generating electricity. In fact, the U.S. Department of Energy announced in 2000 that it was launching a program to develop the rich geothermal energy resources in the western United States. The goal is to have 10 percent of the electricity in the West coming from geothermal energy by 2020.

Although the Bush energy plan doesn't reflect it, the world energy economy is on the verge of a major transformation. Historically, the twentieth century was the century of fossil fuels: first coal, then oil, and finally natural gas were the workhorses of the world economy. But with the advent of the twenty-first century, the sun is setting on the fossil fuel era. The last several decades have shown a steady shift from the most polluting and climate-disrupting fuels toward clean, climate-benign energy sources.

Even the oil companies are now beginning to recognize that the time has come for an energy transition. After years of denying any link between fossil fuel burning and climate change, John Browne, the chief executive officer of BP, announced his new position in a historic speech at Stanford University in May 1997:

> My colleagues and I now take the threat of global warming seriously. The time to consider the policy dimensions of climate change is not when the link between greenhouse gases and climate change is conclusively proven but when the possibility cannot be discounted and is taken seriously by the society of which we are a part. We in BP have reached that point.

At an energy conference in Houston, Texas, in February 1999, Michael Bowlin, CEO of ARCO, said that the beginning of the end of the age of oil was in sight. He went on to discuss the need to shift from a carbon-based to a hydrogen-based energy economy.

The signs of restructuring the global energy economy are unmistakable. Events are moving far faster than would have been expected even a few years ago, driven in part by the mounting evidence that the Earth is indeed warming up and that the burning of fossil fuels is responsible. But can we do what needs to be done fast enough?

We know that social change often takes time. In Eastern Europe, it was fully four decades from the imposition of communism until its demise. Thirty-four years passed between the first U.S. Surgeon General's report on smoking and health and the landmark agreement between the tobacco industry and state governments. Thirty-eight years have passed since biologist Rachel Carson published *Silent Spring,* the wakeup call that gave rise to the modern environmental movement.

Sometimes things move much faster, especially when the magnitude of the threat is understood and the nature of the response is obvious, such as the U.S. response to the attack on Pearl Harbor. Within one year, the U.S. economy had largely been reconstructed. In less than four years, the war was over.

Accelerating the transition to a sustainable future means overcoming the inertia of both individuals and institutions. In some ways, inertia is our worst enemy. As individuals we often resist change. When we are gathered into large organizations, we resist it even more.

At the institutional level, we are looking for massive changes in industry, especially in energy. We are looking for changes in the material economy, shifting from a throwaway mentality to a closed loop/recycle mindset. If future food needs are to be satisfied adequately, we need a worldwide effort to reforest the land, conserve soil, and raise water productivity. Stabilizing population growth means quite literally a revolution in human reproductive behavior—one that recognizes a sustainable future is possible only if we average two children per couple. This isn't a debatable point. It is a mathematical reality.

The big remaining challenge is on the educational front: how can we help literally billions of people in the world understand not only the need for change but how that change can bring a life far better than they have today?

In this connection, I am frequently asked if it is too late. My response is: "Too late for what?" Is it too late to save the Aral Sea? Yes, the Aral Sea is dead; its fish have died, and its fisheries have collapsed. Is it too late to save the glaciers in Glacier National Park in the United States? Most likely. They are already half gone, and it would be virtually impossible now to reverse the rise in temperature in time to save them. Is it too late to avoid a rise in temperature from the buildup of greenhouse gases? Yes. A greenhouse gas-induced rise in temperature is apparently already underway. But is it too late to avoid runaway climate change? Perhaps not, if we quickly restructure the energy economy.

For many specifics, the answer is, yes, it is too late. But there is a broader, more fundamental question: is it too late to reverse the trends that will eventually lead to economic decline? Here I think the answer is no—not if we act quickly.

Perhaps the biggest challenge we face is shifting from a carbon-based to a hydrogen-based energy economy—basically moving from fossil fuels to renewable sources of energy, such as solar, wind, and geothermal. How fast can we make this change? Can it be done before we trigger irreversible damage, such as a disastrous rise in sea level? As I indicated, we know from the United States' response to the attack on Pearl Harbor that economic restructuring can occur at an incredible pace if a society is convinced of the need for it.

We study the archaeological sites of civilizations that moved onto economic paths that were environmentally destructive and could not make the needed course corrections in time. We face the same risk.

There is no middle path. Do we join together to build an economy that is sustainable, or do we stay with our environmentally unsustainable economy until it declines? It isn't a goal that can be compromised. One way or another, the choice will be made by our generation. But what we choose will affect life on Earth for all generations to come.

Lester R. Brown is president and founder of the Earth Policy Institute; the founder and former president of the Worldwatch Institute; a MacArthur Fellow; and the recipient of twenty-two honorary degrees and many awards, including the 1987 UN Environment Prize, the 1989 World Wide Fund for Nature Gold Medal, the 1991 Humanist of the Year Award, and the 1994 Blue Planet Prize for his "exceptional contributions to solving global environmental problems." He has authored or coauthored forty-seven books, nineteen monographs, and countless articles. This article is adapted from Eco-Economy: Building an Economy for the Earth, *which is available online at www.earth-policy.org.*

Originally appeared in *The Humanist,* May/June 2002, pp. 32-34. Excerpted from Chapters 1 & 5 in Lester R. Brown's, *Eco-Economy: Building an Economy for the Earth,* W. W. Norton & Company, NY: 2001. © 2002 by Earth Policy Institute. Reprinted by permission.

Factory Farming in the Developing World

In some critical respects, this is not progress at all.

by Danielle Nierenberg

Walking through Bobby Inocencio's farm in the hills of Rizal province in the Philippines is like taking a step back to a simpler time. Hundreds of chickens (a cross between native Filipino chickens and a French breed) roam around freely in large, fenced pens. They peck at various indigenous plants, they eat bugs, and they fertilize the soil, just as domesticated chickens have for ages.

The scene may be old, but Inocencio's farm is anything but simple. What he has recreated is a complex and successful system of raising chickens that benefits small producers, the environment, and even the chickens. Once a "factory farmer," Inocencio used to raise white chickens for Pure Foods, one of the biggest companies in the Philippines.

Thousands of birds were housed in long, enclosed metal sheds that covered his property. Along with the breed stock and feeds he had to import, Inocencio also found himself dealing with a lot of imported diseases and was forced to buy expensive antibiotics to keep the chickens alive long enough to take them to market. Another trick of the trade Inocencio learned was the use of growth promotants that decrease the time it takes for chickens to mature.

Eventually he noticed that fewer and fewer of his neighbors were raising chickens, which threatened the community's food security by reducing the locally available supply of chickens and eggs. As the community dissolved and farms (and farming methods) that had been around for generations went virtually extinct, Inocencio became convinced that there had to be a different way to raise chickens and still compete in a rapidly globalizing marketplace. "The business of the white chicken," he says, "is controlled by the big guys." Not only do small farmers have to compete with the three big companies that control white chickens in the Philippines, but they must also contend

with pressure from the World Trade Organization (WTO) to open up trade. In the last two decades the Filipino poultry production system has transitioned from mainly backyard farms to a huge industry. In the 1980s the country produced 50 million birds annually. Today that figure has increased some ten-fold. The large poultry producers have benefited from this population explosion, but average farmers have not. So Inocencio decided to go forward by going back and reviving village-level poultry enterprises that supported traditional family farms and rural communities.

Inocencio's farm and others like it show that the Philippines can support indigenous livestock production and stand up to the threat of the factory farming methods now spreading around the world. Since 1997, his Teresa Farms has been raising free range chickens and teaching other farmers how to do the same. He says that the way he used to raise chickens, by concentrating so many of them in a small space, is dangerous. Diseases such as avian flu, leukosis J (avian leukemia), and Newcastle disease are spread from white chickens to the Filipino native chicken populations, in some cases infecting eggs before the chicks are even born. "The white chicken," says Inocencio, "is weak, making the system weak. And if these chickens are weak, why should we be raising them? Limiting their genetic base and using breeds that are not adapted to conditions in the Philippines is like setting up the potential for a potato blight on a global scale." Now Teresa Farms chickens are no longer kept in long, enclosed sheds, but roam freely in large tree-covered areas of his farm that he encloses with recycled fishing nets.

Inocencio's chickens also don't do drugs. Antibiotics, he says, are not only expensive but encourage disease. He found the answer to the problem of preventing diseases in chickens literally in his own back yard. His chickens eat spices and native plants with antibacterial and other medicinal properties. Chili, for instance, is mixed in grain to treat respiratory problems, stimulate appetite during heat stress, de-worm the birds, and to treat Newcastle disease. Native plants growing on the farm, including *ipil-ipil* and *damong maria*, are also used as low-cost alternatives to antibiotics and other drugs.

There was a time when most farms in the Philippines, the United States, and everywhere else functioned much like Bobby Inocencio's. But today the factory model of raising animals in intensive conditions is spreading around the globe.

A New Jungle

Meat once occupied a very different dietary place in most of the world. Beef, pork, and chicken were considered luxuries, and were eaten on special occasions or to enhance the flavor of other foods. But as agriculture became more mechanized, so did animal production. In the United States, livestock raised in the West was herded or transported east to slaughterhouses and packing mills. Upton Sinclair's *The Jungle*, written almost a century ago when the United States lacked many food-safety and labor regulations, described the appalling conditions of slaughterhouses in Chicago in the early 20th century and was a shocking expose of meat production and the conditions inflicted on both animals and humans by the industry. Workers were treated much like animals themselves, forced to labor long hours for very little pay under dangerous conditions, and with no job security.

If *The Jungle* were written today, however, it might not be set in the American Midwest. Today, developing nations like the Philippines are becoming the centers of large-scale livestock production and processing to feed the world's growing appetite for cheap meat and other animal products. But the problems Sinclair pointed to a century ago, including hazardous working conditions, unsanitary processing methods, and environmental contamination, still exist. Many have become even worse. And as environmental regulations in the European Union and the United States become stronger, large agribusinesses are moving their animal production operations to nations with less stringent enforcement of environmental laws.

These intensive and environmentally destructive production methods are spreading all over the globe, to Mexico, India, the former Soviet Union, and most rapidly throughout Asia. Wherever they crop up, they create a web of related food safety, animal welfare, and environmental problems. Philip Lymbery, campaign director of the World Society for the Protection of Animals, describes the growth of industrial animal production this way: Imagine traditional livestock production as a beach and factory farms as a tide. In the United States, the tide has completely covered the beach, swallowing up small farms and concentrating production in the hands of a few large companies. In Taiwan, it is almost as high. In the Philippines, however, the tide is just hitting the beach. The industrial, factory-farm methods of raising and slaughtering animals—methods that were conceived and developed in the United States and Western Europe—have not yet swept over the Philippines, but they are coming fast.

An Appetite for Destruction

Global meat production has increased more than fivefold since 1950, and factory farming is the fastest growing method of animal production worldwide. Feedlots are responsible for 43 percent of the world's beef, and more than half of the world's pork and poultry are raised in factory farms. Industrialized countries dominate production, but developing countries are rapidly expanding and intensifying their production systems. According to the United Nations Food and Agriculture Organization (FAO), Asia (including the Philippines) has the fastest developing livestock sector. On the islands that make up the Philippines, 500 million chickens and 20 million hogs are slaughtered each year.

Despite the fact that many health-conscious people in developed nations are choosing to eat less meat, worldwide meat consumption continues to rise. Consumption is growing fastest in the developing countries. Two-thirds of the gains in meat consumption in 2002 were in the developing world, where urbanization, rising incomes, and globalized trade are changing diets and fueling appetites for meat and animal products. Because eating meat has been perceived as a measure of economic and social development, the Philippines and other poor nations are eager to climb up the animal-protein ladder. People in the Philippines still eat relatively little meat, but their consumption is growing. As recently as 1995, the average Filipino ate 21 kilograms of meat per year. Since then, average consumption has soared to almost 30 kilograms per year, although that is still less than half the amount in Western countries, where per-capita consumption is 80 kilograms per year.

This push to increase both production and consumption in the Philippines and other developing nations is coming from a number of different directions. Since the end of World War II, agricultural development has been considered a part of the foreign aid and assistance given to developing nations. The United States and international development agencies have been leaders in promoting the use of pesticides, artificial fertilizers, and other chemicals to boost agricultural production in these countries, often at the expense of the environment. American corporations like Purina Mills and Tyson Foods are also opening up feed mills and farms so they can expand business in the Philippines.

But Filipinos are also part of the push to industrialize agriculture. "This is not an idea only coming from the West," says Dr. Abe Agulto, president of the Philippine Society for the Protection of Animals, "but also coming from us." Meat equals wealth in much of the world and many Filipino businesspeople have taken up largescale livestock production to supply the growing demand for meat. But small farmers don't get much financial support in the Philippines. It's not farms like Bobby Inocencio's that are likely to get government assistance, but the big production facilities that can crank out thousands of eggs, chicks, or piglets a year.

The world's growing appetite for meat is not without its consequences, however. One of the first indications that meat production can be hazardous arises long before animals ever reach the slaughterhouse. Mountains of smelly and toxic manure are created by the billions of animals raised for human consumption in the world each year. In the United States, people in North Carolina know all too well the effects of this liquid and solid waste. Hog production there has increased faster than anywhere else in the nation, from 2 million hogs per year in 1987 to 10 million hogs per year today. Those hogs produce more than 19 million tons of manure each year and most of it gets stored in lagoons, or large uncovered containment pits. Many of those lagoons flooded and burst when Hurricane Floyd swept through the region in 1999. Hundreds of acres of land and miles of waterway were flooded with excrement, resulting in massive fish kills and millions of dollars in cleanup costs. The lagoons' contents are also known to leak out and seep into groundwater.

Some of the same effects can now be seen in the Philippines. Not far from Teresa Farms sits another, very different, farm that produces the most frequently eaten meat product in the world. Foremost Farms is the largest piggery, or pig farm, in all of Asia. An estimated 100,000 pigs are produced there every year.

High walls surround Foremost and prevent people in the community from getting in or seeing what goes on inside. What they do get a whiff of is the waste. Not only do the neighbors smell the manure created by the 20,000 hogs kept at Foremost or the 10,000 hogs kept at nearby Holly Farms, but their water supply has also been polluted by it. In fact, they've named the river where many of them bathe and get drinking water the River Stink. Apart from the stench, some residents have complained of skin rashes, infections, and other health problems from the water. And instead of keeping the water clean and installing effective waste treatment, the firms are just digging deeper drinking wells and giving residents free access to them. Many in the community are reluctant to complain about the smell because they fear losing their water supply. Even the mayor of Bulacan, the nearby village, has said "we give these farms leeway as much as possible because they provide so much economically."

It would be easy to assume that some exploitive foreign corporation owns Foremost, but in fact the owner is Lucia Tan, a Filipino. Tan is not your average Filipino, however, but the richest man in the Philippines. In addition to Foremost Farms, he owns San Miguel beer and Philippine Airlines. Tan might be increasing his personal wealth, but his farm and others like it are gradually destroying traditional farming methods and threatening indigenous livestock breeds in the Philippines. As a result, many small farmers can no longer afford to produce hogs for sale or for their own consumption, which forces them to become consumers of Tan's pork. Most of the nation's 11 million hogs are still kept in back yards, but because of farms like Foremost, factory farming is growing. Almost one-quarter of the breeding herd is now factory farmed. More than 1 million pigs are raised in factory farms every year in Bulacan alone.

Chicken farms in the Philippines are also becoming more intensive. The history of intensive poultry production in the Philippines is not long. Forty years ago, the nation's entire population was fed on native eggs and chickens produced by family farmers. Now, most of those farmers are out of business. They have lost not only their farms, but livestock diversity and a way of life as well.

The loss of this way of life to the industrialized farm-to-abattoir system has made the process more callous at every stage. Adopting factory farming methods works to diminish farmers' concern for the welfare of their livestock. Chickens often can't walk properly because they have been pumped full of growth-promoting antibiotics to gain weight as quickly as possible. Pigs are confined to gestation crates where they can't turn around. Cattle are crowded together in feedlots that are seas of manure.

Most of the chickens in the country are from imported breed stock and the native Filipino chicken has practically disappeared because of viral diseases spread by foreign breeds. Almost all of the hens farmed commercially for their eggs are confined in wire battery cages that cram three or four hens together, giving each bird an area less than the size of this page to stand on.

Unlike laying hens, chickens raised for meat in the Philippines are not housed in cages. But they're not pecking around in back yards, either. Over 90 percent of the meat chickens raised in the Philippines live in long sheds that house thousands of birds. At this time, most Filipino producers allow fowl to have natural ventilation and lighting and some roaming room, but they are under pressure to adopt more "modern" factory-farm standards to increase production.

The problems of a system that produces a lot of animals in crowded and unsanitary conditions can also be seen off the farm. The *baranguay* (neighborhood) of Tondo in Manila is best known for the infamous "Smoky Mountain" garbage dump that collapsed on scavengers in 2000, killing at least 200 people. But another hazard also sits in the heart of Tondo. Surrounded by tin houses, stores, and bars, the largest government-owned slaughtering facility in the country processes more than 3,000 swine, cattle, and *caraboa* (water buffalo) per day, all brought from farms just outside the city limits. The slaughterhouse does have a waste treatment system where the blood and other waste is supposed to be treated before it is released into the city's sewer system and nearby Manila Bay. Unfortunately, that's not what's going on. Instead, what can't be cut up and sold for human consumption is dumped into the sewer.

Some 60 men are employed at the plant. They stun, bludgeon, and slaughter animals by hand and at a breakneck pace. They wear little protective gear as they slide around on floors slippery with blood, which makes it hard to stun animals on the first try, or sometimes even the second, or to butcher meat without injuring themselves.

The effects of producing meat this way also show up in rising cases of food-borne illness, emerging animal diseases that can spread to humans, and in an increasingly overweight Filipino population that doesn't remember where meat comes from.

There are few data on the incidence of food-borne illness in the Philippines or most other developing nations, and even fewer about how much of it might be related to eating unsafe meat. What food safety experts do know is that food-borne illness is one of the most widespread health problems worldwide.

71

And it could be an astounding 300–350 times more frequent than reported, according to the World Health Organization. Developing nations bear the greatest burden because of the presence of a wide range of parasites, toxins, and biological hazards and the lack of surveillance, prevention, and treatment measures—all of which ensnarl the poor in a chronic cycle of infection. According to the FAO, the trend toward increased commercialization and intensification of livestock production is leading to a variety of food safety problems. Crowded, unsanitary conditions and poor waste treatment in factory farms exacerbate the rapid movement of animal diseases and food-borne infections. *E. coli* 0157:H7, for instance, is spread from animals to humans when people eat food contaminated by manure. Animals raised in intensive conditions often arrive at slaughterhouses covered in feces, thus increasing the chance of contamination during slaughtering and processing.

Cecilia Ambos is one of the meat inspectors at the Tondo slaughterhouse. Cecilia or another inspector is required to be on site at all times, but she says she rarely has to go to the killing floor. Inspections of carcasses only occur, she said, if one of the workers alerts the inspector. That doesn't happen very often, and not because the animals are all perfectly healthy. Consider that the men employed at the plant are paid about $5 per day, which is less than half of the cost of living—and are working as fast as they can to slaughter a thousand animals per shift. It's unlikely that they have the time or the knowledge to notice problems with the meat.

Since the 1960s, farm-animal health in the United States has depended not on humane farming practices but on the use of antibiotics. Many of the same drugs used to treat human illnesses are also used in animal production, thus reducing the arsenal of drugs available to fight food-borne illnesses and other health problems. Because antibiotics are given to livestock to prevent disease from spreading in crowded conditions and to increase growth, antibiotic resistance has become a global threat. In the Philippines, chicken, egg, and hog producers use antibiotics not because their birds or hogs are sick, but because drug companies and agricultural extension agents have convinced them that these antibiotics will ensure the health of their birds or pigs and increase their weight.

Livestock raised intensively can also spread diseases to humans. Outbreaks of avian flu in Hong Kong during the past five years have led to massive culls of thousands of chickens. When the disease jumped the species barrier for the first time in 1997, six of the eighteen people infected died. Avian flu spread to people living in Hong Kong again this February, killing two. Dr. Gary Smith, of the University of Pennsylvania School of Veterinary Medicine, also warns that "it is not high densities [of animals] that matter, but the increased potential for transmission between firms that we should be concerned about. The nature of the farming nowadays is such that there is much more movement of animals between farms than there used to be, and much more transport of associated materials between farms taking place rapidly. The problem is that the livestock industry is operating on a global, national, and county level." The foot-and-mouth disease epidemic in the United Kingdom is a perfect example of how just a few cows can spread a disease across an entire nation.

Modern Methods, Modern Policies?

The expansion of factory farming methods in the Philippines is raising the probability that it will become another fast food nation. Factory farms are supplying much of the pork and chicken preferred by fast food restaurants there. American-style fast food was unknown in the Philippines until the 1970s, when Jollibee, the Filipino version of McDonald's, opened its doors. Now, thanks to fast food giants like McDonald's, Kentucky Fried Chicken, Burger King, and others, the traditional diet of rice, vegetables, and a little meat or fish is changing—and so are rates of heart disease, diabetes, and stroke, which have risen to numbers similar to those in the United States and other western nations.

The Filipino government doesn't see factory farming as a threat. To the contrary, many officials hope it will be a solution to their country's economic woes, and they're making it easier for large farms to dominate livestock production. For instance, the Department of Agriculture appears to have turned a blind eye when many farms have violated environmental and animal welfare regulations. The government has also encouraged big farms to expand by giving them loans. But as the farms get bigger and produce more, domestic prices for chicken and pork fail, forcing more farmers to scale up their production methods. And because the Philippines (and many other nations) are prevented by the Global Agreement on Tariffs and Trade and the WTO from imposing tariffs on imported products, the Philippines is forced to allow cheap, factory-farmed American pork and poultry into the country. These products are then sold at lower prices than domestic meat.

Rafael Mariano, a leader in the Peasant Movement for the Philippines (KMP), has not turned away from the problems caused by factory farming in the Philippines. He and the 800,000 farmers he works with believe that "factory farming is not acceptable, we have our own farming." But farmers, he says, are told by big agribusiness companies that their methods are old fashioned, and that to compete in the global market they must forget what they have learned from generations of farming. Rafael and KMP are working to promote traditional methods of livestock production that benefit small farmers and increase local food security. This means doing what farmers used to do: raising both crops and animals. In mixed crop–livestock farms, animals and crops are parts of a self-sustaining system. Some farmers in the Philippines raise hogs, chickens, tilapia, and rice on the same farm. The manure from the hogs and chickens is used to fertilize the algae in ponds needed for both tilapia and rice to grow. These farms produce little waste, provide a variety of food for the farm, and give farmers social security when prices for poultry, pork, and rice go down.

The Philippines is not the only country at risk from the spread of factory farms. Argentina, Brazil, Canada, China, India, Mexico, Pakistan, South Africa, Taiwan, and Thailand are all seeing growth in industrial animal production. As regu-

The Return of the Native... Chicken

Bobby Inocencio believes that the happier his chickens are, the healthier they will be, both on the farm and at the table. Native Filipino chickens are a tough sell commercially, he says, because they typically weigh in at only one kilogram apiece. But Inocencio's chickens are part native and part SASSO (a French breed), and grow to two kilos in just 63 days in a free-range system. They are also better adapted to the climate of the Philippines, unlike white chickens that are more vulnerable to heat. As a result, Inocencio's chickens not only are nutritious, but taste good. Raising white chickens, he says, forced small farmers to become "consumers of a chicken that doesn't taste like anything." Further, his chickens don't contain any antibiotics and are just 5 percent fat, compared to 35 percent in the white chicken. Because they are not raised in the very high densities of factory farms, these chickens actually enrich the environment with their manure. They also provide a reliable source of income for local farmers and give Filipinos a taste of how things used to be.

lations controlling air and water pollution from such farms are strengthened in one country, companies simply pack up and move to countries with more lenient rules. Western European nations now have among the strongest environmental regulations in the world; farmers can only apply manure during certain times of the year and they must follow strict controls on how much ammonia is released from their farms. As a result, a number of companies in the Netherlands and Germany are moving their factory farms—but to the United States, not to developing countries. According to a recent report in the *Dayton Daily News*, cheap land and less restrictive environmental regulations in Ohio are luring European livestock producers to the Midwest. There, dairies with fewer than 700 cows are not required to obtain permits, which would regulate how they control manure. But 700 cows can produce a lot of manure. In 2001, five Dutch-owned dairies were cited by the Ohio Environmental Protection Agency for manure spills. "Until there are international regulations controlling waste from factory farms," says William Weida, director of the Global Reaction Center for the Environment/Spira Factory Farm project, "it is impossible to prevent farms from moving to places with less regulation."

Mauricio Rosales of FAO's Livestock, Environment, and Development Project also stresses the need for siting farms where they will benefit both people and the environment. "Zoning," he says, "is necessary to produce livestock in the most economically viable places, but with the least impact." For instance, when livestock live in urban or peri-urban areas, the potential for nutrient imbalances is high. In rural areas manure can be a valuable resource because it contains nitrogen and phosphorous, which fertilize the soil. In cities, however, manure is a toxic, polluting nuisance.

The triumph of factory farming is not inevitable. In 2001, the World Bank released a new livestock strategy which, in a surprising reversal of its previous commitment to funding of large-scale livestock projects in developing nations, said that as the livestock sector grows "there is a significant danger that the poor are being crowded out, the environment eroded, and global food safety and security threatened." It promised to use a "people-centered approach" to livestock development projects that will reduce poverty, protect environmental sustainability, ensure food security and welfare, and promote animal welfare. This turnaround happened not because of pressure from environmental or animal welfare activists, but because the large-scale, intensive animal production methods the Bank once advocated are simply too costly. Past policies drove out smallholders because economies of scale for large units do not internalize the environmental costs of producing meat. The Bank's new strategy includes integrating livestock–environment interactions in to environmental impact assessments, correcting regulatory distortions that favor large producers, and promoting and developing markets for organic products. These measures are steps in the right direction, but more needs to be done by lending agencies, governments, non-governmental organizations, and individual consumers. Changing the meat economy will require a rethinking of our relationship with livestock and the price we're willing to pay for safe, sustainable, humanely-raised food.

Meat is more than a dietary element, it's a symbol of wealth and prosperity. Reversing the factory farm tide will require thinking about farming systems as more than a source of economic wealth. Preserving prosperous family farms and their landscapes and raising healthy, humanely treated animals, should also be viewed as a form of affluence.

Further Reading:
World Society for the Protection of Animals,
www.wspa.org.uk

Danielle Nierenberg is a Staff Researcher at the Worldwatch Institute.

COMMON GROUND FOR
Farmers and Forests

ALARMED BY SIGNS OF EXTENSIVE DEFORESTATION OVER THE PAST DECADES, GROUPS IN COSTA RICA ARE DEVELOPING PROGRAMS THAT COMBINE ECOLOGICAL AWARENESS AND SUSTAINABLE AGRICULTURE

by Joyce Gregory Wyels

With more than 25 percent of its land protected in national parks and private reserves, Costa Rica enjoys an enviable reputation as an ecological paradise. Green-hued travel posters tout cone-shaped volcanoes blanketed by forests, home to a variety of flora and fauna all out of proportion to the diminutive size of this Central American nation. Rare toucans, resplendent quetzals, and scarlet macaws delight bird watchers, while adventure-travel enthusiasts wax lyrical about kayaking mangrove swamps, hiking verdant rain-forest trails, and rafting white-water rivers fringed with tropical foliage. Costa Rica's forests have even spawned a new sport: the ingenious canopy tour, in which participants don rappeling gear to zip through the treetops on cables attached to elevated platforms. From this vantage point, the travel posters have it right: broad vistas of undulating green spread toward the horizon, underscoring the tourism slogan, "All Natural Ingredients."

But a look at the landscape from an even loftier perspective suggests trouble in paradise. NASA (U.S. National Aeronautics and Space Administration) has compiled nearly thirty years' worth of photographs taken from space, which

document the tracts of land covered in forest and the growing settlements, farms, and pasturage that continually chew at their edges. In 1993, student interns from EARTH University (Escuela de Agricultura de la Región del Trópico Húmedo), with the support of NASA scientists, began interpreting the images. Last summer the results were released in *Costa Rica desde el espacio* (Costa Rica from Space), a compendium of 125 photographs published in book form with bilingual text by NASA and EARTH, with the financial support of UNESCO (United Nations Educational, Scientific, and Cultural Organization) and Banco San José.

> COSTA RICA'S FORESTS HAVE EVEN SPAWNED A NEW SPORT: THE CANOPY TOUR, IN WHICH PARTICIPANTS RAPPEL THROUGH THE TREETOPS ON CABLES ATTACHED TO ELEVATED PLATFORMS

The images, which include aerial and field shots as well as photographs from

space, cast doubt on Costa Rica's green image. Whereas forest once covered 99 percent of the country, by 1983 it was down to 17 percent. The photographs show traces of fire along both coasts, and great swaths of pastureland replacing old-growth trees. Forests that once spread from the slopes of the Turrialba Volcano in the Cordillera Central to the flatlands of Tortuguero on the east coast have disappeared.

NASA astronauts, including Costa Rica's own Franklin Chang-Díaz, presented the book at press conferences and other public gatherings last summer. Newspaper editorials warned of dire consequences if the trend continued, one going so far as to predict that if San José and the surrounding Central Valley are transformed into a megalopolis, "within twenty-five years we would go from the 'Switzerland of Central America' to the 'Calcutta of the Caribbean'."

The loss of forest lands could decimate the carefully nurtured tourism industry, which has surpassed coffee and bananas as Costa Rica's number-one income earner. But reasons for concern go far beyond the loss of tourism revenue. Forest environments prove their worth in more ways than maintaining habitat for ecotourists or even for the wildlife that dwindles as forests become fragmented.

On a global level, tropical rain forests absorb excess atmospheric carbon, thereby reducing global warming. Locally, even the simple act of providing shade helps to regulate temperatures and safeguard organisms that affect plant and animal life all along the food chain. Moreover, trees and their root systems anchor the soil, keeping erosion in check.

One compelling reason for preserving the rain forest stems from our ignorance about the extent of its resources. In Costa Rica, where 0.035 percent of the earth's surface supports 5 percent of the world's biodiversity, the next medical breakthrough or nutritional supplement may be awaiting discovery among the endemic plants and organisms of a little-understood ecosystem.

Some of the most serious damage comes from commercial logging and roads cut into the forest. The heavy equipment that accompanies logging erodes the soil and suffocates streambeds, killing fish and the animals that feed on them. Roads establish new channels for runoff, thereby altering drainage patterns and aggravating erosion.

In Costa Rica, however, forest loss stems less from commercial logging than from people simply clearing land for farms and pasturage as a source of livelihood. "The possibility to own small farms has been something available in Costa Rica for quite a while," says agroecologist Dr. Stephen Gliessman, who has studied farming systems in Costa Rica for the past thirty years. "It's different from other Central American countries in which much less land is available to small farmers."

But even when the trees disappear as a result of farmers clearing land for cultivation, the resulting erosion and water contamination contribute to environmental degradation. Especially in the tropics, with its torrential rains, deforestation triggers far-reaching consequences. Agroecologist Dr. Reinhold Muschler cites the cropland on the shoulders of the Irazú Volcano as an example.

"We have very steep hills worked in ways that keep them completely denuded at times," he says. "So the rain washes tons of fertile topsoil away, and it's irreversibly gone. This is one of the main problems linked to deforestation. With the loss of forest cover or a permanent ground cover you have this exposure of the soil, and with that you have tremendous erosion, leaving that landscape denuded and causing problems downstream."

Among the problems caused by deforestation, explains Muschler, are siltation in hydroelectric power plants and in marine coastal areas. In Cahuita National Park, for example, clay particles in the eroded material change the composition of the water. This in turn changes the way the water transmits light, and therefore the coral reef suffers. Says Muschler, "We have seen serious negative effects in many marine environments that are due to deforestation in the hinterlands, sometimes hundreds of kilometers away."

The good news is that even as rural poverty and the demand for farmland push more people into the forest, concepts of ecology and sustainable agriculture have emerged that promise benefits for both farmers and forests. One group that espouses a holistic approach to conservation and rural development is the Tropical Agricultural Research and Higher Education Center (Centro Agronomico Tropical de Investigación y Enseñanza—CATIE) on the outskirts of Turrialba.

CATIE grew out of an earlier organization established in 1942 to conduct research and train personnel in tropical agriculture. The center began by building collections of coffee, cocoa, and fruit trees, followed by other crops and forest species. Today the twenty-five-hundred-acre campus contains more than three hundred species of trees and crops, distributed among fields, forests, swamps, and lake. The diversity of its ecological niches attracts some 200 of the 850 bird species in Costa Rica, plus numerous migrant birds.

Another organization that educates scientists regarding conservation issues in tropical regions is the Organization for Tropical Studies (OTS). A consortium of fifty-six universities from the U.S. and Central America operating at three Costa Rican field stations, OTS also embraces a philosophy of reaching out to surrounding communities.

But when Stephen Gliessman first studied tropical biology and ecology at OTS as a graduate student in the early 1970s, he encountered a paradox. "I was thinking about how ecology tells us a lot about how nature works," says Gliessman. "I was confounded by the fact that it didn't look to me like agriculture was using the same knowledge. People were clearing forests and farming for a short period of time, then having to abandon their farms and clear more forest. It just didn't seem like that should be happening—there ought to be a way to apply ecology to agriculture so that agriculture could stay permanently on a cleared piece of land and not continually move into the forest to find more after what they cleared had degraded."

After earning his Ph.D., Gliessman went to work for a small private farm in southern Costa Rica, applying his knowledge of ecology to help make Finca Loma Linda work as a sustainable farm and ecosystem. Gliessman presented the results of his research at the First International Congress of Ecology in Tel Aviv, Israel, in 1974. The author of *Agroecology: Ecological Processes in Sustainable Agriculture*, he later organized the agroecology course for OTS and started the Agroecology Program in the Department of Environmental Studies at the University of California Santa Cruz.

Last summer Gliessman joined Reinhold Muschler of CATIE to coordinate the International Course on Tropical Agroecology and Agroforestry, the first time the course had been offered in Spanish and in Latin America. Twenty-six researchers—agronomists and agroforestry specialists from universities, government agencies, and nongovernmental organizations in twelve countries—came together at the CATIE campus to enhance their knowledge of agroecological principles and to explore ways of helping farmers move toward environmentally friendly farming practices.

Participants in the CATIE course also shared their own innovations, describing coffee-tasting rooms in Matagalpa,

Nicaragua, micro-credit programs in Huatusco, Veracruz, Mexico, and a homestay internship program for college students at Costa Rica's Finca Loma Linda that offers learning experiences for both farmers and students.

Juan José Jiménez Osorio and Rosana Gutierrez Jiménez, from Yucatán, described a one-year training program in Maní in which young Maya farmers learn about Maya history along with sustainable farming practices. "The school is unique," says Roberta Jaffe, Gliessman's partner and a science educator who applies her skills to supporting sustainable communities. "Young farmers, usually eighteen to twenty-five years old, come from the surrounding pueblos to learn how to enhance their farming practices, and they are taught about basic Maya history." Jaffe expressed surprise that these farmers would need such instruction, inasmuch as they were living the culture. "But," she was told, "most of them don't have an education beyond primary grades. Many of them have never even been to the nearby pyramids."

It was the school in Maní, Escuela de Agricultura Ecológica, that inspired Gliessman and Jaffe to start a networking group from among the researchers present at CATIE last summer. They call it "CAN," for Community Agroecology Network. "These groups were working in isolation," says Jaffe. "We saw that they could benefit tremendously from learning from each other."

Many of the researchers at CATIE reported that as they spread the gospel of sustainable agriculture, a common response from farmers was, "That's the way my grandparents used to farm." For centuries, indigenous farmers and forest dwellers had intuitively followed sounder ecological principles than the mechanized methods that came into vogue in the 1960s.

According to Muschler, the traditional combination of maize, beans, and squash is an example of crops that complement one another. The beans convert atmospheric nitrogen into a form of nitrogen fertilizer that is useful to the plants and enriches the soil. The corn produces a crop, and once it is harvested the stalks can be used as climbing poles for the beans. The squash provides additional soil cover and produces a nutritious vegetable.

"This is a traditional system that is not a monoculture," says Muschler. "This is an intermediate step of polycultures. As you go toward further complicated systems you end up with systems that have annual crops and some perennial crops—like maybe some fruit trees or bushes of different kinds—and ultimately you will have some timber species and maybe other tree species that provide a series of ecological functions."

He lists as examples of those functions microclimate improvement, reduction of wind, and balancing of the water availability in a watershed. That's why there's such a big difference between a forested watershed and a deforested watershed, says Muschler: "It's the presence of the trees and their deep roots that helps sustain the soil and helps the rainwater to infiltrate into the soil and to be retained in the soil." Organic matter from dead leaves and plant roots functions as a sponge in retaining the water. At the same time, certain trees can help to fertilize the crops by converting the nitrogen in the atmosphere to a form that nourishes the plants.

The vaunted "Green Revolution" of the 1960s, however, focused on increasing crop yields without concern for environmental or health problems. Massive applications of chemical fertilizers, herbicides, and pesticides depleted soil fertility and contaminated water, while single crops reduced genetic diversity. Far from today's understanding of "green," the revolution led to an industrial and technological approach to agriculture.

"The Green Revolution with its overemphasis on high yields undermines all of our understanding of applying ecology to agriculture," says Gliessman. "Technology did increase yields, but it created imbalances in other areas. Agroecology gives us a framework for understanding not just the ecology, but how we connect to the health and viability of the communities in which people live."

Although agroecology applies to crops in general, the topic that kept coming up at CATIE was the global coffee crisis. A glut on the world market has caused a steep drop in prices, so that the grower now earns only 5 percent of the retail market value—or less than production costs. Coffee, an important crop in other Central American economies, is Costa Rica's largest export. One out of four agricultural workers derives a livelihood from coffee growing, and 75 percent come from small farms. "They're at the stage where they're pulling out plants," says Jaffe. Worse yet, some workers and small farmers have been forced off their land and into urban centers.

For those growers affiliated with CATIE, the center has set a minimum fair price paid directly to the farmers, who receive training in farm diversification with native trees, soil conservation, and biological pest and disease management.

As coffee connoisseurs have learned, shade-grown, organic berries produce a better quality coffee than berries exposed to the sun. Several coffee companies that promote shade-grown coffee subscribe to fair-trade practices as well, in which coffee farmers receive a fair share of profits regardless of world price fluctuations. The problem is that so far, in the United States, only two-tenths of 1 percent (0.2 percent) of the coffee sold is fair-trade coffee.

FARMERS OUGHT TO BE ABLE TO GET A DIFFERENT PRICE BASED ON A SYSTEM THAT PROTECTS RESOURCES, KEEPS PEOPLE ON THE LAND, ... AND OPENS UP THOSE FORESTS FOR REFORESTATION

The solution, some believe, lies in paying farmers for the environmental services they provide. Muschler offers the example of members of a village at the bottom of a hill paying fees for the drinking water and irrigation water that's dependent on the watershed farther uphill. Payment would go to those farmers who maintain their land intact in such a way that it can provide the hydrological services of a forest.

Muschler wants to make every consumer of water aware that water comes ultimately from a spring. In addition, everybody who turns on a television should understand that "hydroelectric power (in Costa Rica we have 90 percent of electricity from hydropower) can only function as long as the dam is intact and the lake that feeds the turbine is not subject to siltation from all the soil that is eroded on the hillsides around it."

"We're talking about forest structure and the services a forest can provide in terms of protecting the soil, producing water, tying up carbon dioxide, and protecting biodiversity," says Gliessman. "The *bosque de café* [coffee forest] can do a lot of things a forest does, yet provide a livelihood for the farmers who manage it. When farmers sell their coffee they ought to be able to get a different price based on coffee from a system that protects natural resources, keeps people on the land, reduces the need to cut down any more forest, and maybe even opens up those forests for a certain degree of reforestation."

That's precisely the idea behind a program started two years ago at Finca Loma Linda. Speaking for the farmers in four cooperatives that have banded together as Programa Pueblos, Darryl Cole-Christensen, who owns Loma Linda, puts it this way: "Treat us fairly in your purchase of our coffee, and we will manage more competently your environment and ours." But despite the high quality of their coffee, the farmers of Programa Pueblos are struggling economically.

The missing link, says Cole-Christensen, is a marketing organization in the U.S. Programa Pueblos has enlisted former student interns to help with direct marketing, as well as faculty, family members, and friends. "From month to month the number of people involved in this marketing structure is increasing," says Cole-Christensen.

Jaffe sees CAN, too, as a way of cutting out some of the middle-man profits and "taking back the concept of global village—where global village has been co-opted by multinational businesses—but maybe we can make global village a way of connecting consumer and producer to the benefit of both."

> WE ARE CREATING A MODEL FOR HOW LESSER DEVELOPED AREAS OF THE WORLD CAN ACHIEVE ECONOMIC STABILITY AND WELL-BEING WHILE ALSO MAINTAINING ENVIRONMENTAL INTEGRITY

Programa Pueblos is currently building a community center where interns work with families on community and environmental policy. The interns have launched a market study to enable farm families to earn income through diversified, small-scale produce production. They plan to welcome visitors to demonstration plots to showcase pilot research projects that can later be extended to co-op members' farms. Cole-Christensen envisions creating links between producers and consumers by way of a network of e-mail communication and video and multimedia. The hope is to extend "reforestation of our watersheds and the high summits of our mountains to an entire region of Costa Rica." Ultimately, says Cole-Christensen, "we are creating a model that demonstrates how lesser developed areas of the world can achieve economic stability and well-being while also maintaining environmental integrity."

Even now, Muschler sees reason for optimism: "While some ten or fifteen years ago the prognosis for the forest cover of Costa Rica was rather negative and there was a series of satellite images and aerial photographs showing rampant deforestation, over the past five or six years there has been a lot of effort to reforest. I think we are through that trough of deforestation. We have seen a reverse trend, and now Costa Rica has more forest cover than ten years ago."

Joyce Gregory Wyels is a California-based travel writer and frequent contributor to Américas. *Aerial and space photographs are from* Costa Rica from Space *and appear courtesy of Heliconia Press, San José, Costa Rica. Readers interested in further information on agroecology in the tropics should refer to the website www.agroecology.org.*

WHERE THE SIDEWALKS END

One out of every seven people now lives in a slum—or at least that's the UN's best estimate. More and more slum residents are organizing to improve their lot, as their numbers swell in cities all over the world.

by Molly O'Meara Sheehan

SQUINTING IN THE SUNLIGHT, George Ng'ang'a leads me up a mound of dirt and rubbish on the edge of his Nairobi neighborhood to take in the view. To the south unfolds a safari scene of grassy plains dotted with acacia bushes as far as I can see. To the north stands a dense gathering of gangly shacks cobbled together with cloth, mud, tin, rocks, and sheets of plastic. There are about 800 homes in all crowded onto some 5 to 6 hectares, says Ng'ang'a.

On city maps, the location of this settlement—called "Mtumba" by the 6,000 people who live there—shows up as prime habitat for rhino and giraffe. That's because this unsanctioned community lies on the edge of Nairobi National Park. Mtumba is only one of the many slums around Nairobi. In fact, more than half of the residents of Kenya's capital city cannot afford to live in "formal" housing, and have been forced to find shelter in slums like this one.

Ng'ang'a turns to me and tells me to call him "Castro," which, he says, is his nickname. He has the physique of a bear and is clean shaven, but he insists he was thin and bearded in his youth. I'm not sure if he's joking about the physical resemblance, but it's clear that he's passionate and politically active. For several years in a row the people of Mtumba have chosen Castro to be the leader of the community's governing council in informal elections—informal because the city government does not serve slums, so the people of Mtumba have found their own ways to organize and police themselves.

"We can't depend on the government for anything," says Castro as we walk through the settlement. One of his neighbors, a solemn man named Tom Werunga, joins in our stroll. Werunga, who carries a Bible, tells me that he's a pastor. He points out a water tap—one of two small spigots that supply water for the entire settlement. But no city water is piped here. Instead, these taps are fed by pri-

vate companies that truck in tanks. And they sell their water at a premium. As of yet, no company has seen fit to establish any sort of business setting up toilets or sewers. Instead the 6,000 people who live here share three flimsy pit latrines. "Flying toilets," I learn, are baggies of human excrement that are flung atop roofs or into rubbish piles.

NAIROBI

"When you wake up in the morning, the important thing to do first is to find out where are your shoes. Why shoes are useful: when you walk without them your legs can get injured by anything dangerous like bones, thorns and many others. So I will suggest that shoes are the most useful object in our home."—Serah Waithera, a 15-year old girl living in Nairobi's Mathare slum. Reprinted from: *Shootback: Photos by Kids from the Nairobi Slums* (London: Booth-Clibborn Editions, 1999).

I am scribbling notes, trying to pay attention to the latrines Castro is showing me, but my eyes are stinging in the acrid air. Cinders and fumes from untended piles of burning trash mingle with ash and smoke from charcoal cooking fires where women prepare meals. At night, kerosene fumes from lanterns join the stew. More than 80 percent of Nairobi's households use charcoal for cooking, but the air is worst in neighborhoods such as this, which lack both electricity and trash removal.

Everything in Mtumba, it seems, is insecure and informal. There is no land ownership. There is no public infrastructure. And there is no protection provided by the law. Mtumba's families have moved together twice before, says Castro. They landed in this location in 1992. Since then Nairobi officials have threatened to evict the community several times. And on one occasion, he says, officials sent in bulldozers to completely demolish the settlement. Some families have seen their homes destroyed as many as 10 times. "Every day we are waiting for the demolition squad," says Castro. "We are refugees in our own country."

IT IS NEIGHBORHOODS LIKE MTUMBA—not Greenwich Village in Manhattan or the Rive Gauche in Paris—that are setting the trends for modern urban living. The UN estimates that somewhere between 835 million and 2 billion people now live in some type of slum, whether in a *kampung* in Indonesia, a *favela* in Brazil, a *gecekondu* in Turkey, or a *katchi abadi* in Pakistan. The population of slum dwellers in some of the world's largest cities—Bombay, Bogotá, and Cairo, for example—now outnumbers the population of people living in formal housing.

NAIROBI

Families living in slums often have no access to state-sponsored education. Lacking government support, the Mtumba slum in Nairobi pooled resources for a schoolhouse. Community members built the structure, and now support four teachers.

In many cities—particularly in sub-Saharan Africa and South Asia—explosive urban growth is combining with the world's worst poverty to fuel the proliferation of slums. The world's population increased by 2.4 billion in the past 30 years, and half of that growth was in cities. Over the next three decades, global population is expected to increase by another 2 billion. Demographers expect that nearly *all* of that population increase will end up in developing-country cities, due to urban migration and high birth rates (see graph, next page).

While most poor people still live in rural areas, poverty is rapidly urbanizing. As of 1998, more than 1.2 billion people were living in extreme poverty (on less than the equivalent of about $1 a day), unable to meet even basic food needs. Martin Ravallion of the World Bank estimates that the urban share of the world's extreme poverty is currently 25 percent. He projects that it is likely to reach 50 percent by 2035.

A number of factors are driving the growth of cities worldwide. Rural economies in many regions have been hard hit by environmental degradation, military or ethnic conflicts, and the mechanization of agriculture, which has curbed the number of rural jobs. The prospect of better-paying jobs has drawn many people to cities.

Latin America is by far the most urbanized region of the developing world. About 75 percent of people in Latin America live in cities—along with 75 percent of the poor. While only 37 and 38 percent of Asians and Africans live in cities respectively, a number of nations in these regions are beginning to see poverty shift to urban centers. For instance, the proportion of people living below the poverty line in rural Kenya between 1992 and 1996 increased from 48 to 53 percent, while the share of people living below the poverty line in Nairobi doubled from 25 to 50 percent.

Castro tells me that his family's land was taken by the colonial Kenyan government in 1952 to build a golf course. "My father was a businessman," he says, "so we went to different places, like nomads." Castro continued the itinerant lifestyle as a young man, but then he got married and began looking for a better life for his family. Eventually, he says, "we came to the Nairobi slums, even though I have an education."

IN GENERAL, THE "OFF-THE-BOOKS" NATURE OF MTUMBA and other informal communities confers certain advantages. Rents are lower than in formal housing. There are no property taxes. Residents can skirt cumbersome zoning laws that separate housing from businesses, and set up shop inside their homes or just outside. Mtumba's commercial strip boasts rows of brightly painted storefronts, each about 1 meter wide. There are produce stands, coffee shops, a "movie house" showing videos, a barber shop, and an outfit that collects old newspapers. But the short-term benefits of living and working outside the formal economy rarely outweigh the long-term costs to residents—and to the cities that have failed to address their needs.

Slums are often located in a city's least-desirable locations—situated on steep hillsides, in floodplains, or downstream from industrial polluters—leaving residents vulnerable to disease and natural disasters. Another long-term cost is the premium residents pay for basic services. The African Population and Health Research Center recently released a report showing that Nairobi's slum dwellers pay more than residents of wealthy housing estates for water—and, as a result, use less than is adequate to meet health needs. "A family needs 100 liters per day for drinking and cleaning," says Mtumba's Tom Werunga. As that much water costs 25 Kenyan shillings (30 cents), it could easily eat up half the income of people who, on average, make about 50 to 60 shillings (60 to 75 cents) per day.

Landlords operating in slums can easily gouge their tenants without fear of legal recourse. And the proportion of renters in slums is higher than commonly thought, as vacant land close to employment opportunities tends to be quickly developed by enterprising landlords. In fact, four out of five slum residents in Nairobi are renters, according to a study done by the Kenyan government and UN-HABITAT, the UN agency for human settlements, which happens to be headquartered in Nairobi. The

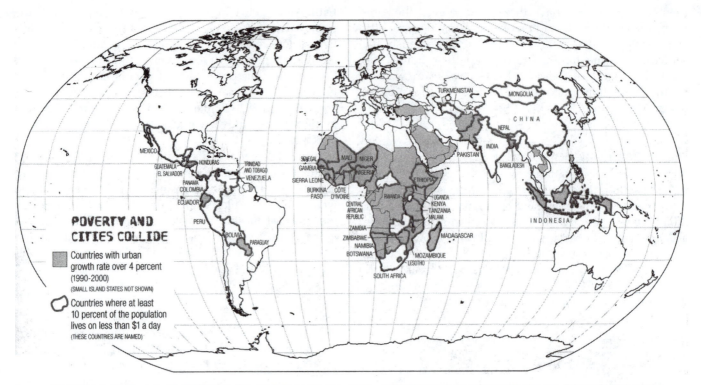

Sources: *World Urbanization Prospects: The 1999 Revision* (New York: United Nations, 2001); and *World Development Indicators* (Washington, DC: World Bank, 2001)

shacks are lucrative investments, finds the survey, yielding a return in less than two years (compared to 10 to 15 years in the formal property market). Yet landlords do not typically reinvest their profits in the shacks by repairing them or hooking them up to electricity or water, and tenants have no way to hold landlords accountable.

Lacking adequate access to water, toilets, and trash removal, crowded slums also breed diseases that threaten the public health of entire cities. More than half of Nairobi's 3 million people live in slums, squeezed into just 5 percent of the city's land area. In urban centers throughout the developing world, the AIDS virus is facilitating outbreaks of tuberculosis—and both diseases are spreading rapidly. In the Nairobi slums, the mortality rate of children under five years of age is 151 per 1,000 births, far higher than the average of 61 per 1,000 for the city as a whole.

Economic inequalities may significantly hamper public health, according to several new studies. *The Society and Population Health Reader* has brought together journal articles showing that economic inequality in the United States and parts of Europe correlates with reduced public health. In Nairobi, where slums occasionally abut posh, gated enclaves, the economic disparities are as glaring as the public health nightmare.

The growth of slums in an era of unprecedented economic prosperity may also contribute to tensions that threaten local, national, and even global security. "Poor urban settlements are breeding grounds for disease, crime, and terrorism," warned Anna Tibaijuka, the head of UN-HABITAT, in April 2002. While desperate situa-

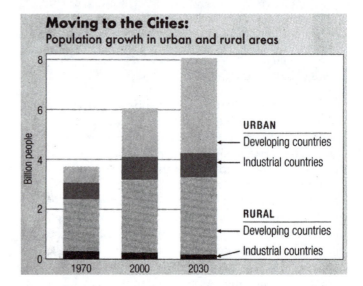

URBAN POVERTY By 2007 more than half of the world's population will reside in cities. In the next 30 years the global population is expected to grow by 2 billion, and demographers project that developing-country cities—many of which are already faced with dire poverty—will be the locations of the most growth.

tions may foster problems, it is the poor who are disproportionately the victims of crime. Some slums are crime ridden and other are nearly crime free, but those that lack municipal or community policing are usually more dangerous.

Following the September 11, 2001 attacks on the United States, *New York Times* columnist Tom Friedman wrote that in an increasingly interconnected world, it will be impossible to ignore the problems of people living in desperate conditions at home or abroad: "if you don't visit a bad neighborhood, a bad neighborhood will visit you."

WALKING AGAIN WITH CASTRO, I AM BEING PURSUED by a friendly, giggling swarm of small children, none taller than my waist. They want to hold my hands. My tour guide is talking about the three vehicles owned by various people in Mtumba—one old car and two bicycles—but my attention is drawn to the children. Many of them have no shoes, yet are following us over sharp rocks, human and animal waste, and all sorts of garbage.

I looked at the kids' feet and I cringed. It is impossible to watch bright-eyed children play in toxic trash and human waste, and listen to their articulate parents describe their efforts and their hopes to build a better life, and not feel obliged to help somehow. This well-intentioned impulse to help slum dwellers into better housing, however, has been carried out with rather disastrous consequences throughout history.

In the United States, for example, the 1949 Housing Act paved the way for cities to raze blighted neighborhoods and build giant public housing projects to house the newly displaced inhabitants. Brazil, Colombia, Egypt, and South Korea were among the developing nations that launched huge public housing campaigns in the 1960s. These costly efforts destroyed the networks of family and friends that poor people had used to survive. Communities often had to move from inner-city locations to outlying areas with fewer job prospects. Added transportation costs meant less could be spent on food. In many cases, the people whose homes were destroyed could not afford the new public projects, which ended up housing wealthier residents. "Urban renewal" projects often had the perverse effect of worsening living conditions for the people they were intended to help.

A major shift began to occur in the 1970s, as city planners were faced with the fact that poor people had been improving their neighborhoods more effectively and with less money than many government projects. Drawing on his experiences working in the slums of Lima, Peru in the 1960s, British architect John F. C. Turner challenged the prevailing orthodoxy with his influential 1972 book, *Freedom to Build*, warning that officials should stop doing more harm than good.

Lacking city services, some communities have managed to close the gap themselves. One of the trailblazers was Akhter Hameed Khan, who in 1980 began mobilizing the community of Orangi, the largest squatter settlement in Karachi, Pakistan. He started a research institute called the Orangi Pilot Project to help residents organize and build a sewer system. Each block collected money and began construction of their own sewers, which served some 90 percent of Orangi's residents by the late 1990s. Between 1982 and 1991, infant mortality rates in the settlement dropped from 130 per thousand to 37 per thousand.

IN THE SLUMS OF NAIROBI, COMMUNITIES LONG NEGLECTED by the government are just beginning to gain some level of political effectiveness. In Mtumba, for instance, residents have begun to organize. "On our own," says Tom Werunga, "we have built a school." Four teachers juggle morning and afternoon shifts to teach more than 400 children in three classrooms. The classroom I saw boasted a small chalkboard, and about 30 to 40 small children, who jumped up smiling from their desks as we passed.

RIO DE JANEIRO

One out of every six of Rio's 5.9 million residents lives in a favela, or slum, according to government estimates, but citizen's groups say the share is as high as one-third of the metro region's 12 million people. Entire communities have been wiped out by mudslides, as erosion has claimed land under makeshift houses on Rio's hillsides. "It raises doubts in people's minds if you give an address here—you have to give a false address to be treated fairly," said a woman in the Nova Brasilia section of Rio to U.S. scholar Janice Perlman. A city-wide program, Favela-Bairro, aims to integrate these neighborhoods into the city with pavement, water, sewers, and electricity.

With the help of a local nongovernmental organization, the Pamoja Trust, Mtumba has started a savings scheme and opened a bank account to pool funds. They hope to save up enough to purchase land at a better location. So far, they have saved about 300,000 Kenyan shillings ($3,800) altogether. According to Pamoja Trust's Jack Makau, his organization would like to match the savings accrued by the Mtumba families, shilling for shilling, and help them invest it, to speed the time necessary to reach the 5 million or so shillings that will be needed.

The residents of Mtumba and Nairobi's other slums are starting to flex some political muscle, bolstered by a city-wide federation, *Muungano wa Wanavijiji*. "Unity is strength," says Jane Weru, the head of Pamoja Trust, which is supporting the federation in 40 of Nairobi's more than 100 slums. *Muungano* members are setting up savings groups, which help build trust and can be turned into revolving loan funds. They are also collecting data on their neighborhoods and sharing experiences to help build coalitions that will help sway government policies in their favor.

Slum residents in Nairobi are also learning from their counterparts around world, loosely organized by Slum/Shack Dwellers International (SDI). The Group was founded in 1996 when the Asian Coalition for Housing Rights joined forces with the South African Homeless People's Federation. Today, the group boasts members from Argentina, Cambodia, Colombia, India, Kenya, Madagascar, Namibia, Nepal, the Philippines, South Af-

TOILET POWER

Jockin Arputham is President of the National Slum Dwellers Federation of India and leader of Slum Dwellers International, a network of grassroots organizations in Cambodia, India, Kenya, South Africa and Thailand, among other countries. He has led the struggle for housing rights in Mumbai, India, since the 1970s, where he started the National Slum Dwellers Federation, which now spans 34 Indian cities. Rasna Warah, Editor of Habitat Debate, *the journal of UN-HABITAT, interviewed him during the World Urban Forum in Nairobi in April 2002. (Printed with permission.)*

Rasna Warah: Tell me a bit about your background.

Jockin Arputham: I am a slum dweller who has lived in Mumbai's Mankhurd Janata Colony for the last 35 years. I was not always a slum dweller. I belonged to an upper middle class family, which lost everything in 1963. That is when I left my home in Kolargold Field for Bangalore and then for Mumbai in 1964. I had never lived in a slum before that, so it was a new culture and new concept for me.

My first experience with activism was when I brought together children from the slum and organized them to carry uncollected garbage in the slum and dump it in front of the Mumbai Municipality's offices. Then in 1974, I invited officers from the Municipality to visit our slum. The idea was to lock one of the officers in the public toilet for a whole day, which we did. He was there for eight hours. In the end, the police had to take him out. But we had made our point. That incident changed people's attitudes towards cleanliness in public toilets.

RW: One of the main issues your organization agitates around is that of toilets. Why is this such an important issue for you?

JA: In India, a public toilet is not simply a toilet. Public toilets are community centers where people meet to exchange news about what is happening in the community. When you go to a public toilet, you get all the news about the settlement. In fact, today in India, if you want to mobilize people, you first go to the public toilets.

In India, 90 per cent of slum dwellers use public toilets maintained by the municipality. But I call these toilets "monuments" because the minute they become the responsibility of the municipality, the service becomes defunct. The National Slum Dwellers Federation has been urging the government to allow slum communities to plan, design, construct, and maintain public toilets. As a result, in the city of Mumbai, we are now constructing toilet blocks—10,000 toilet seats in total.

The Prime Minister has also given a grant of 10 billion rupees to build toilets all over India. These toilets will be planned, designed and maintained by the communities themselves.

RW: What is Slum Dwellers International?

JA: SDI is not a political movement or a social service organization. It is a platform through which urban poor communities take responsibility for improving their lives. It is a voice of the urban poor. When the poor don't take responsibility for their lives, everyone from NGOs to the World Bank will take responsibility away from them. They will tell you how to live, how to eat, how to dress. They will even tell you how to use a toilet. What nonsense! I am an adult, why should anyone control my life? However, instead of merely protesting against these institutions, we are going a step forward by saying, "involve us."

Institutions such as the World Bank have been unsuccessful in many cases because their projects lacked one key ingredient—the involvement of the community. For instance, the World Bank once took seven years to build one toilet in a slum in Mumbai. Seven years! Why? Because they spent much of the time "studying" Indian culture, Indian values, even Indian ways of shitting. Tell me, how can a consultant from London know these things?

But since the World Bank started to work with the National Slum Dwellers Federation, we have managed to build 100 toilet blocks, or 2,200 toilets, within one year. At this rate, for the first time in the history of independent India, no one in Mumbai will have to squat on the streets because there will be one public toilet for every 50 people.

RW: But don't governments have a responsibility towards the urban poor? Shouldn't they be providing these services?

JA: Certainly governments have this responsibility. But they are not doing it. We used to have the handout approach. But now we are saying that we need to split responsibility. In the case of toilets, we are saying to governments, "You pay the capital costs, but we will plan, design and maintain the toilets."

RW: You still live in a slum, even though your economic conditions have improved. Why?

JA: I think I continue to live there because my whole life's work began there. Everything I learned, all the knowledge I gained, was from the slum. Besides, both my children are married and it is just me and my wife. Why would we need a bigger place? I've lived in my current home for 15 years, why not another 25? If I move out, I'll feel as if I exploited the very people who made me who I am. I am very proud to be a slum dweller. I don't see my work as a job; it is my life.

rica, Swaziland, Thailand, Zambia, and Zimbabwe. "A lot of what we do in Nairobi," says Pamoja Trust's Jack Makau, "has been tried out in other cities by the SID network."

IN RECENT YEARS, THESE NEW COALITIONS HAVE ARTICULATED ground-breaking strategies for urban development, where governments engage slum dwellers as equal partners in efforts to improve communities. "We are not coming here to beg," declared Jockin Arputham, the head of Slum Dwellers International, at UN-HABITAT's World Urban Forum in Nairobi in May 2002. "We can sit together with you—national governments, city authorities, and bilateral aid agencies—to plan the city" (see "Toilet Power").

Where local and national governments have been willing to seriously engage those living in urban slums, the partnership has often produced significant results. But for the most part, governments still have a long way to go to help address the problems faced by people living in slums. In general, slum leaders like Arputham have identified three key obstacles that governments must surmount in order to become more effective partners:

1) HOME SECURITY

"Land is the key to implement any project for development," says a Mtumba woman who is involved in the community's self-run school. She explains that the people of her community have difficulty convincing themselves—let alone anyone else—to invest in water, toilets, or any sort of improvement. Why bother if the neighborhood could be bulldozed the next day? Indeed, a central obstacle to any sort of "self-help" in many slums is that the residents do not belong on the land where they live in the eyes of the law.

If governments were to grant people in informal settlements legal recognition or titles to the property where they live, it could open up new opportunities for development, and even credit. Buildings without titles are "dead capital," says Peruvian economist Hernando de Soto. They are useful only for whatever shelter they provide. Buildings with titles, in contrast, can have a second "life" in capital markets, where their owners can leverage them.

> "In India if you want to mobilize people, you first go to the public toilets," says Jockin Arputham, president of National Slum Dwellers Federation of India. In May 2002, his organization built a toilet on the front lawn of the UN headquarters in Nairobi to publicize the lack of public toilets and sewers in slums.

De Soto was instrumental in prompting Peru to undertake a massive titling program, which formalized some 1 million urban land parcels between 1996 and 2000, first in the *pueblos jovenes* of Lima, and then in other cities. In his recent book, *The Mystery of Capital,* de Soto suggests that titling programs could have a huge global impact. He estimates that the value of real estate not legally owned in the developing world and former Soviet bloc nations is $9.3 trillion.

Granting titles to residents in much of Lima and some other Latin American cities has been fairly straightforward, as a number of informal settlements arose after groups of settlers planned "invasions" of unused public lands. But in places like Kenya, many slums are on private land or on public land given—often under the table—to large-scale shack builders, who rent out their tenement housing. Sorting out ownership can be further complicated by a confusing mix of English land laws and African customary laws. One new innovation in Kenya is a "community land trust," which allows a neighborhood to collectively own its property, while each household retains some individual property rights.

The issue of secure land tenure is gaining in prominence. Heads of state meeting in New York for the UN's Millenium Summit in 2000 pledged to improve the lives of 100 million slum dwellers by 2020. The two measures of "improvement" are to be access to sanitation and security of tenure. When asked how improved security will be measured, Billy Cobbett of the Cities Alliance acknowledges that "it's tricky." Many governments don't count slum dwellers in their censuses, let alone measure their sense of security.

2) EMPLOYMENT OPPORTUNITIES

Most people come to cities seeking jobs. And the slums that many of these people end up living in—with rickety homes, mounds of refuse, and inadequate water supplies—could become key sources of employment. At little cost, municipal authorities could employ slum dwellers to build sewers, collect trash, compost organic waste, or otherwise improve their communities. If organic waste is composted, it can be used to nourish urban agriculture, which can provide both food and jobs. Cities could also revamp their policies on transportation, land use, and small-scale credit to improve the ability of poor people to make a living.

In 2000, the Kenyan government committed itself to working with the slum dwellers federation, local authorities, and the UN on a seven year slum-upgrading initiative. This program aims to make physical improvements—to extend roads and services into slums to connect them to the rest of the city. "We're looking at all possible sources of job generation," says UN-HABITAT's Chris Williams, including providing housing, water, electricity, and other services.

Schemes to collect and compost organic waste—such as paper, food scraps, and even human excrement—can

EVEN IN A SLUM, LIFE

While no two slums in the world are alike, and conditions vary greatly, getting by in the "informal" world is a daily struggle.

MONEY
Without property to leverage, you are hard pressed to get a reasonable loan. But a number of slums have started "revolving loan funds," where residents pool their savings for local use in small-scale loans.

HOME
The perpetual threat of eviction—because you do not have legal claim to the land where you live—reduces your incentive to improve your surroundings. Peru has begun to address this dilemma by issuing 1 million property titles to slum residents.

COOKING
Unless you have access to propane or electricity, you must cook your meals over a smoky fire pit burning charcoal or scraps of wood.

GARBAGE
Instead of curbside garbage pickup, you live with curbside garbage dumps. Burning the garbage—and creating noxious fumes—is how you keep levels of trash down.

TRANSPORTATION
Your main mode of transportation is your feet, so hopefully you can afford a pair of shoes. Bicycles, buses, and jitneys can greatly expand your employment opportunities by getting you closer to workplaces.

EDUCATION
If you are lucky, your community has cobbled together a school for your children to attend, and collectively supports a teacher.

WATER
Getting clean water can be difficult or expensive for you. Local waterways often double as sewers, and water piped in or brought in tanks by truck can eat up much of your day's meager earnings.

TOILETS
If you think sharing a toilet with your family is frustrating, try sharing with 1,000 families. Infectious diseases spread quickly when there are few sewers, and open waterways are often fouled by waste.

help nurture urban gardens and reduce the problems and costs of waste management while producing food and money. The UN Development Programme estimates that 800 million urban farmers harvest 15 percent of the world's food supply—and the share could grow if governments promoted, rather than discouraged, the practice. Agriculture provides the highest self-employment

earnings in small-scale enterprises in Nairobi, and the third highest in all of urban Kenya.

High transportation costs limit poor people's access to jobs. Zoning laws that separate homes from businesses discriminate against the poor, as do decisions to invest in infrastructure for private cars, rather than dedicated bus lanes, cheap transit, safe pedestrian walkways, or bicycle paths. "More than 95 percent of money that is meant to tackle transport issues in Kenya goes to motorization, while less than 5 percent of Kenyans actually own cars," says Jeff Maganya of the Nairobi office of the global Intermediate Technology Development Group (ITDG). Today more than 40 percent of Nairobi's residents can't afford to pay bus fares.

Most people would benefit if governments were to shift their priorities towards cheaper forms of transportation, including information jitneys (small buses called *matatus* in Nairobi) and bicycles. For many years, high luxury taxes on bicycles and a large fee for registering bicycles prevented poor people from buying and keeping them in Nairobi. Isaac Mburu, a bicycle mechanic who lives in Mtumba, had his bicycle confiscated by local authorities because he could not pay the fee. When Kenya reduced its tax on bicycles from 80 percent to 20 percent between 1986 and 1989, bicycle sales surged by 1,500 percent.

Governments can also take steps to open up lines of credit in informal communities, not only for home improvement, but for small-business development. Even in the poorest neighborhoods, there are buildings and money-making activities that could be leveraged to increase economic opportunities and strengthen communities. Nairobi's *jua kali,* or "hot sun," workers—street hawkers selling vegetables, motor parts, and all manner of goods and services—act as a crucial source of income for many poor people.

3) GOVERNMENT REPRESENTATION

A number of factors can contribute to silencing the voices of the poor and limiting public scrutiny of key decisions about how resources are allocated: collusion between politicians and real estate developers; government influence over or control of the press; or a weak civil society, for example. The wealthy, even if a small minority, simply have greater political power.

MANILA

More than one-half of Asia's urban residents live in slums, according to a recent article in the Harvard International Review. In the Philippines, poor people have settled in slum dwellings along the banks of Manila's Pasig River, where they risk being washed away by floods.

Government corruption also takes a disproportionate toll on slum residents. "When you take a complaint to a local authority employed by the government," says Isaac Mburu, who lives in Nairobi's Mtumba slum, "if you go without cash, you won't be served." While 67 percent of all Kenyans surveyed recently by Transparency International-Kenya said that interactions with public officials required bribes, 75 percent of the poorest and least educated said they were forced to pay bribes. An independent fact-finding team visited Kenya in March 2000 and concluded that "the land and housing situation is characterized by forced evictions, misallocation of public land, and rampant land grabbing through bureaucratic and political corruption." According to Transparency International's Michael Lippe, "corruption is a tax on the poor."

In some parts of the world, however, corruption is being thwarted by community organizers and committed leaders. Porto Alegre, Brazil has become famous for a municipal budgeting experiment started in 1989 that invites citizens to engage in setting public priorities and shows people how funds are allocated. A survey done after the first year of participatory budgeting in Porto Alegre revealed that the process had amplified the voices of the city's poor. Most of that city's slum population had indicated that clean water and toilets were their highest priority, whereas the government previously assumed that public transport was at the top of their list.

Today, more than 200 cities in Latin America have introduced participatory budgeting. In July 2001, Brazil enacted a national "City Statute" that requires municipalities to include citizens in urban planning and management, through participatory budgeting, among other measures. While only a small share of a city budget is usually up for grabs, the process does get important issues on the agenda and helps thwart corruption.

In Bombay, both the municipality and poor neighborhoods have gained as a result of the evolving partnership between local authorities and the national slum dwellers federation. "Fifteen years ago, we were just trying to get poor people to be part of the city," said Sheela Patel, director of the India-based Society for the Promotion of Area Resource Centres. "Now there's a realization that this is a key component of good governance." For example, she says, "when hawking is illegal, the municipality loses 170 million rupees ($3.5 million) per month by not giving the hawkers licenses."

In Nairobi, citizens convened the first ever Nairobi Civic Assembly in January 2002 to demand that the government open itself up to all citizens, including the poor majority. "We have a city without citizens because most of them have no voice," said Davinder Lamba, the head of the local human rights group, Manzigira Institute. Participants discussed how they might tackle a number of specific problems, from the city council's failure to provide water in poor neighborhoods to corrupt "land-grabbing" by public officials.

NEIGHBORHOOD BY NEIGHBORHOOD, THINGS ARE beginning to change. For years, whenever residents of a Nairobi slum called Huruma Ghetto tried to repair their homes, the city council blocked them, forcing them to pay bribes or forbidding their efforts on the grounds that they were squatters on public land. The community's initial efforts to organize themselves to overcome these obstacles met with failure. Once, when the community collectively refused to pay the tribes, their houses were set ablaze.

Banding together, and fortified by allies, Huruma Ghetto's residents are getting local authorities to work with them, rather than against them. In May 2002, I watched as the Huruma Ghetto held a ground-breaking ceremony for a model home paid for by its locally organized savings group and approved for construction by the Nairobi slums, as well as activist friends from all over the world (including Jockin Arputham of Slum Dwellers International), came to Huruma Ghetto to take part.

"With the savings scheme, we are not only collecting money, we are collecting people," says David Mwaniki, a 37-year old father of five who makes a living hawking utensils. He also serves as the assistant to the secretary of Huruma's community council, which organized the savings group. "We want to eradicate poverty, and we want people living in informal settlements all over the world to join us, so we can wipe out slums."

Molly O'Meara Sheehan *is a senior researcher at the Worldwatch Institute.*

FOR FURTHER INFORMATION

Jorge Hardoy, Diana Mitlin, and David Satterthwaite, *Environmental Problems in an Urbanizing World* (London: Earthscan, 2001)

Cities in a Globalizing World: Global Report on Human Settlements 2001 (Nairobi: UN-HABITAT, 2001)

Environment & Urbanization journal: www.iied.org/eandu

Habitat Debate journal: www.unhabitat.org/hd

World Bank, "Upgrading Urban Communities," www.worldbank.org/html/fpd/urban//urb_pov/up_body.htm

For information about Shack/Slum Dwellers International (SDI) go to: www.homeless-international.org

For information on Nairobi's Pamoja Trust, contact landrite@wananchi.com

From *World Watch*, November/December 2002, pp. 20-32. © 2002 by Worldwatch Institute, www.worldwatch.org.

UNIT 3

Energy: Present and Future Problems

Unit Selections

11. **Beyond Oil: The Future of Energy**, Fred Guterl, Adam Piore, and William Underhill
12. **Powder Keg**, Keith Kloor
13. **Living Without Oil**, Marianne Lavelle
14. **Renewable Energy: A Viable Choice**, Antonia V. Herzog, Timothy E. Lipman, Jennifer L. Edwards, and Daniel M. Kammen
15. **Fossil Fuels and Energy Independence**, B. Samuel Tanenbaum

Key Points to Consider

- According to recent and still controversial projections on oil supplies, when will the world begin to run out of oil (that is, reach peak production after which the supply of oil will no longer be sufficient to meet the demand)? How can the world adjust to this potential shortfall within such a short period of time?

- What are some of the connections between military conflict in the Middle East and the United States' supply of oil? Are there reasons for the failure to develop strong alternative energy policies in the face of potential disruption of energy flows as a consequence of war?

- What are some of the major benefits of such alternate energy sources as solar power and wind power? Do these energy alternatives really have a chance of competing with fossil fuels for a share of the global energy market?

- What are the benefits in combining energy conservation strategies with new technologies? Will such combinations really work to reduce the overall demand for fossil fuels, as some proponents suggest?

 Links: www.dushkin.com/online/
These sites are annotated in the World Wide Web pages.

Alliance for Global Sustainability (AGS)
http://globalsustainability.org/

Alternative Energy Institute, Inc.
http://www.altenergy.org

Communications for a Sustainable Future
http://csf.colorado.edu

Energy and the Environment: Resources for a Networked World
http://zebu.uoregon.edu/energy.html

Institute for Global Communication/EcoNet
http://www.igc.org/

Nuclear Power Introduction
http://library.thinkquest.org/17658/pdfs/nucintro.pdf

U.S. Department of Energy
http://www.energy.gov

There has been a tendency, particularly in the developed nations of the world, to view the present high standards of living as exclusively the benefit of a high-technology society. In the "techno-optimism" of the post–World War II years, prominent scientists described the technical-industrial civilization of the future as being limited only by a lack of enough trained engineers and scientists to build and maintain it. This euphoria reached its climax in July 1969 when American astronauts walked upon the surface of the Moon, an accomplishment brought about solely by American technology—or so it was supposed. It cannot be denied that technology has been important in raising standards of living and permitting Moon landings, but how much of the growth in living standards and how many outstanding and dramatic feats of space exploration have been the result of technology alone? The answer is few—for in many of humankind's recent successes, the contributions of technology to growth have been no more important than the availability of incredibly cheap energy resources, particularly petroleum, natural gas, and coal.

As the world's supply of recoverable (inexpensive) fossil fuels dwindles and becomes more important as a factor in international conflict, it becomes increasingly clear that the energy dilemma is the most serious economic and environmental threat facing the Western world and its high standard of living. With the exception of the specter of global climate change, the scarcity and cost of conventional (fossil fuel) energy is probably the most serious threat to economic growth and stability in the rest of the world as well. The economic dimensions of the energy problem are rooted in the instabilities of monetary systems produced by and dependent on inexpensive energy. The environmental dimensions of the problem are even more complex, ranging from the hazards posed by the development of such alternative sources as nuclear power to the inability of developing world farmers to purchase necessary fertilizer produced from petroleum, which has suddenly become very costly, and to the enhanced greenhouse effect created by fossil fuel consumption. The only answers to the problems of dwindling and geographically vulnerable, inexpensive energy supplies are conservation and sustainable energy technology. Both require a massive readjustment of thinking, away from the exuberant notion that technology can solve any problem. The difficulty with conservation, of course, is a philosophical one that grows out of the still-prevailing optimism about high technology. Conservation is not as exciting as putting a man on the Moon. Its tactical applications—caulking windows and insulating attics—are dog-paddle technologies to people accustomed to the crawl stroke. Does a solution to this problem entail the technological fixes of which many are so enamored? Probably not, as it appears that the accelerating energy demands of the world's developing nations will most likely be met first by increased reliance on the traditional (and still relatively cheap) fossil fuels. Although there is a need to reduce this reliance, there are few ready alternatives available to the poorer developing countries. It would appear that conservation is the only option.

But conservation is probably not enough to solve the environmental problems related to fossil fuel use. In "Beyond Oil: The Future of Energy," a team of science writers from *Newsweek* magazine suggests that controversial projections, which show global oil production peaking in the next couple of years, may well be accurate, making alternative energy strategies more than just attractive—they become necessary. Even if we adopted all the available alternative energy sources immediately, there would still probably be significant shortfalls. Such a relatively pessimistic outlook could be tempered if promising new experimental strategies such as the hydrogen fuel cell prove useful in tests now being carried out in, among other places, Iceland.

Two alternate approaches to the possibility of reducing dependence on foreign oil—and the problems that might be created thereby—are found in the unit's next two selections. In "Powder Keg" science writer Keith Kloor discusses some of the environmental and social consequences of increased natural gas development in Wyoming's Powder River Basin. Here, large energy companies own the mineral rights under the land owned and operated by livestock ranchers. The disjunctions between what the energy companies are allowed to do at the surface because they own the subsurface and the best uses of the surface of the land by ranchers are profound. A three-way struggle between big energy, the federal and state governments, and environmental advocates puts ranchers in the unusual position of siding with environmentalists. In "Living Without Oil" journalist Marianne Lavelle takes on the problem of large SUVs—sports utility vehicles—and their high levels of gasoline consumption. She notes that a large number of energy-saving devices could be used to make these vehicles more energy-efficient without sacrificing their power and size. Development of these alternatives has become more necessary in light of military developments in the Middle East. Economists suggest, however, that the primary solution to the problem of America's dependence on foreign oil would be to do what European countries do: tax oil products such as gasoline and heating oil more heavily.Then, "Renewable Energy: A Viable Choice" by a research team from the University at Berkeley's Renewable and Appropriate Energy Laboratory, focuses on the public policies that are necessary to implement new energy systems.

The final article in this unit on energy stresses the importance of the need to combine conservation strategies with fuel-saving technologies. In "Fossil Fuels and Energy Independence," engineering professor B. Samuel Tanenbaum notes that ultimately we will probably derive most of our energy from solar sources. But until we learn to harness that energy efficiently and cost-effectively, we will need to rely on a number of other alternatives to fossil fuels. These other alternatives already exist and most of them—such as wind power—are already being used, albeit not widely enough. But Tanenbaum reminds us that enormous savings in energy expenditures could be accomplished by the simplest method of all: conservation.

Beyond Oil:
THE FUTURE OF ENERGY
When Wells Go Dry

ENERGY: The rate of global oil production will start to fall in just a few years, says a controversial geologist. And alternative technologies aren't ready yet.

BY FRED GUTERL

As KENNETH DEFFEYES walks the five blocks from the Princeton University campus to his home, he veers sharply through a parking lot and then without warning takes a diagonal path across a side street. He doesn't seem to be paying any particular attention to where his sneaker-clad feet are taking him. His hands are tucked firmly in his parka, his eyes are looking up at a cloudless blue sky and his mind is where it usually is: on the world's supply of oil. In particular, Deffeyes is trying to explain why anybody should believe that the entire human enterprise of oil exploration—the search for reserves, the drilling of wells, the extraction of crude and all the attendant calculations of supply and demand—why this whole messy business should obey a simple but elegant piece of mathematics.

Deffeyes has reached a conclusion with far-reaching consequences for the entire industrialized world. So far-reaching that many of his colleagues in the field of petroleum geology dismiss it. The conclusion is this: in somewhere between two and six years from now, worldwide oil production will peak. After that, chronic shortages will become a way of life. The 100-year reign of King Oil will be over. And there will be nothing that President George W. Bush or Saudi princes or the invisible hand of the marketplace will be able to do about it. "There's nothing we could conceivably do now that would have much of an effect on the oil supply for at least 10 years," says Deffeyes.

This news isn't all bad. For the past decade, climate scientists have lobbied for drastic reductions in the use of fossil fuels, which release carbon dioxide into the air when burned, creating the "greenhouse effect." And

since September 11, some political leaders have called for reducing Western dependence on oil from the Middle East. At the same time, alternative technologies have advanced considerably in recent years. Farms of slender windmills have sprung up in California and Europe, generating electrical power at prices nearly competitive with conventional fuels. Photovoltaic panels that convert sunlight to electricity have fallen in price and are being built into roofing shingles. Hydrogen-fuel cells, another "clean" energy technology, offers an increasingly attractive long-term alternative not only to electrical power but to replace the internal combustion engine in automobiles; President Bush has made fuel cells the centerpiece of his long-term energy plan.

> "There's nothing we could do now that would have much effect on the oil supply for at least 10 years."
>
> –KENNETH S. DEFFEYES, Princeton geologist

Despite the progress in alternative fuels, however, the world will have to depend on hydrocarbons such as oil, coal and natural gas for a while yet. Without the political will to pursue alternatives with the wartime vigor of those who created the Manhattan Project to build the atom bomb, it would take decades to replace old power plants and phase out the ubiquitous gas station. Deffeyes, by conjuring up the bad years of the 1970s, when the Organization of Petroleum Exporting Countries restricted oil supplies and sent Western economies reeling, is injecting a large dose of urgency to the debate.

Future Fuels

Many experts believe that over the next 50 years fossil-fuel supplies will diminish. Renewable hydrogen, wind and solar power will emerge to provide most of the world's energy.

In 50 Years ...

The age of fossil fuels is over. Wind, solar and hydrogen power meet most of the world's energy needs.

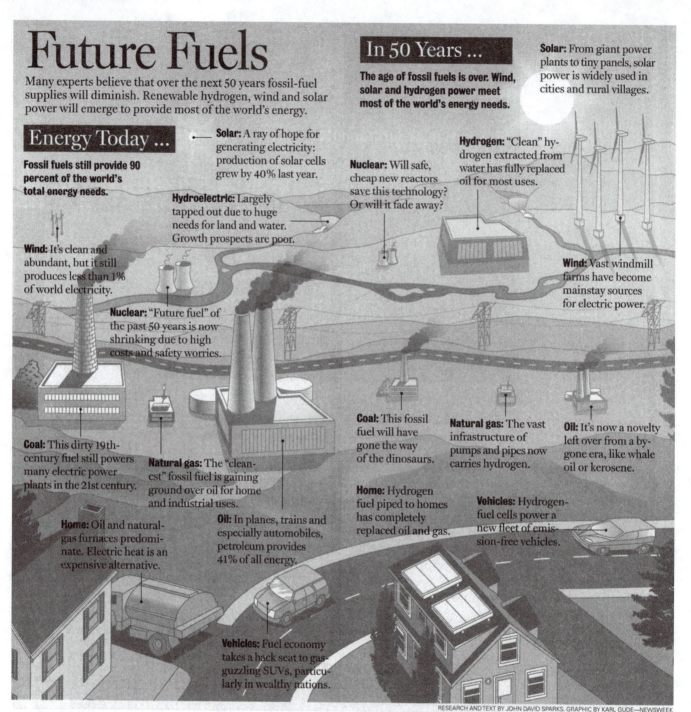

Energy Today ...

Fossil fuels still provide 90 percent of the world's total energy needs.

Solar: A ray of hope for generating electricity: production of solar cells grew by 40% last year.

Hydroelectric: Largely tapped out due to huge needs for land and water. Growth prospects are poor.

Wind: It's clean and abundant, but it still produces less than 1% of world electricity.

Nuclear: "Future fuel" of the past 50 years is now shrinking due to high costs and safety worries.

Coal: This dirty 19th-century fuel still powers many electric power plants in the 21st century.

Natural gas: The "cleanest" fossil fuel is gaining ground over oil for home and industrial uses.

Home: Oil and natural-gas furnaces predominate. Electric heat is an expensive alternative.

Oil: In planes, trains and especially automobiles, petroleum provides 41% of all energy.

Vehicles: Fuel economy takes a back seat to gas-guzzling SUVs, particularly in wealthy nations.

Solar: From giant power plants to tiny panels, solar power is widely used in cities and rural villages.

Hydrogen: "Clean" hydrogen extracted from water has fully replaced oil for most uses.

Nuclear: Will safe, cheap new reactors save this technology? Or will it fade away?

Wind: Vast windmill farms have become mainstay sources for electric power.

Coal: This fossil fuel will have gone the way of the dinosaurs.

Natural gas: The vast infrastructure of pumps and pipes now carries hydrogen.

Oil: It's now a novelty left over from a by-gone era, like whale oil or kerosene.

Home: Hydrogen fuel piped to homes has completely replaced oil and gas.

Vehicles: Hydrogen-fuel cells power a new fleet of emission-free vehicles.

RESEARCH AND TEXT BY JOHN DAVID SPARKS. GRAPHIC BY KARL GUDE—NEWSWEEK

How did Deffeyes arrive at his gloomy forecast? He likes to say that he grew up in the Oklahoma oil patch: his father was a petroleum engineer, and during summers off from high school and college the younger Deffeyes worked odd jobs in the oil fields. But the real genesis of his thinking was a brief stint he did at Shell Oil's research labs in the late 1950s, at the very start of his career. There he worked under the late geophysicist M. King Hubbert, who in 1956 made a startling prediction. At the time, oil reserves were being discovered at an accelerating pace, and the oil business was booming, with no apparent end in sight. But Hubbert had done groundbreaking work by applying a combination of geophysics and statistical analysis to the base of U.S. oil reserves, and from there projecting domestic production rates for years to come. The result was a bell curve: production was then going up, but it would inevitably peak and then start to decline. The math told Hubbert that the peak would come in the early 1970s, which clashed with the conventional wisdom of the time. In fact, that is more or less what happened: U.S. oil production hit its high in 1970, then started down just in time for the OPEC-induced oil crisis. But at the time Deffeyes was one of the few people who took Hubbert's prediction seriously. "I realized that a contracting oil industry was not a good career prospect," he says, "so I decided to get out and go into academia."

After spells at the University of Minnesota and Oregon State, Deffeyes pitched up at Princeton (and served as John McPhee's guide and mentor for his memorable writings on geology, collected as "Annals of the Former World"). There he applied Hubbert's methodology to global oil supplies. The results shocked him: the bell curve would peak sometime between 2004 and 2008, depending on how you crunched the numbers.

But why should oil production follow a bell curve? Deffeyes offers the analogy of a hunter shooting at a target. Each time he's going to miss by a certain amount due to wind or air temperature or faulty aim or hand tremors. If he fires 1,000 shots, you would measure how far each shot falls from the center of the target and plot how many shots fall at each interval of distance. The resulting graph will take the shape of a bell. The fat part of the graph will sit over the target, where most of the shots landed, and the two downward slopes will correspond to those distances on either side of the bull's-eye. Oil exploration is a similarly hit-and-miss affair. Despite all the high-tech methods geologists use nowadays, nine out of every 10 exploration wells turn out to be dry. It stands to reason, Deffeyes argues, that as more of the world's oil is discovered, the smaller the new finds are going to get. Deffeyes's bell curve is a plot of the volume of oil discovered each year for most of the 20th century. The curve rises slowly at first, as geologists picked off the easy-to-find fields—the ones that announced themselves with telltale tar pits and oil slicks. The curve shoots up during the '50s and '60s, when geologists discovered many of the bigger, deeper reserves, such as those in the North Sea, the Bass Strait and Saudi Arabia. As the century draws to a close, the pace of discovery actually accelerates, but the finds are smaller. The curve begins to flatten out.

Already, Saudi Arabia is the only oil-producing country that doesn't sell as much as it can pump. That fact has allowed the Saudis to dominate global markets, but they can't offset the shortages yet to come. Pumping at maximum capacity, they would add less than 4 percent to the world supply. Nor is technology likely to offer a way out. Energy companies, says Deffeyes, invested billions of dollars in the past 30 years to improve their ability to discover and extract oil, and it's unlikely that any breakthroughs in the offing would significantly change the equation. Technology would help eke out the hardest-to-reach reserves, but the curve would still fall. "What would it take to get another hump in the curve, with a peak farther out?" says Deffeyes. "There would have to be another kind of oil field altogether. And there's no evidence that such a thing exists."

Deffeyes, who reported his findings last October in a book, "Hubbert's Peak," has his detractors. Ronald Charpentier, a geologist at the U.S. Geological Survey, wrote recently in the journal Science that Deffeyes's estimates are based on a "questionable methodology" that is attractive in large part because it requires "modest data and human resources." The USGS's 2000 survey of world oil supply—which took 100 man-years to prepare, as opposed to the six months Deffeyes spent running his numbers—shows a worldwide oil supply of more than 3 trillion barrels (a trillion more than Deffeyes estimates). And Charpentier argues that additional discoveries, such as deep-water oil now being extracted in the South Atlantic, as well as untapped reserves in the Caspian Sea, Siberia and Africa, could change the outlook considerably. "There's a lot of oil and gas out there," he says, "but it's not necessarily where you want it to be and at a price you want it to be." Deffeyes says he doesn't dispute the possibility of finding new reserves, but he insists they're likely to be less fruitful than the USGS thinks.

What if Deffeyes is right? How will the world satisfy its growing demand for energy? Carmakers (not to mention President Bush) have pinned their hopes on hydrogen-fuel cells, which would emit no carbon and wouldn't entail drilling in unfavorable parts of the world. General Motors has stepped up its research budget for fuel cells from $1 million a year in 1990 to $100 million this year. "We believe hydrogen is the long-term answer," says Matt Fronk, GM's chief engineer for fuel cells. Photovoltaics (solar panels) have been hobbled by the high cost of semiconductors used to make them, but cheaper alternatives are in the offing. The National Renewable Energy Laboratory in Colorado, for instance, is developing an inexpensive way to deposit ultrathin layers of semiconductors on glass, steel or flexible plastic sheets. Says Christoph Frei of the World Economic Forum: "Everyone expects the energy mix to change."

But developments such as these, as exciting as they are, will not be of much use in the next decade. In that time, natural gas will probably be the most attractive hydrocarbon substitute for oil, at least for electricity generation and heating. It is relatively clean and, by all estimates, plentiful. Using it in cars, though, would entail a whole new fuel infrastructure that couldn't be built in a hurry. The only other real alternative would be conservation. According to conservation advocate Amory Lovins, improving the average fuel efficiency of vehicles in the United States by 2.7 miles per gallon would equal all U.S. oil imports from the Persian Gulf.

There's always the chance, of course, that the next decade will come and go without so much as a wobble in the supply of oil. Deffeyes allows for the possibility that he is wrong. "You've got to give Hubbert credit for getting it right," he says. "Then you ask yourself if he was lucky or if he really knew." Deffeyes sighs. "And you just don't know." With global warming, that brings to two the number of urgent but tentative reasons to pursue alternatives to fossil fuels.

With ADAM PIORE *and* SANDY EDRY

Hot Springs Eternal

HYDROGEN POWER: People mocked Bragi Arnason's vision of producing energy from the H in H_2O. Now the first test is about to be launched in his native Iceland. Next: the world?

By Adam Piore

IT'S A LITTLE BEFORE NOON, AND an anemic winter sun rises slowly over an Icelandic wasteland of jagged lava rocks. At a steaming pool of milky blue water, Bragi Arnason is talking about the future. A ruddy-faced chemistry professor with a white pompadour and a heavy maroon coat, Arnason has a staccato voice that barely rises above the hiss of steam surging up from the earth's core. For almost three decades, it has been Arnason's dream to harness this volcanic power to split the H from H_2O, and create a society free of fossil fuels: the first "hydrogen economy." "I will see the first steps," says Arnason, raising a gloved hand to the horizon, where a geothermal power plant gushes white steam. "My children, they will watch the whole transformation. My grandchildren, they will live with this new energy economy."

Oblivious to the frigid wind whipping across the plain just west of Reykjavik, Arnason spells out the vision that earned him the moniker "Professor Hydrogen." First, take all of Iceland's cars and fishing trawlers and gradually replace their internal combustion engines with electric motors that run on hydrogen-fuel cells, just like American space shuttles. Meanwhile, harness the power of Iceland's active volcanoes and raging rivers to begin producing pure hydrogen gas on a mass scale. Arnason has even inspired some compatriots to imagine Iceland as the "Kuwait of the North," a major source of energy in a world where all nations have decided to replace hydrocarbons with hydrogen. "There were not many people who really listened to him" at first, says Valgerdur Sverrisdottir, Iceland's minister of Industry and Commerce, throwing back her head and laughing. "It was so far away—to use water for cars?"

It's not far out anymore. The roster of experts who see hydrogen as the most likely replacement for oil in the long term now includes not only futurists and environmentalists but also the oilmen of the Bush administration and researchers at General Motors and Ford. Iceland's plan is now backed by DaimlerChrysler, Shell and the European Union, which plan to spend tens of millions of euros to create the first societal lab test of a hydrogen economy. In the coming months, Iceland will roll out three hydrogen-powered buses and begin constructing a filling station where hydrogen gas will be produced on-site. If all goes according to plan, this demonstration will expand to cars and fishing vessels in 2005, and all vehicles within 30 to 40 years. Other nations are likely to follow. The only question is when, says Margaret Mann, an engineer at the U.S. National Renewable Energy Laboratory. "In the long term we have to move to hydrogen. It's the only way to really divorce ourselves from fossil fuels."

Even now, the vision of a hydrogen economy sounds too good to be true. Hydrogen occurs naturally in water, so the supply is as close to endless as the oceans. And pure hydrogen is a harmless gas, not a toxic liquid, so spills would dissipate in the air. Hydrogen-fuel cells emit only water vapor, and the electric motors make nary a sound. In the case of Iceland, the process will rely on the island nation's abundant steam and water to generate the power to make hydrogen fuel. The process involves bombarding H_2O with ions to split off the hydrogen, which is recombined with oxygen in a fuel cell that makes charged ions to run a motor, and water molecules as a byproduct. In short, Arnason plans to use natural energy to make a powerful hydrogen fuel that emits only water mist as waste, promising a bottomless well of clean energy that produces no greenhouse gases and no threat of global warming.

ARNASON THOUGHT it was an obvious idea. In the 1970s, he was living on top of a glacier and mapping Iceland's reservoirs of hot groundwater as part of his doctoral thesis in chemistry. The reservoirs were no secret in a nation where cooks have been known to bake bread by burying boxes of dough in the ground, and farmers drilling in the backyard are likely to set off geysers of boiling-hot vapor. But Arnason was the first to map them, revealing huge reserves of natural geothermal energy.

It was a portentous discovery. Iceland grew up energy-poor, a harsh reality that rankled an independent-minded nation founded in the 900s by Norwegian Vikings. With no fossil fuels of its own, and huge energy needs due to the cold weather and the economy's reliance on gas-guzzling fishing vessels, Iceland was heavily dependent on foreign oil. It reeled from recurrent oil shocks, and its inflation rate averaged a staggering 17.6 percent between 1944 and 1995. Driving down the frigid North Atlantic coast, Arnason recently recalled that it was "quite natural" for a nation in this state of energy servitude to ask, "How can we change this?"

Clean Energy: Electricity With No Waste

Like a battery, a fuel cell passes electrical energy between positive and negative poles. Unlike a battery, a fuel cell generates current rather than simply storing it. Chemical reactions between platinum catalysts and the highly charged poles start the juice flowing.

HOW A FUEL CELL WORKS

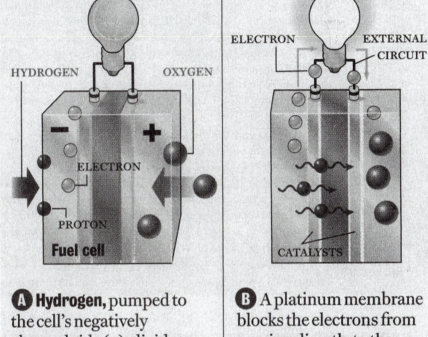

A **Hydrogen,** pumped to the cell's negatively charged side (–), divides into protons and electrons. Oxygen is pumped into the positively charged side (+), which attracts electrons.

B A platinum membrane blocks the electrons from moving directly to the positive side of the cell, forcing them through an **external circuit,** where they form a usable current.

C **Protons** pass through the membrane between the two sides. They join with the oxygen to form water. When the water is expelled, oxygen refills the cell, and the process begins again.

TEXT BY JOHN D. SPARKS, GRAPHIC BY TONIA COWAN

From the beginning, Arnason saw hydrogen as the answer, knowing full well it appeared a bit nuts. In the 1970s and 1980s, hydrogen-fuel cells were huge and expensive, and the fact that they were used by the U.S. space program only made Arnason's vision seem farther out there. Arnason delivered his first paper on the subject in 1978 with trepidation. "I was not quite sure whether I should publish it. I thought that maybe people would think I was crazy," Arnason recalls. "[My mentor] read it and said, 'If you really believe this could be realized in 20 to 30 years, you should start talking about it now'."

He did, mostly to Rotary clubs, his classes and other small groups in Iceland, where he soon gained a reputation. "He was always called a guru, not by the general public but by the energy people," says Johann Mar Mariusson, vice president of Landsvirkjun, the National Power Company, and a longtime friend. "He didn't like the name Professor Hydrogen. He thought people were mocking him."

Then the world caught up. Arnason's work began to pay off in the 1990s, when global oil and auto companies started to take seriously the idea of hydrogen as the "next oil." The first breakthrough came in 1992, when Ballard Power Systems of Vancouver, British Columbia, demonstrated the first hydrogen-powered bus—the precursor to the models that will roll onto the streets of Reykjavik. The big surprise: Ballard's bus produced 150 kilowatts, or 15 times more power than most engineers expected from a hydrogen motor. "People kept looking around and asking, 'Where's the rest of it hidden?'" recalls Firoz Rasul, Ballard's CEO. "They never thought it could be done." Daimler-Benz stepped up and invested $250 million, and together they turned to Iceland.

Iceland just made sense. It has extreme weather conditions to test the durability of new automotive models, and a small population of 280,000, for whom high energy costs were reason enough to experiment with alternative energies. Moreover, Iceland's Professor Hydrogen was already well acquainted with the head of fuel-cell research at Daimler, Ferdinand Panik, who had been seeing Arnason at obscure conferences for years. The support of this big-shot scientist from a major multinational silenced the skeptics. In 1998, when officials from Daimler (now merged with Chrysler) arrived in Reykjavik to cut the deal, it was such a big local story that talks were moved to a secret site outside town. "We had to hide," recalls parliamentarian Hjalmar Arnason (no relation to Professor Hydrogen), a leader of Iceland's negotiating team. "Everything went crazy."

The interest of an auto giant drew other global players. The result is Icelandic New Energy, owned 51 percent by Iceland and 49 percent by private interests including DaimlerChrysler, Shell Oil and Norway's Norsk Hydro. The European business executives were stunned at how fast business got done in Reykjavik, a tiny capital where everyone knows everyone. It also didn't take long to interest the European Union, which so far has allocated €2.85 million to the bus phase of the project. The EU plans to follow with similar bus projects in Britain, Germany, Spain and at least four other nations. "If it works in Iceland, it'll work in other countries," says Panik. "It's the perfect place to start."

ICELAND IS ALSO UNUSUALLY AMBITIOUS. Cities from Vancouver to Palm Springs have deployed hydrogen buses, but Iceland represents the first attempt to phase out fossil fuels entirely, and thus become the first society to eliminate greenhouse emissions. Icelandic New Energy will be doing extensive market research to refine the design of hydrogen stations and cars, and to make the whole idea of the hydrogen economy "socially acceptable." One key hurdle is cost: even under the most favorable estimates, Icelandic New Energy expects the hydrogen to eventually cost about twice as much as gasoline, though it would generate twice as many miles per gallon.

Iceland is a model in the making. The International Code Council in Virginia is holding a public hearing in April on draft fire, electrical and industrial-fuel codes to govern a hydrogen economy. This June the International Organization for Standardization will consider safety guidelines for tanks, containers and fueling stations that hold hydrogen, which is about as flammable as gasoline. "Iceland is small enough that you can actually test and validate codes and standards and so on," says Rasul. "Iceland is going to be able to show where the rest of the world can go."

The question is when, and at what cost. Shell Hydrogen figures it would cost at least $19 billion to build hydrogen fuel plants and stations in the United States, $1.5 billion in Britain and $6 billion in Japan. That's compared with a "matter of millions of dollars in tiny Iceland," says CEO Don Huberts. "Also, in Iceland people do not drive their cars off an island, so we won't have to wait for the infrastructure to develop elsewhere."

Iceland may prove unique in other ways, too. Its huge reserves of natural geothermal energy are unrivaled in Europe, and some dream of exporting the hydrogen to the continent and creating a booming new industry (though first they'll have to figure out a way to get it there). Whatever happens, when the new energy age dawns, Arnason will have the satisfaction of knowing that the first seeds of the hydrogen economy grew in his Viking land of snowy moonscapes and steaming water holes. And Professor Hydrogen will no longer have reason to hate his nickname.

With STEFAN THEIL in Berlin

Taking the Breeze

ELECTRICITY: New technologies make Europe take another look at wind as a power source

BY WILLIAM UNDERHILL

THE NATIVES OF LEWIS KNOW wind—sometimes too well. Every winter the Atlantic gales come blasting across the northern tip of Scotland's Outer Hebrides. The wind hardly slows down after striking land; in the island's marshy interior, gusts regularly exceed 100 miles an hour. Everyone stays indoors but the sheep. Tourists arrive in summer, lured by mild temperatures and wide expanses of unspoiled countryside; even so, there's rarely a calm day. "The weather here is changeable," says Nigel Scott, a spokesman for the local government. "But the wind is constant."

The brutal climate could finally be Lewis's salvation. The place has been growing poorer and more desolate for generations, as young people seek sunnier prospects elsewhere. But now the energy industry has discovered the storm-swept island. The multinational engineering and construction giant AMEC and the electricity generator British Energy are talking about plans to erect some 300 outsize wind turbines across a few thousand acres of moorland and peat bog. If the $700 million project goes through, the array will be Europe's largest wind farm, capable of churning out roughly 1 percent of Britain's total electrical needs—and generating some badly needed jobs and cash for the people of Lewis. "We have been slowly bleeding to death," says Iain McIver of the Stornoway Trust, the proposed site's owners. "The benefits from this project will continue to flow for as long as the wind blows over the Western Isles."

It sounds like the answer to a lot of prayers—and not only on Lewis. Enthusiasts around the world call wind a perfect alternative to fossil fuels and nuclear power: safe, inexhaustible and free. "This is simply one of the cheapest ways of reducing our output of greenhouse gases," says Christian Kjaer of the Brussels-based European Wind Energy Association. Still, not everyone is such a fan. "I find it incredible that organizations which describe themselves as 'Green' or 'Friends of the Earth' can contemplate the ravage of our hills with these industrial installations," says Margaret Thatcher's former press secretary Sir Bernard Ingham, a champion of nuclear power. Even environmentalists confess to a few reservations. "The wind industry is as capable of environmental insensitivity as any other," says Roger Higman, a senior campaigner for Friends of the Earth.

The energy industry isn't bothered by such quibbles. For the last seven years the world market for wind turbines has grown by an average of 40 percent annually. Last year alone, generating capacity worldwide jumped by almost a third. The more wind plants you build, the cheaper and more powerful you can make them. Turbine makers are now mass-producing giant machines—the rotors' 211-foot diameter is wider than the wingspan of a jumbo jet—that once existed only in theory. Today one standard-issue turbine can produce at least 1 megawatt of power, more than double the typical model's output of 20 years ago and enough to provide electricity for as many as 800 modern households. The next generation, capable of more than twice that output, is already emerging.

The new turbines are not just bigger; they're smarter. The basic design, a triple-vaned rotor atop a vertical shaft, hasn't changed much in 50 years. The big difference is that the towers are taller; the new ones rise almost 300 feet above the ground. The higher they go, the stronger and steadier are the winds they catch. But scientists keep tweaking the specs in subtler ways, too. Tough, light structural materials have been borrowed from the aeronautics industry. Likewise for "vortex generators," tiny fins added to the surfaces of wind rotors and aircraft wings. They induce turbulence that helps prevent stalling at low speeds. Best of all for people who live nearby, im-

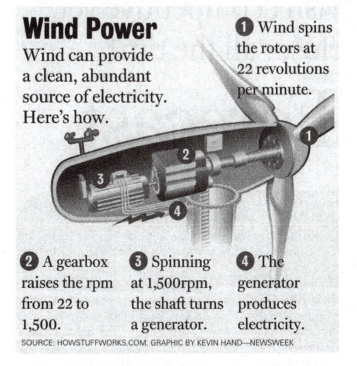

Wind Power
Wind can provide a clean, abundant source of electricity. Here's how.

❶ Wind spins the rotors at 22 revolutions per minute.

❷ A gearbox raises the rpm from 22 to 1,500.

❸ Spinning at 1,500rpm, the shaft turns a generator.

❹ The generator produces electricity.

SOURCE: HOWSTUFFWORKS.COM. GRAPHIC BY KEVIN HAND—NEWSWEEK

proved design on the latest models has cut noise to a relative whisper.

Still, some nature lovers hate wind power. Turbines seem to hold the same fatal attraction for birds that bug zappers have for mosquitoes, although no one is quite sure why. And the best sites for wind farms are often previously unspoiled hilltops. "We don't think esthetics are an ecological criterion," says Sven Tiske of Greenpeace. "If we opposed a nuclear power station just because it didn't look good, everyone would laugh." Not everyone. Ask Robert Woodward, a British art historian who campaigns against the spread of wind farming. His holiday residence is in Wales, at the edge of the Cambrian Mountains, and since the early 1990s the view from the hillside above his house has encompassed more than 100 turbines, all flailing out of sync. "A staggeringly beautiful landscape is being devastated," says Woodward, who used to support environmentalist groups—until wind power blew him away. "This is the first time in the history of the conservation movement that concern for the landscape has become a dirty concept."

Such gripes are one reason the industry is experimenting with offshore facilities. Viewed from land, even the tallest offshore farm is a tiny thing. Besides, sea breezes tend to be stiff, and steady enough to raise a turbine's output at least 20 percent. Denmark is already proving the idea can work. Take a walk on the beach in Copenhagen's northern suburbs, and you'll notice that the horizon is broken by a distant line of 20 turbines. They're built on a submerged limestone reef, almost two miles out to sea. Billed as the world's largest offshore wind farm when it opened last year, the installation generates enough power for 30,000 city households. Now vast tracts of open water are being staked out, from western Ireland to the Baltic. Nick Goodall, director of the British Wind Energy Association, says: "The great thing about offshore wind power is that you are limited only by your imagination and the amount of money that the banks will lend you."

Bankers aren't getting too swept away. Building and running a farm at sea costs up to 40 percent more than onshore. Consider just the price of laying undersea cable to deliver electricity from turbine to grid. Construction is tricky, usually requiring a pile driver to sink the turbine's concrete foundation deep into the seabed, and maintenance is inconvenient and expensive. "Offshore is a luxury," says Per Krogsgaard of the Danish consultancy BTM. "And it will be for a very, very long time."

Which leaves one question: can the wind industry undersell its competitors? The answer seems easy. After all, the fuel is literally as free as the wind. The chief expense is setting up a turbine farm. That's still too high a price to drive fossil-fuel plants out of business. "It is still cheaper to put more coal into an existing power station than to build a new wind farm," admits Kjaer. If you built a new coal-fired plant today, wind enthusiasts insist they could put up their own power installation and equal your rates. But no one is staging any such contest: unlike America, Europe has a chronic surplus of generating capacity.

Even so, the question is not academic. Sometimes there's no choice but to build new generating facilities. Conventional plants get old and obsolete, and environmental laws keep getting tougher. A recent British government report on energy policy includes a projection that by 2020, wind is likely to beat nuclear's prices and roughly match those of natural-gas power stations. But the wind industry seems unable to offer a solid cost-comparison against other power sources right now. Why?

"That's really a political question," says Andrew Garrad, of the wind-power consultants Garrad Hassan. "Scratch anywhere beneath the surface of energy economics, and you will find politics." It's officeholders, not engineers, who set subsidy levels for different energy sources, and their decisions can be far more inscrutable than the wind. Not that wind advocates are complaining. So far wind power is thriving only in countries—notably Denmark, Spain and Germany—where governments force utility companies to pay their suppliers premium rates for wind energy.

Wind still has one big drawback: sometimes it refuses to blow. Users need a dependable backup power supply for days when the turbines won't turn. Hard-core enviros say wind is only a stopgap; in the long run, they argue, there's no alternative to the tough conservation measures that are too painful for any politician to espouse. At present, though, wind has no eco-OK rivals. Environmentalists have fallen out of love with big hydroelectric plants, and Europe has few suitable sites left anyway. Solar technology remains prohibitively expensive and space-consuming. And biomass—burning fuel crops such as wood chips or agricultural waste in steam-powered generators—is a messy business that requires the right equipment on the spot. You can't just burn the stuff in a conventional coal-fired plant, for example.

The people of Lewis are putting their money on wind. "This is all about preserving the environment," says Nigel Scott, who moved here seven years ago. "If we don't go down the wind-energy road, in the long run there won't be any habitats to protect." No one seems too worried about turbines' spoiling the peat bog's vistas—or about what might happen if the winter gales should ever turn gentle.

POWDER KEG

THE GAS INDUSTRY HAS BEEN BUSY IN WYOMING'S PRAIRIES AND GRASSLANDS, BUILDING THOUSANDS OF MILES OF ROADS AND SINKING MORE THAN 10,000 WELLS IN THE PAST THREE YEARS. BUT IN THE POWDER RIVER BASIN, RANCHERS ARE JOINING ENVIRONMENTALISTS TO TRY TO STILL THE DRILLS.

BY KEITH KLOOR

Ed swartz does not seem like the kind of guy you would threaten with bodily harm. He has spent almost all of his 62 years in a windswept corner of northeastern Wyoming, herding cattle, baling hay, and building waterlines to keep his ranch from going dry. He still works as long as the daylight lasts, even after suffering a near-fatal heart attack in 1995. But shortly after his ranch started dying in 1999, he stood up on a bus full of state officials from Wyoming and Montana and started fuming about his dried-up meadows and polluted creekbeds. Most audiences would have sympathized with Swartz. But this particular bus carried coal-bed methane executives and various state officials, including the governors of Wyoming and Montana. They were touring drilling sites in the Powder River basin; the Wyoming officials wanted to impress their counterparts in neighboring Montana with how smoothly the booming development was going.

Ed Swartz's property, however, which sits in the coal-rich basin, has been inundated with coal-bed methane discharge water from drilling sites near his ranch, contaminating his creek and preventing him from irrigating his alfalfa, the cattle ranch's lifeline. On this day he just wanted to make sure everyone on the bus was aware of it. And that he was not the only rancher with this problem. John Kennedy, a local coal-bed methane operator, was furious. "You're a liar!" he yelled at Swartz, cutting him off. Swartz, a strapping, leathery, third-generation Wyomingite whose grandfather homesteaded the family ranch in 1904, kept talking, unperturbed. "I'm going to hurt you!" Kennedy warned. Swartz said to go ahead and try. Inflamed, Kennedy again thundered, "I'm going to hurt you!" But the blustery driller never left his seat. Finally, Montana's governor, Judy Martz, settled things down, so the tour could continue.

Sitting around his kitchen table, chain-smoking low-tar cigarettes, Swartz tells me the story over a pot of turbo-charged coffee, a glinty grin on his face. Kennedy, I learn, has a reputation for harassing similarly aggrieved ranchers at public meetings and over the phone. Swartz, who has long been an active, bedrock Republican in what is a virtual one-party—Republican—state, is now considered a pariah for rocking the boat. "People think I'm just a rabble-rousing rancher around here," he says defiantly. That, undoubtedly, is because of the Clean Water Act lawsuit he has filed against Wyoming state officials and the drilling company (Denver-based Redstone Inc.) he asserts is responsible for the damages to his land.

Coal-bed methane development is a relatively new form of natural gas extraction that has exploded in the Powder River basin since 1997, with more than 10,000 gas wells already sunk on private and state lands. It is like mother's milk to state officials, because it produces both tax revenues and campaign contributions from the energy companies. "They are so oriented to the energy industry that they could give a red rat's ass about a rancher," Swartz bristles. "I'm as serious as I can be about that. There's not one of them that is concerned about a rancher around here."

ALONG THIS STORIED AND BLOODIED FRONTIER, INDIAN BATTLES RAGED, RANGE WARS WERE FOUGHT, AND BUFFALO BILL AND BUTCH CASSIDY SEALED THEIR MYTHS.

The 8-million-acre Powder River basin straddles the Wyoming–Montana border, offering a snapshot of the quintessential Old West, with its rolling hills and prairies nestled between the Bighorn Mountains and the Black Hills, 100 miles to the east. Along this storied and bloodied frontier, Indian battles raged, range wars first erupted between homesteaders and cattle barons, and Buffalo Bill and Butch Cassidy sealed their myths. It is the place, many claim, where the American cowboy was born.

The Powder River region is also where the Great Plains meet the Rocky Mountains, a bare, reddish, lunarlike landscape that skips between parched, stumpy buttes, green meadows, and unspoiled streams. It is a vast, mixed ecosystem of grassland and sagebrush, where, in that timeless Darwinian race, prairie dogs burrow deep and antelope hurtle fast and high, coyotes and mountain lions quick on their heels. The Powder River—described by settlers as "a mile wide and an inch deep, too thin to plow and too thick to drink"—is a 375-mile tributary of the Yellowstone River. Plying its shallow, muddy waters are the globally imperiled sturgeon chub, the channel catfish, and 23 other native fish.

During the 20th century the long boom-and-bust cycles of strip mines and oilfields left their own indelible imprint on the landscape, in the form of sawed-off hilltops and abandoned "orphan" wells. Even so, ranchers and environmentalists have always taken solace that the deep and large gashes scarring the land were mostly confined to a few areas.

No longer. The latest boom to hit the Powder River basin has spread out in a chaotic patchwork, pockmarking the historic landscape with thousands of miles of powerlines, pipelines, roads, compressor stations, and wellheads. Methane is a natural gas found in the region's plentiful coal seams. Water pressure holds the gas in the coal; pumping the water out in large volumes releases the gas. The process also produces wastewater laced with sodium, calcium, and magnesium—too saline to be used for irrigation, too tainted to be dumped in waterways. So in a semi-arid region where water is precious, energy companies are forced to store the methane water in "containment" pits, from which it often runs into water wells and into the tributaries of the Powder River.

"IT'S SO DAMN DISCOURAGING. EVERYTHING I WORKED FOR, THAT MY GRANDFATHER WORKED FOR, AND THAT MY SON IS WORKING FOR IS BEING WIPED OUT."

The resulting environmental damage in the Powder River basin has hit ranchers and the land equally hard. "It's so damn discouraging," Swartz tells me in a craggy voice tinged with resignation. "Everything I worked for, that my grandfather worked for, and that my son is working for is being wiped out." Over the years the family has endured many droughts (including the one the region is suffering today), its share of machinery breakdowns, and several diseases afflicting their cattle. But nothing compares with the poisonous runoff that is killing the ranch's vegetation. "They're [Redstone] using my place as a garbage dump," Swartz fumes.

About a year and a half ago, at the boom's peak, drillers were pumping 55 million gallons of water to the surface every day. Underground aquifers were being depleted and cottonwood trees flooded. The massive runoff of the methane-tainted water

has become so alarming—polluting creeks and streams and altering natural river flows—that earlier this year the conservation group American Rivers named Wyoming's Powder River as one of America's Most Endangered Rivers for 2002. "There could be 139,000 coal-bed methane wells in the Powder River basin by the end of the decade," says Rebecca Wodder, president of American Rivers, referring to energy-industry and government estimates. "Despite this, federal and state agencies have yet to formulate an adequate plan for minimizing the environmental consequences of drilling in the Powder River basin."

And they weren't about to until the U.S. Environmental Protection Agency (EPA) issued a report last May, slamming the Bureau of Land Management (BLM) for failing to assess the fallout from coal-bed methane development. At the time, drillers were already having their way on Wyoming's private and public lands, owing to lax state and federal environmental safeguards. They had just set their sights on a mother lode of rich coal-seam deposits on BLM lands in the Powder River region. Then, with the bureau poised to give the operators a quick go-ahead, the EPA's report stopped them in their tracks, throwing into doubt the development of gas wells on 8 million public acres in Wyoming. In particular, the EPA cited concerns about air quality from dust and compressor emissions, and the impact on wildlife and water quality. The EPA report forced the BLM to redo its environmental-impact statement on 51,000 new gas wells slated for development on federal lands.

The reassessment, scheduled to be released in January, stands to reverberate through out the Rocky Mountain West, where a gold rush mentality has taken hold. Energy officials have called the area the "Persian Gulf of natural gas." Gas companies have already struck hard and fast in Colorado's San Juan basin; now they're waiting for the green light on federal lands to expand there and across Montana and Wyoming.

Moreover, coal-bed methane development in the West is the cornerstone of President Bush's proposed domestic energy plan, which claims the area has enough natural gas to supply the energy needs of the United States for seven years. "The region is enormously rich in minerals," Ray Thomasson, a Colorado-based energy consultant, told a recent Denver gathering of energy experts and industry officials. "We just have to find out where the sweet spots are."

IN WYOMING, AS IN MOST OF THE WEST, SUBSURFACE RIGHTS SUPERSEDE SURFACE RIGHTS, SO UNLESS A LANDOWNER OWNS BOTH—FEW DO—WHOEVER OWNS THE MINERAL RIGHTS UNDER THE LAND HOLDS THE TRUMP CARD.

DESPITE THEIR PROUD BEARING, NANCY AND Robert Sorenson can't mask the sorrow and anguish in their voices as they describe what it's like to live in the first "sweet spot" of the coal-bed methane boom. Both natives of Wyoming, the Sorensons have spent the past 30 years—almost half their lives—on a 3,500-acre cattle ranch in the Powder River basin, 20 miles north of Ed Swartz's spread. They, too, have had methane-contaminated water run off onto their property, flooding their soil and boxwood elder trees.

The Sorensons have graciously invited me into their home for a lunch of stir-fried chicken and homemade nut bread and a discussion of their unwanted quandary. Robert is taciturn yet direct, and casts a sharp gaze behind his full mustache; Nancy has a warm, open smile but speaks softly and deliberately, always searching for the right words. Recently, rich coal seams have been discovered under their land. But since the Sorensons own only partial mineral rights, when the drillers came knocking, they had no power to turn them away. In Wyoming, as in most of the West, subsurface rights supersede surface rights, so unless a landowner owns both—few do—whoever owns the mineral rights under the land holds the trump card.

Though the Sorensons stand to collect royalties, they are agonizing over the repercussions. "I worry about our neighbors downstream, who will be hit hard by this," says Nancy, a retired schoolteacher and an active board member of the Powder River Basin Resource Council, a grassroots group that has united ranchers and local environmentalists. She also worries about the fate of the wildlife she sees every day on her morning walks around the ranch, such as antelope, mule deer, and foxes. (" I've seen a mountain lion twice, which was quite a thrill!") Living on the land all their lives, the Sorensons have become attuned to its natural rhythms, mindful of even the subtlest changes in predator-prey relationships. Over the years the couple has watched, with admiration, as ecological forces exerted their own balance. "When the rabbits become too much of problem, the bobcats take care of it," says Nancy. But, she adds in a doleful whisper, "I hate to see what happens to all these animals once this drilling starts, because you are fragmenting their environment."

Overall, the Powder River basin is home to more than 157,000 mule deer, 108,000 pronghorn antelope, and almost 12,000 elk. Even the BLM admitted in its first—albeit inadequate—environmental assessment that the proposed coal-bed methane development "may result in loss of viability of federal lands… and may result in trends toward federal listing" under the Endangered Species Act for 16 species, including the white-tailed prairie dog, the burrowing owl, and the Brewer's sparrow.

As clouds darken the early afternoon and a light, intermittent drizzle begins to fall, the Sorensons mourn the transformation—almost overnight—of their rural community into an industrial zone. "It's a total change of a way of life," says Robert, whose family homesteaded in the Powder River basin in 1881.

He's right. For two days I have zigzagged hundreds of miles west from the city of Gillette—the satellite home base of energy companies—to Sheridan, another energy outpost. In between antelope sightings, it seemed that for every 20 miles I rode along Highway 14-16, new dirt roads were being bulldozed in the rolling foothills for pipelines, and open pits were being dug for wastewater containment ponds. What's more, this was a quiet period, because the unseasonably wet, cold weather hampered drilling. In warmer temperatures, for instance, coal-bed methane operators use an atomizer to spray methane water at

high pressure, so that most of it will evaporate—although the salt and minerals still coat the ground.

"IMAGINE IF YOU LIVED ACROSS THE STREET FROM A POWER PLANT. THAT'S WHAT IT'S LIKE FOR PEOPLE THAT LIVE NEAR A COAL-BED METHANE FIELD." SOME RANCHERS LIKEN THE SOUND TO THAT OF 747S TAKING OFF—CONSTANTLY.

During construction of a particular site, it's not uncommon, the Sorensons say, to have a hundred trucks trundling in 24 hours a day. Once the wellheads are sunk and the compressor stations built alongside—sheltered in houselike structures—the noisy, whirring process of methane gas extraction runs all day and all night. "Imagine if you lived across the street from a power plant," says Robert. "That's what it's like for people that live near a coal-bed methane field." Some ranchers liken the sound to that of 747s taking off—constantly.

Yet as Nancy admits, somewhat awkwardly, not everyone in the community is put off by the development. Though Wyomingites are deeply proud of their ties to the land and the ranching tradition, it's not an easy life. "I'm aware of neighbors who welcomed this [the gas development] with open arms because they were so far at the end of their rope," she says. "And because this is going to keep them on the land a little while longer, I'm absolutely understanding of their position." Her voice trails off, a faraway look in her eyes. "You know," she continues, a few seconds later, "we've been able to make a living off this place, which is quite unusual. Robert has made some smart moves, and we've both worked hard, but we've also been lucky in that we haven't had a major illness. So I can't blame other people for how they feel."

Like Ed Swartz, the Sorensons direct their anger at state officials, who they believe have permitted the coal-bed methane operators to run roughshod over ranchers and the land. For their ranch, the Sorensons signed 13 separate lease deals, like pipeline and powerline rights-of-ways, with energy companies. "Not once did we have a choice," says Nancy. What's more, Wyoming, unlike neighboring Montana and most states, doesn't have laws mandating that proposed industrial development—such as gas extraction—on nonfederal lands be assessed beforehand for environmental impact.

Shortly after the widespread environmental damages became evident several years ago, the Sorensons, Ed Swartz, and many other ranchers met with their state representatives to plead for tighter regulations on the drilling and for a set of rancher rights—to no avail.

"They've all just been bought and sold," says Robert. "Really, they all have." (Nearly 70 percent of all campaign contributions to Wyoming's state legislators come from the oil and gas industry.)

"They all know where the paychecks come from," Nancy chimes in. "There isn't anybody in local government that is going to fight this. And neither am I. We just need some tougher regulations."

The ranchers' biggest concern is the depletion of their underground aquifers. In Colorado, where coal-bed methane development has taken off in a number of areas, drillers are required by state law to clean the discharged methane water of all pollutants and "reinject" it into the ground. Not so in Wyoming. Drillers say that requirement would make their operations less cost-effective. Many ranchers, joining forces with the formidable Powder River Basin Resource Council, have also been petitioning state and federal lawmakers for a mandatory Surface Owner Agreement—which would allow landowners to have a say about where pipelines and roads are built. (The government broke its promise that powerlines would be buried and kept to minimum.) Above all, what pains the ranching community is the lack of adequate bonding—money the energy companies pay to cover land damages and reclamation. Just look around, Nancy says, at all the abandoned wells from previous booms. Undoubtedly, history will repeat itself, she asserts, if energy companies aren't required to pony up much more than the paltry $25,000 bond they pay for unlimited wellheads on Wyoming's federal lands (it's $75,000 for state and private lands).

Given this history, I'm surprised to hear the Sorensons say they are not against coal-bed methane development in principle (and that goes for Ed Swartz and the other Powder River basin ranchers I spoke with).

"No," Robert answers resolutely. "We just want to see it done right."

A FEW MONTHS AFTER MY TRIP TO THE Powder River basin last May, the bottom fell out of the natural gas market, dropping prices to less than a dollar for every thousand cubic feet, from a high of $12 two years earlier, a price that fueled the drilling frenzy. Some industry experts and plenty of environmentalists attribute the Powder River boom to the manipulation of energy prices in California by Enron and other energy companies. Whatever the reason, the plunging price has slowed development in the Powder River basin and given ranchers—and wildlife—some breathing room.

"We have a window of opportunity to get the problems addressed," says Jill Morrison, an organizer with the Powder River Basin Resource Council. "It's good we have this slowdown. Maybe now we can get better planning and development." Morrison cautions it could merely be a lull in a volatile energy market, and that in the meantime, drillers might use the depressed prices as an excuse to pay less money to repair damages to the land.

No matter the outcome, Ed Swartz, the Sorensons, and other Powder River ranchers under siege aren't counting on their politicians for help. In August, three months after the EPA released its critical report of the BLM's environmental assessment, Wyoming's governor, Jim Geringer, lambasted the EPA for its "aggressive and ill-informed approach" to coal-bed methane development in the Powder River region. And in September, perhaps to shore up their flagging spirits, Montana Governor Judy Martz told a roomful of oil and gas lobbyists that they were "the true environmentalists"—as opposed to, as one meeting attendee said, the "radical" critics of energy development.

After I heard both comments, I figured these officials had never seen Ed Swartz's polluted stream or had lunch with the Sorensons. "We are very environmentally conscious," Nancy told me during my visit, "and most ranchers we know are." Her tone was gentle but firm. "I'm not opposed to having development, but there should be some equilibrium."

It never occurred to me then, or months later, when Governor Martz and the gas-industry proponents made their barbs, that the Sorensons or Ed Swartz were "radicals." But I sure do know who the "true environmentalists" are in Wyoming.

WHAT YOU CAN DO

For more information, call the Powder River Basin Resource Council at 307-672-5809 or log on to www.powderriverbasin.org. You can also call or e-mail the Wyoming state office of the BLM at 307-775-6256; state_wyomail@blm.gov.

LIVING WITHOUT OIL

As war looms, the search for new energy alternatives is all the more urgent.

By Marianne Lavelle

Grant Goodman wanted to do his part to reduce U.S. dependence on foreign oil. So two years ago, the Phoenix concrete producer began using biodiesel—made from refined soybean oil—to fuel his fleet of 130 diesel-powered cement mixers and excavators. For his efforts, Goodman in 2001 won a local entrepreneur of the year award and plaudits from the Environmental Protection Agency. But protecting the Earth was not Goodman's only concern. "Let's start with national security—the billions and billions we waste dancing around the issue, protecting those pipelines, invading Iraq, doing whatever else we're doing in the Middle East. It all gets down to continuing the flow of oil to this country."

Goodman's stance hasn't been easy. Biodiesel fuel sold for 70 cents per gallon more than regular diesel fuel, giving competitors of his Rockland Materials a decided edge. "It cost me a few hundred grand," says Goodman. Those harsh economics forced him last year to resort to a petroleum mix including 40 percent or less of biodiesel. But don't count him out. He plans to build his own soybean oil refinery this year to help him return to 100 percent biodiesel. Goodman has urged other local businesses to make the switch, but as long as petroleum is cheaper, he says, "I'm this guy screaming in the wind."

Sure, in theory, everyone agrees the nation should break its 20 million-barrel-a-day oil habit, 58 percent of it imported. Last week, President Bush noted that "sometimes we import from countries that don't particularly like us. It jeopardizes our national security." Antiwar protesters, who argue that Iraq's massive oil reserves have made it a U.S. target, use sharper rhetoric. "No blood for oil!" they shouted at demonstrations at gasoline stations around the country last week. At the other end of the political spectrum, Martin Feldstein, who headed former President Reagan's panel of economic advisers, has argued that the United States should set a goal of complete oil independence by the year 2020. "Otherwise, we will

---

BIODIESEL

Biodiesel, processed from any vegetable oil, can power most diesel-engine vehicles without modification.

STATUS: Sales tripled in 2002 to 15 million gallons. Gained exposure in an episode of *The West Wing*. Used in many government vehicles, especially in national parks.

PROS: Burns cleaner than petroleum diesel. Could boost rural economy, especially soybean farming.

CONS: Double the cost of petroleum diesel; fueling stations are scattered. Does not eliminate pollution when blended with petroleum diesel.

---

continue to be hostage to the policies of the current and future rulers of Saudi Arabia, Iraq, Iran, and their neighbors." And indeed, the jitters of potential war in the Middle East and political upheaval in Venezuela, the nation's fourth-largest oil supplier, have pushed up the price of gasoline for eight consecutive weeks. If global events turn awry, an oil price shock could, as has happened repeatedly in the past, tip the struggling economy back into recession.

Within reach. But has anyone found a reasonable alternative to the black gold that fuels the U.S. economy? Some answers seem tantalizingly close, especially for transportation, which consumes the vast majority of our oil. Hundreds of truck fleets and bus systems already run on two diesel-fuel alternatives, biodiesel and natural gas. Meanwhile, biotechnology has made it possible to extract fuels from farm products like corn husks, long discarded as waste. And, of course, there are the many recent advances in the harnessing of energy from the world's most abundant element, hydrogen—the science for which Bush pledged $1.2 billion support in his State of the Union message.

101

Redesigning the SUV

SUVs could be more fuel efficient, a National Academy of Sciences study found. Installing all these new technologies, some already in use, could improve gas mileage by 25 percent to 50 percent.

The cost: $1,219 to $2,923 per vehicle.

Smarter cylinders

A system that automatically deactivates some cylinders on easy drives, conserves extra power when it isn't needed.

Fuel-economy gain:
3% to 6%

Cost:
$112 to $252

Enhanced aerodynamics

Sleeker front ends and reduced tire-rolling resistance can cut wind drag.

Fuel-economy gain:
2% to 4%

Cost:
$14 to $196

Lighten up

High-strength, lighter-weight metals can replace some steel.

Fuel-economy gain:
3% to 4%

Cost:
$210 to $350

Better engines

"Variable valve timing," which adjusts the intake of the air that mixes with fuel, along with other systems that reduce friction, can improve efficiency.

Fuel-economy gain:
7% to 18%

Cost:
$253 to $641

Improved transmissions

"Continuously variable transmissions" can keep the engine running in the most efficient range at all times.

Fuel-economy gain:
4% to 9%

Cost:
$140 to $350

End idling

An integrated, 42-volt starter-generator essentially "turns off" the vehicle briefly at traffic lights, conserving fuel.

Fuel-economy gain:
5% to 10%

Cost:
$280 to $630

Source: National Academy of Sciences, 2002

But much more money and an even broader government commitment will be needed to reverse the current U.S. trajectory toward greater oil addiction. After all, largely because of the popularity of gas-guzzling sport utility vehicles, the average fuel economy of the 2003 fleet of cars sank 6 percent below the peak set 15 years ago. Critics say that until the new technology is ready to help the nation kick the oil habit, the Bush administration should focus on breaking the addiction step by step. Fuel-economy regulations, they argue, could force greater use of the breakthrough hybrid gas-electric engine and other lesser-known innovations that can squeeze more miles out of every gallon of gasoline.

Japan's government, for example, vows to put 10 million "ecofriendly" cars on its roads by 2010, a number it hopes will include not only 50,000 hydrogen fuel cell cars but also natural gas vehicles, electric autos, and hybrids. Japan's auto industry views that as an attainable goal, given the tax incentives and subsidies that support it. Stephen Tang, president of Millennium Cell, an Eatontown, N.J., firm that has developed a hydrogen fueling system, is hopeful that a similar commitment will catch fire here. "If we can get the oil man to say the word 'hydrogen,' that's significant progress," says Tang.

In his so-called FreedomFUEL initiative, the president zeroed in on what is unquestionably the most promising alternative fuel. Hydrogen is everywhere, and when used to power a special battery called a fuel cell, its only waste product is water. It's an alluring option, but slippery. Hydrogen is extremely difficult to harness, store, and distribute. And many people are most familiar with hydrogen for its darkest moment: the 1937 Hindenburg dirigible disaster. However, scientists reported in 1998 that the zeppelin's flammable coating, not its fuel, ignited this deadly blaze.

Lip service? All of the major oil firms have investments in hydrogen; in fact, BP's new motto is "Beyond Petroleum." But energy analyst Fadel Gheit of Fahnestock & Co. says the corporate commitment is "minuscule." The Royal Dutch/Shell Group's promised $1 billion for renewable energy over the next four years fades beside its $24.6 billion capital investment, mostly in oil and natural gas, in 2002 alone. "These companies don't want to be left out in the event that some of these ventures come to fruition," Gheit says, but "they're not holding their breath."

How soon will cars that run on hydrogen be on the market? "My answer has always been 'four years after we figure out how to have hydrogen at the corner gas sta-

ETHANOL

An alcohol fuel fermented from corn mash. But new biotech methods would use farm waste, such as cornstalks, as feedstock.

STATUS: Sales rose 20 percent in 2002 to 2.13 billion gallons.

PROS: The new biotech method, if successfully commercialized, would cut more pollution than the corn process. Boosts fuel combustion. Made domestically; aids farm economy.

CONS: Provides fewer mpg and costs more than gasoline. Usually mixed with gasoline, a practice that trims environmental benefits. Traditional production, which uses diesel fuel and electricity, diminishes energy and antipollution gains.

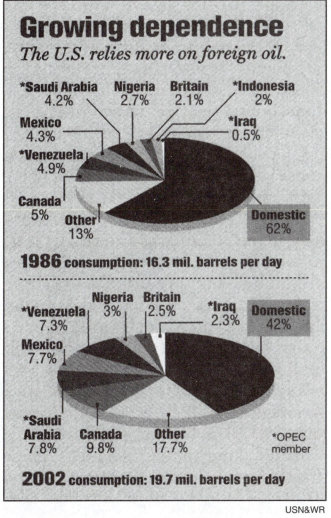

Growing dependence
The U.S. relies more on foreign oil.

*Saudi Arabia 4.2%   **Nigeria** 2.7%   **Britain** 2.1%   ***Indonesia** 2%
Mexico 4.3%   ***Iraq** 0.5%
*Venezuela 4.9%
Canada 5%   **Other** 13%   **Domestic** 62%

1986 consumption: 16.3 mil. barrels per day

*Venezuela 7.3%   **Nigeria** 3%   **Britain** 2.5%   ***Iraq** 2.3%   **Domestic** 42%
Mexico 7.7%
*Saudi Arabia 7.8%   **Canada** 9.8%   **Other** 17.7%   ***OPEC member**

2002 consumption: 19.7 mil. barrels per day

USN&WR

Source: U. S. Energy Information Administration

tion,' " says Thomas Moore, vice president of DaimlerChrysler's advanced car division. Perhaps scores of firms are working on solutions. DaimlerChrysler, which has earmarked $1.4 billion for fuel cell research from 2001 to 2004, has worked with Millennium Cell on a concept car, the Natrium, named after the Latin word for sodium. It is fueled with a water solution of the compound sodium borohydride, and a chemical reaction releases hydrogen as needed. Energy Conversion Devices, a Rochester Hills, Mich., company chaired by former General Motors CEO Robert Stempel and developer of the nickel metal hydride battery that now powers hybrid cars, has worked with ChevronTexaco to convert that same technology into a hydrogen storage and delivery system. A metal hydride element aboard the car would absorb hydrogen, like a sponge, then release it as needed into the fuel cell to power the vehicle.

Although President Bush called hydrogen a "pollution free" technology, that isn't necessarily the case. Extracting hydrogen from its most common source, water, requires electricity. Energy Secretary Spencer Abraham says that electricity could come from coal, a domestic but dirty source, or from nuclear energy, an option whose expansion the U.S. public has not welcomed. Hydrogen also can be gleaned from gasoline, an idea that has garnered notable support from Big Oil. Environmentalists want to see large amounts of new wind and solar power deployed to help generate the fuel, and the Bush plan would put most funding toward that goal. But wind power, although competitive and growing at a rate of 28 percent last year, still accounts for less than 1 percent of U.S. electricity. Expensive solar's footprint is even smaller. Until renewable energy is more widespread, many suspect that hydrogen will be manufactured out of a clean, though not ideal, alternative fuel, natural gas.

That's the hope of H2Gen of Alexandria, Va., which plans to roll out its first on-site hydrogen generation stations using natural gas later this year. The company hopes to silence critics who say distributing hydrogen would be prohibitively expensive, requiring either tanker trucks of liquid hydrogen or construction of a new nationwide pipeline system. H2Gen's idea is to hook its 6-by-7-foot fueling stations to existing natural gas lines and, through an on-site chemical process, extract hydrogen at a cost competitive with that of gasoline. "We see ourselves as a transition to the renewable hydrogen future," when there's enough wind and solar energy to produce hydrogen from water, says Sandy Thomas, company president.

When consumers begin to see hydrogen cars in showrooms, which General Motors Vice President Larry Burns thinks will be by 2015, they may not be recognizable. GM's version, the Hy-wire, has no hood, steering wheel, or pedals. The driver uses handgrips to steer, accelerate, and brake while looking out through a floor-to-ceiling windshield. "We think we can truly reinvent the automobile and the industry around the fuel cell and make good money doing that," says Burns.

High stakes. But with terrorism a national concern, many observers see 12 years as too long to wait for the

KEEP ON GUZZLIN'

At the Detroit auto show earlier this year, the world's car companies tried to one-up each other with proclamations about future SUVs that will get the gas mileage of a pedicab, with lower emissions than a banana. But there's an important prerequisite: Consumers must not be asked to sacrifice a hint of horsepower or a sliver of size. "We want to produce vehicles where you can be proud of your contribution to the environment," said General Motors President Rick Wagoner, "but don't really have to make any trade-offs."

An ad campaign targets SUVs.

SUVs have come into the cross hairs in a series of high-profile assaults, including the "What Would Jesus Drive?" campaign promoted last fall by religious leaders as well as commercials dreamed up by pundit Ariana Huffington, which equate SUV owners with terrorist sympathizers. Even federal safety czar Jeffrey Runge has piled on, saying in a recent speech that he wouldn't ride in SUVs that score below average on government rollover tests, including the Ford Explorer and the Jeep Liberty.

Consumers appear unmoved. Automakers sold 8.6 million light trucks in 2002, as trucks outsold cars for the first time. This year, Ford expects to sell nearly 200,000 of the gas hog Expedition. The conventional wisdom in Detroit is that big, powerful vehicles will carry the day as long as fuel stays below $2 a gallon. "The price of gas is a killer," says one executive.

Few incentives. With Americans wedded to their Explorers and TrailBlazers, Detroit and Washington are offering little to lure them away. The only hybrid gas-electric vehicles are small sedans from Toyota and Honda. They'll haul the soccer balls but not the team. By next year, some new offerings will begin to test whether sport utilitarians have a conscience. Ford plans to sell a hybrid version of its popular Escape SUV in early 2004 that will get about 40 miles per gallon, some 50 percent more than a conventional Escape. But with two drive trains instead of one, it will cost $2,000 to $3,000 more than the standard Escape. Later in '04, Toyota plans a hybrid version of the Lexus RX330 luxury SUV.

How aggressively the automakers market such new products will both determine their success and indicate whether companies are sincere about boosting fuel economy. "They have created a demand for huge SUVs through advertising and zero-percent financing," says Huffington. "How much has Detroit spent advertising hybrid cars?"

The biggest factor could be General Motors. The world's largest automaker last month said it would put 1 million hybrid vehicles on the road by 2008, many of them the biggest trucks and SUVs in its fleet. That's if demand materializes. Another GM technology, displacement on demand, metes out fuel only to those cylinders where it is needed and could cut fuel use by 10 percent or more. But GM has lobbied aggressively against any increases in federal mileage standards, and critics say its hybrid technology barely breaks new ground. Besides, car nuts at the Detroit show were far less impressed with GM's hybrid announcement than with the conceptual "Sixteen"—a mock Cadillac with a 16-cylinder gasoline engine.—*Richard J. Newman*

country to wean itself from Middle East oil. Given the high stakes, is any alternative ready *now*? U.S. farms and fields have yielded some homegrown energy choices, like biodiesel, natural gas, and ethanol, but it has been hard for them to challenge the entrenched oil industry with its relatively low prices and robust infrastructure. Biodiesel would seem to have the inside track; it can be pumped into nearly any diesel engine tank with no modification. In fact, Rudolph Diesel used peanut oil to power the engine he debuted at the 1900 World's Fair. Biodiesel can be made from any fat or vegetable oil—even used and purified kitchen grease—although it is usually made from the

nation's second-biggest crop, soybeans. Environmental benefits are impressive; 100 percent biodiesel eliminates sulfur emissions and cuts particulate matter and some other pollutants by about 50 percent.

Oddly, U.S. government policies have served as both boon and restraint to biodiesel. The 1992 Energy Policy Act, which encourages a percentage of federal and state government vehicles to run on alternative fuels, was amended in 1998 to give credit for biodiesel use. Sales have exploded 30-fold since 1999 to 15 million gallons. But to get credit, the federal government requires only that the fleets run on a mix of 20 percent biodiesel, 80 percent petroleum. With biodiesel as much as double the cost, and taxed at the same rate as petroleum diesel, governments seldom buy more than the 20 percent mix.

Naturally, this practice reduces environmental benefits. Biodiesel also increases emissions of one smog-producing pollutant, nitrogen oxide, or NOx. Although technical solutions, such as adjusting engine timing, appear to be available, some environmentalists remain lukewarm. Daniel Becker, head of the Sierra Club's energy program, says the "french fry grease hustlers" are not competing with petroleum at all but are vying for market share against an alternative fuel his organization prefers: natural gas.

Gassing up. Big-city residents have seen natural gas buses popping up in their mass transit systems. City governments have found they can move closer to meeting tough new federal antismog standards by converting some of their fleet to CNG, or compressed natural gas. Natural gas is not environmentally benign; it emits greenhouse gases, and exploration and drilling often prove controversial. But CNG engines emit virtually no particulate matter, toxic chemicals, or sulfur and 50 percent less NOx. What's more, 85 percent of natural gas consumed in the United States is produced domestically; nearly all the rest is from Canada. The 130,000 CNG vehicles on U.S. roads last year displaced 124 million gallons of gasoline, and the sector is growing 10 percent a year. Natural gas passenger cars and pickup trucks are now available.

HYDROGEN

Although extremely common, hydrogen must be extracted from water or fossil fuels using electricity or chemical processing.
STATUS: High-tech solutions are under development to produce, store, and distribute hydrogen.
PROS: When burned or fed into a fuel cell battery, its only waste output is water. Could power homes and businesses as well as cars. Would be produced domestically. Extends the possibilities for wind and solar power.
CONS: Some extraction methods—for example, coal-fired electricity—pollute. Hydrogen is a flammable gas; safe delivery must be perfected. Dogged by chicken-and-egg problem: Lacking fuel stations, no hydrogen vehicles will be built. Without vehicles, no fuel network can survive.

However, cost and inconvenience can be stumbling blocks. The suggested starting price of a Honda Civic GX natural gas car is $20,510, nearly 60 percent more than a gasoline-run sedan and $1,000 more than Honda's hybrid gas-electric Civic. The price tag of the heavy-duty CNG engines is still $20,000 to $50,000 more than that of a traditional diesel engine. Finding a fueling station that pumps out the pressurized gas can be a challenge. When Wash-

ington, D.C.'s transit agency, Metro, decided last year to add 250 new CNG buses to its fleet, opponents complained the agency could buy twice as many diesel buses for the $105 million tab of the vehicles and their fueling station. But advocates noted that over time, Metro would save money on fuel. Natural gas is the only oil alternative that has been selling for less than gasoline; the discount last fall was about 25 percent. Also, tax incentives from the federal government and some states help offset capital costs. New solutions also are on the horizon for the fueling problem. FuelMaker, a Canadian company in which Honda has a 20 percent stake, plans this year to begin sales of its "Phill" appliance, which will allow drivers to fill up from natural gas lines at home; expected price: about $1,000.

NATURAL GAS

Natural gas, drilled out of underground supplies in the United States and Canada, is compressed for use in specially designed engines.
STATUS: In addition to the trillions of cubic feet used for electricity and heating, sales for transportation are growing 10 percent annually and reached the equivalent of 124 million gallons of gasoline in 2002.
PROS: Prices fluctuate, but it now costs 25 percent less than gasoline; eliminates particulate matter, toxic chemicals, and sulfur; reduces smog.
CONS: Prices spike when demand is high, as in winter; its growing use in electricity and as a hydrogen source could pressure prices; drilling disrupts the environment.

Second chance. After the 1970s oil crises, policymakers turned to what seemed at the time to have the most potential: ethanol. Corn alcohol hasn't exactly lived up to its promise. The 2.13 billion gallons of ethanol produced last year, up 20 percent from the previous year, still amounted to less than 2.6 percent of U.S. oil imports. More accurately called an additive than a replacement fuel, ethanol is typically mixed with 90 percent petroleum; an oxygenate, it boosts combustion and reduces tailpipe emissions.

But ethanol's green image has faded of late. Diesel tractors plant, fertilize, and harvest the corn used to make ethanol, and substantial coal-fired electricity is used to process the grain. Cornell University scientist David Pimentel, author of a study showing ethanol consumes more energy than it produces, calls it "unsustainable, subsidized food burning." Federal and state governments spend about $1 billion a year to support ethanol, most of which goes to agribusiness giants like Archer Daniels Midland, which owns 35 percent of the market.

But the ethanol industry could be transformed by biotechnology. Researchers can now unlock the sugars found in tough agricultural waste products—corn husks, rice hulls, saw grass, and wood chips—which can then be fermented into an alcohol that can fuel vehicles. This so-called cellulosic, or biomass, ethanol would require less energy to produce and could be manufactured from material that is now burned or buried. Getting biomass ethanol from the laboratory to the highway has been slow. BC International of Dedham, Mass., which plans to build a plant in Louisiana to convert sugar cane waste into fuel, is having a hard time getting $90 million to build the refinery. "It's a combination of the economy and the fact that it's the first of its kind," says Vice President John Doyle. "Bankers and investors love to say, 'Where is one of these running?' "

The Department of Energy has supported biomass ethanol research and development, announcing $75 million in grants in December. Recipients include Cargill Dow and DuPont, which have successfully used biotech to convert corn into packaging materials, plastics, and synthetic fibers like those now made from petroleum. "The government has a role in helping to defray some of the risk" if the nation wants faster development of cellulosic ethanol, says Pat Gruber, vice president of Cargill Dow. "Many plants never work or take years to work. It's a scary thing from an investment standpoint," says Gruber. But the Bush administration's new budget would reduce biomass ethanol funding, leaving the private sector to lead future development.

Until true oil replacements can make a dent in the market, many argue that the government should focus on reducing consumption by requiring automakers to build more fuel-efficient cars. The Sierra Club calculates that if the vehicle fuel economy average of 20.8 mpg were raised to 40 mpg, it would save upwards of 3 million barrels of oil a day. "The technology is out there, but the government needs to give the automakers a very clear signal," says David Friedman of the Union of Concerned Scientists. Still, carmakers say that consumers don't want the smaller cars that higher standards would entail. And at an Alliance to Save Energy conclave last fall, Karen Knutson, a top aide to Vice President Cheney, cited studies showing that smaller cars are more hazardous: "Every mile per gallon saved kills lives," she said, eliciting hissing from the audience. Fuel-efficiency advocates say carmakers would not have to downsize but could install new technologies that could make even SUVs, which have their own safety concerns, more efficient. Greater savings can be reached if more manufacturers produced, and more consumers drove, the hybrid gasoline-electric engines pioneered by Honda and Toyota.

High hopes. The hybrid's popularity, coupled with mounting criticism of the SUV's gas-guzzling tendency, is forcing the market to respond (see box, Keep on Guzzlin'). But GM's Burns says that consumers will not pay additional costs that are inevitable with high-efficiency vehicles unless they get greater performance and better value. "Just having a strong preference for fuel economy—we don't think that's going to happen at $1.50 a gallon," says Burns. He may have a point. Statistics show that as fuel economy has grown, the number of miles that Americans drive each year has risen.

That's why many economists say the more efficient way to help Americans give up oil would be to do what the Europeans do—tax gasoline more heavily. "Some would argue that the price of oil ought to include some of the cost of the defense establishment, since we seem periodically to have to send the military over to defend oil producers," says oil economist Philip Verleger of the Council on Foreign Relations. It's an idea, however, that has been a nonstarter among antitax Republicans and populist Democrats alike, since it would hit low-income and rural Americans the hardest. Some creative solutions have been suggested, including an offsetting cut in Social Security taxes for working Americans and an elaborate system of vouchers. But Verleger believes the U.S. government passed up its one chance to enact such a tax. "If George Bush had pointed to the wreck of the World Trade Center, and said, 'We must correct this problem,' and the only way is by raising the cost of gasoline on a phased-in basis, it would have worked," he says. "It was the golden opportunity missed."

RENEWABLE ENERGY: A VIABLE CHOICE

By Antonia V. Herzog, Timothy E. Lipman, Jennifer L. Edwards, and Daniel M. Kammen

Renewable energy systems—notably solar, wind, and biomass—are poised to play a major role in the energy economy and in improving the environmental quality of the United States. California's energy crisis focused attention on and raised fundamental questions about regional and national energy strategies. Prior to the crisis in California, there had been too little attention given to appropriate power plant siting issues and to bottlenecks in transmission and distribution. A strong national energy policy is now needed. Renewable technologies have become both economically viable and environmentally preferable alternatives to fossil fuels. Last year the United States spent more than $600 billion on energy, with U.S. oil imports climbing to $120 billion, or nearly $440 of imported oil for every American. In the long term, even a natural gas–based strategy will not be adequate to prevent a buildup of unacceptably high levels of carbon dioxide (CO_2) in the atmosphere. Both the Intergovernmental Panel on Climate Change's (IPCC) recent Third Assessment Report and the National Academy of Sciences' recent analysis of climate change science concluded that climate change is real and must be addressed immediately—and that U.S. policy needs to be directed toward implementing clean energy solutions.[1]

Renewable energy technologies have made important and dramatic technical, economic, and operational advances during the past decade. A national energy policy and climate change strategy should be formulated around these advances. Despite dramatic technical and economic advances in clean energy systems, the United States has seen far too little research and development (R&D) and too few incentives and sustained programs to build markets for renewable energy technologies and energy efficiency programs.[2] Not since the late 1970s has there been a more compelling and conducive environment for an integrated, large-scale approach to renewable energy innovation and market expansion.[3] Clean, low-carbon energy choices now make both economic and environmental sense, and they

provide the domestic basis for our energy supply that will provide security, not dependence on unpredictable overseas fossil fuels.

Energy issues in the United States have created "quick fix" solutions that, while politically expedient, will ultimately do the country more harm than good. It is critical to examine all energy options, and never before have so many technological solutions been available to address energy needs. In the near term, some expansion of the nation's fossil fuel (particularly natural gas) supply is warranted to keep pace with rising demand, but that expansion should be balanced with measures to develop cleaner energy solutions for the future. The best short-term options for the United States are energy efficiency, conservation, and expanded markets for renewable energy.

Traditional power plants based on fossil fuels emit pollutants that contribute substantially to climate change. Renewable energy sources are becoming economically viable and environmentally preferable alternatives.

For many years, renewables were seen as energy options that—while environmentally and socially attractive—occupied niche markets at best, due to barriers of cost and available infrastructure. In the last decade, however, the case for renewable energy has become economically compelling as well. There has been a true revolution in technological innovation, cost improvements, and our understanding and analysis of appropriate applications of renewable energy resources and technologies—notably solar, wind, small-scale hydro, and biomass-based

Figure 1. Capital cost forecasts for renewable energy technologies

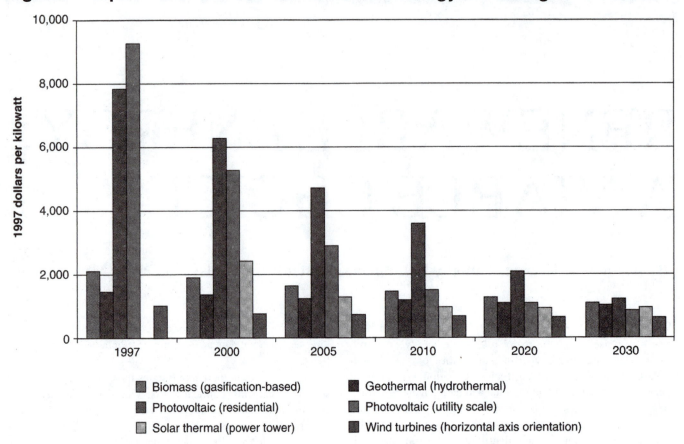

Legend:
- Biomass (gasification-based)
- Photovoltaic (residential)
- Solar thermal (power tower)
- Geothermal (hydrothermal)
- Photovoltaic (utility scale)
- Wind turbines (horizontal axis orientation)

SOURCE: U.S. Department of Energy (DOE), *Renewable Energy Technology Characterizations*, Topical Report prepared by DOE Office of Utility Technologies and EPRI, TR-109496, December 1997.

energy, as well as advanced energy conversion devices such as fuel cells.[4] There are now a number of energy sources, conversion technologies, and applications that make renewable energy options either equal or better in price and services provided than the prevailing fossil-fuel technologies. For example, in a growing number of settings in industrialized nations, wind energy is now the least expensive option among all energy technologies—with the added benefit of being modular and quick to install and bring on-line. In fact, some farmers, notably in the U.S. Midwest, have found that they can generate more income per hectare from the electricity generated by a wind turbine than from their crop or ranching proceeds.[5] Also, photovoltaic (solar) panels and solar hot water heaters placed on buildings across the United States can help reduce energy costs, dramatically shave peak-power demands, produce a healthier living environment, and increase the overall energy supply.

The United States has lagged in its commitment to maintain leadership in key technological and industrial areas, many of which are related to the energy sector.[6] The United States has fallen behind Japan and Germany in the production of photovoltaic systems, behind Denmark in wind and cogeneration system deployment, and behind Japan, Germany, and Canada in the development of fuel-cell systems. Developing these indus-tries within the United States is vital to the country's international competitiveness, commercial strength, and ability to provide for its own energy needs.

Renewable Energy Technologies

Conventional energy sources based on oil, coal, and natural gas have proven to be highly effective drivers of economic progress, but at the same time, they are highly damaging to the environment and human health. These traditional energy sources are facing increasing pressure on a host of environmental fronts, with perhaps the most serious being the looming threat of climate change and a needed reduction in greenhouse gas (GHG) emissions. It is now clear that efforts to maintain atmospheric CO_2 concentrations below even double the pre-industrial level cannot be accomplished in an oil- and coal-dominated global economy.

Theoretically, renewable energy sources can meet many times the world's energy demand. More important, renewable energy technologies can now be considered major components of local and regional energy systems. Solar, biomass, and wind energy resources, combined with new efficiency measures

Figure 2. Actual electricity costs in 2000

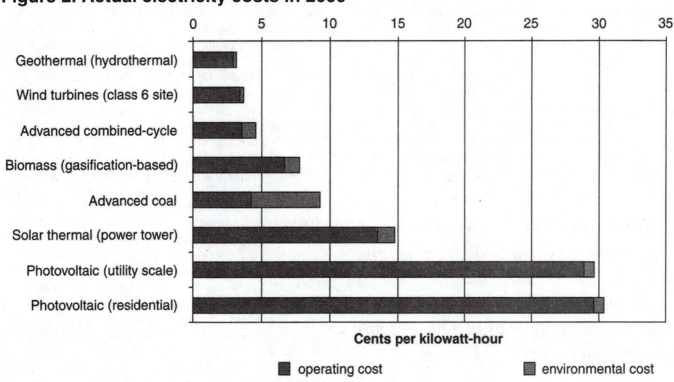

Cents per kilowatt-hour

■ operating cost   ■ environmental cost

SOURCE: R.L. Ottinger et al., *Environmental Costs of Electricity* (New York: Oceana Publications, Inc., 1991); and U.S. Department of Energy, *Annual Energy Outlook 2000*, DOE/EIA-0383, Energy Information Administration, Washington, D.C., December 2000.

available for deployment in California today, could supply half of the state's total energy needs. As an alternative to centralized power plants, renewable energy systems are ideally suited to provide a decentralized power supply that could help to lower capital infrastructure costs. Renewable systems based on photovoltaic arrays, windmills, biomass, or small hydropower can serve as mass-produced "energy appliances" that can be manufactured at low cost and tailored to meet specific energy loads and service conditions. These systems have less of an impact on the environment, and the impact they do have is more widely dispersed than that of centralized power plants, which in some cases contribute significantly to ambient air pollution and acid rain.

There has been significant progress in cost reductions made by renewable technologies (see Figure 1).[7] In general, renewable energy systems are characterized by low or no fuel costs, although operation and maintenance costs can be considerable. Systems such as photovoltaics contain far fewer mechanically active parts than comparable fossil fuel combustion systems, and are therefore likely to be less costly to maintain in the long term.

Costs of solar and wind power systems have dropped substantially in the past 30 years and continue to decline. For decades, the prices of oil and natural gas have been, as one research group noted, "predictably unpredictable."[8] Recent analyses have shown that generating capacity from wind and solar energy can be added at low incremental costs relative to

additions of fossil fuel–based generation. Geothermal and wind can be competitive with modern combined-cycle power plants—and geothermal, wind, and biomass all have lower total costs than advanced coal-fired plants, once approximate environmental costs are included (see Figure 2).[9] Environmental costs are based, conservatively, on the direct damage to the terrestrial and river systems from mining and pollutant emissions, as well as the impacts on crop yields and urban areas. The costs would be considerably higher if the damage caused by global warming were to be estimated and included.

The push to develop renewable and other clean energy technologies is no longer being driven solely by environmental concerns; these technologies are becoming economically competitive. According to Merrill Lynch's Robin Batchelor, the traditional energy sector has lacked appeal to investors in recent years because of heavy regulation, low growth, and a tendency to be cyclical.[10] The United States' lack of support for innovative new companies sends a signal that U.S. energy markets are biased against new entrants. The clean energy industry could, however, become a world-leading industry akin to that of U.S. semi-conductors and computer systems.

Renewable energy sources have historically had a difficult time breaking into markets that have been dominated by traditional, large-scale, fossil fuel–based systems. This is partly because renewable and other new energy technologies are only now being mass produced and have previously had high capital costs relative to more conventional systems, but also because coal-, oil-, and gas-powered systems have benefited from a range of subsidies over the years. These include military expenditures to protect oil exploration and production interests overseas, the costs of railway construction to enable economical delivery of coal to power plants, and a wide range of tax breaks.

One disadvantage of renewable energy systems has been the intermittent nature of some sources, such as wind and solar. A solution to this problem is to develop diversified systems that maximize the contribution of renewable energy sources but that also use clean natural gas and/or biomass-based power generation to provide base-load power (energy to meet the daily needs of society, leaving aside the peak in energy use associated, for example, with afternoon and evening air-conditioner or heating demands).

Solar energy can be harnessed by both industrial-scale and smaller residential solar power systems. Solar and other renewable energy sources could take over for fossil fuel–based power plants while saving energy and money.

Renewable energy systems face a situation confronting any new technology that attempts to dislodge an entrenched technology. For many years, the United States has been locked in to nuclear- and fossil fuel–based technologies, and many of its secondary systems and networks have been designed to accommodate only these sources. The U.S. administration's recent National Energy Policy plan focused on expanding the natural gas supply, without any attention to the benefits of building a diverse energy system.[11] The plan would add one to two new power plants each week for the next several years. The majority of these plants would be fired by natural gas, making the country far more dependent on natural gas than it ever was on oil—even at the height of the OPEC crisis in the 1970s.

Renewable energy technologies are characterized by low environmental costs, but many of these environmental costs are termed "externalities" and are not reflected in market prices. Only in certain areas and for certain pollutants do these environmental costs enter the picture. The international effort to limit the growth of GHG emissions through the Kyoto Protocol may lead to some form of carbon-based tax, which continues to face stiff political opposition in the United States. It is perhaps more likely that concern about emissions of particulate matter and ozone formation from fossil-fuel power plants will lead to expensive mitigation efforts by the plant operators, and this would help to tip the balance toward cleaner renewable systems.

There are two principal rationales for government support of R&D to develop clean energy technologies. First, conventional energy prices generally do not reflect the social and environmental cost of pollution. Second, private firms are generally unable to appropriate all the benefits of their R&D investments. The social rate of return for R&D exceeds the returns captured by individual firms, so they do not invest enough in R&D to maximize social benefits.[12] Public investment, however, would help spread innovation among clean energy companies, which would benefit the public.

Publicly funded market transformation programs (MTPs) for desirable clean energy technologies would provide an initial subsidy and incentive for market growth, thus stimulating long-term demand. A principal reason for considering MTPs is inherent in the production process. When a new technology is first introduced, it is more expensive than established substitutes. The unit cost of manufactured goods then tends to fall as a function of cumulative production experience. Cost reductions are typically very rapid at first and then taper off as the industry matures—resulting in an "experience curve". Gas turbines, photovoltaic cells and wind turbines have all exhibited this expected price-production relationship, with costs falling roughly 20 percent for each doubling of the number of units produced.[13]

If producers of clean energy consider the experience-curve effect when deciding how much to produce, they will "forward-price," producing at a loss initially to bring down their costs and thereby maximize profit over the entire production period. In practice, however, the benefits of production experience often spill over to competitor producers, and this potential problem discourages private firms from investing in bringing new products down the experience curve. Publicly funded MTPs can help correct the output shortfall associated with these experience effects.[14]

This suggests an important role for MTPs in national and international technology policies. MTPs are most effective with emerging technologies that have steep industry experience curves and a high probability of major long-term market penetration once subsidies are removed. The condition that these technologies be clean mitigates the risk of poor MTP performance, because the investments will alleviate environmental problems whose costs were not taken into account for the older, dirtier energy technologies. Renewable energy products are ideal candidates for support through MTPs, via federal policies that reward the early production of clean energy technologies.

Energy Efficiency

Energy efficiency improvements have contributed a great deal to economic growth and increased standard of living in the United States over the past 30 years, and there is much potential for further improvements in the decades to come. According to the U.S. Department of Energy (DOE), increasing energy efficiency could cut national energy use by 10 percent or more by 2010 and about 20 percent by 2020. The recent Interlaboratory Working Group study Scenarios for a Clean Energy Future estimates that cost-effective end-use technologies could reduce electricity consumption by about 1,000 billion kilowatt-hours (kWh) by 2020, almost entirely offsetting the projected growth in electricity use.[15] This level of energy savings would reduce U.S. carbon emissions by approximately 300 million metric tons, and many of these changes can actually be accomplished with an increase in profits. Still more benefits can be had for investments of only a few cents per kilowatt-hour, far less than the energy cost of new power plants.

Energy efficiency is the single greatest way to improve the U.S. energy economy. Based on data published by the Energy Information Administration (EIA), the American Council for an Energy Efficient Economy (ACEEE) estimates that total energy use per capita in the United States in 2000 was almost identical to that in 1973, while over the same period, economic output (measured by Gross Domestic Product or GDP) per capita increased 74 percent. Furthermore, national energy intensity (energy use per unit of GDP) fell 42 percent between 1973 and 2000, with about 60 percent of this decline attributable to real improvements in energy efficiency and about one-quarter due to structural changes and fuel switching. Between 1996 and 2000, GDP increased 19 percent while primary energy use increased just 5 percent. These statistics clearly indicate that energy use and GDP do not have to grow or decline in lock step with each other, but GDP can, in fact, increase while energy use does not.[16]

The federal government's energy efficiency programs have been a resounding success. Last year, DOE documented the results of 20 of its most successful energy efficiency and renewable energy technology initiatives over the past two decades.[17] These programs have already saved the nation 5.5 quadrillion Btu (British Thermal Units) of energy, equivalent to the amount of energy needed to heat every household in the United States for about a year, and worth about $30 billion in avoided energy costs. Over the last decade, the cost to taxpayers for these 20 programs has been $712 million, less than 3 percent of the energy bill savings that the programs created.[18]

In 1997, the President's Committee of Advisors on Science and Technology (PCAST), a panel that consisted mainly of distinguished academics and private-sector executives, conducted a detailed review of DOE's energy efficiency R&D programs. PCAST concluded, "R&D investments in energy efficiency are the most cost-effective way to simultaneously reduce the risks of climate change, oil import interruption, and local air pollution, and to improve the productivity of the economy." PCAST recommended that the DOE energy efficiency budget be doubled between the fiscal years of 1998 and 2003. They estimated that this could produce a 40-to-1 return on investment for the nation, including reductions in fuel costs of $15 billion to $30 billion by 2005 and $30 billion to $45 billion by 2010.[19] Funding for these DOE programs in the last several years has fallen far short of the PCAST recommendations.

Increasing the efficiency of homes, appliances, vehicles, businesses, and industries must be an important part of a sound national energy and climate-change policy. Increasing energy efficiency reduces energy waste (without forcing consumers to cut back on energy services or amenities), lowers GHG emissions, saves consumers and businesses money (because the energy savings more than pay for any increase in initial cost), protects against energy shortages, reduces energy imports, and reduces air pollution. Furthermore, increasing energy efficiency does not create a conflict between enhancing national security and energy reliability, on the one hand, and protecting the environment on the other.

Climate Change

The threat of global climate change is finally producing a growing understanding and acknowledgement by some in U.S. industry and government that a responsible national energy policy must include a sound global climate-change mitigation strategy. President George W. Bush has rejected the Kyoto Protocol, but the U.S. Congress, in particular the Senate, appears poised to take action to reduce domestic GHG emissions. For example, Senators Jim Jeffords (I-Vt.) and Joe Lieberman (D-Conn.) and Representatives Sherwood Boehlert (R-N.Y.) and Henry Waxman (D-Calif.) recently introduced legislation in Congress to reduce the emission of four pollutants from electricity generation. The legislation puts a national cap on power plants' emissions of nitrogen oxides, sulfur dioxide, mercury, and carbon dioxide, and requires every power plant to meet the most recent emission control standards. It allows market-oriented mechanisms such as the trading of emissions credits, which is widely seen as a way to control pollution and stimulate innovation at the lowest cost. In emissions trading, total emissions are capped and then a market is created involving those firms that have excess credits to sell (resulting from decreased emissions due to efficiency and other improvements) and those firms needing to purchase credits due to emissions exceeding an allocated baseline. In the United States, nitrogen oxides and sulfur dioxide markets have been highly successful. The CO_2 reductions required by the legislation would bring emissions back to 1990 levels by 2007, and the costs of implementing such measures would likely be dwarfed by the resulting benefits of industrial innovation.[20]

A tax credit that rewards fuel-efficient vehicles, including hybrid vehicles now on the market, could be used along with a tax penalty for inefficient vehicles to encourage purchases of fuel-efficient vehicles and the development of new technologies.

Legislation that controls the four major pollutants from power plants in an integrated package will help reduce regulatory uncertainties for electric generators and will be less costly than separate programs for each pollutant. Although voluntary action by companies is an attractive idea, in the last 10 years, voluntary actions have failed to reduce carbon dioxide emissions in the United States. Instead, emissions have increased by 15.5 percent since 1990, with an annual average increase of 1.5 percent since 1990, and they continue to increase.[21] EIA recently released data showing an increase of 2.7 percent in U.S. carbon dioxide emissions from 1999 to 2000. Solutions will become more costly and difficult if mandatory emissions reductions are not enacted now.

111

Policy Options

The ultimate solutions to meeting the nation's energy needs cost-effectively and reducing GHG emissions must be based on private-sector investment bolstered by well-targeted government R&D and incentives for emerging clean energy technologies. The United States now has the opportunity to build a sustainable energy future by engaging and stimulating the tremendous innovative and entrepreneurial capacity of the private sector. Advancing clean energy technologies requires a stable and predictable economic environment.

Research and Development Funding

Federal funding and leadership for renewable energy and energy efficiency projects has resulted in several notable successes, such as the U.S. Environmental Protection Agency's (EPA) Energy Star and Green Lights Programs, which have been emulated in a number of countries. Fifteen percent of the public-sector building space in the country has now signed up for the Energy Star buildings program, saving more than 21 billion kWh of energy in 1999 and reducing carbon emissions by about 4.4 million metric tons, which according to EPA, has resulted in $1.6 billion in energy bill savings. Despite these achievements, funding in this area has been scant and so uneven as to discourage private sector involvement. By increasing funding for these EPA programs, their scope could be considerably expanded.

The Bush administration's proposed cuts in its 2002 fiscal year budget for DOE's renewable energy and energy efficiency programs would harm existing public-private partnerships as well as R&D. This budgetary roller coaster harms all investments and sends mixed signals to industry.[22] Steadily increasing funding would transform the clean energy sector from a good idea to a pillar of the new economy.

Tax Incentives

The R&D tax credit, which goes to companies based on their R&D expenditures, has proven remarkably effective and popular with private industry, so much so that there is a strong consensus in both Congress and the administration to make this credit permanent. To complement this support of private-sector R&D, tax incentives directed toward those who use the technologies would provide the "demand pull" needed to accelerate the technology transfer process and the rate of market development.

Currently, non-R&D federal tax expenditures aimed at the production and use of energy have an unequal distribution across primary energy sources, distorting the market in favor of conventional energy technologies. Renewable fuels make up 4 percent of the United States' energy supply, yet they receive only 1 percent of federal tax expenditures and direct fiscal spending combined (see Table 1).[23] The largest single tax credit in 1999 was the Alternative Fuel Production Credit, which totaled more than $1 billion.[24] This income tax credit, which has gone primarily to the natural gas industry, was designed to reduce dependence on foreign energy imports by encouraging the production of gas, coal, and oil from unconventional sources (such as tight gas formations and coalbed methane) within the United States. Support for the production and further development of renewable fuels, all found domestically, would have a greater long-term effect on the energy system than any expansion of fossil-fuel capacity.

A production tax credit (PTC) of 1.7 cents/kWh now exists for electricity generated from wind power and "closed loop" biomass (biomass from dedicated energy crops and chicken litter). The wind power credit, in particular, has proven successful in encouraging strong growth of U.S. wind energy over the last several years—with a 30-percent increase in 1998 and a 40-percent increase in 1999. Approximately 2,000 megawatts (MW) of wind energy will be under development or proposed for completion before the end of 2001 (a 40-percent increase from 2000), when the federal wind energy PTC is scheduled to expire. Currently, Germany has twice the U.S. installed wind energy capacity, and the major wind-turbine manufacturers are now in Europe.[25]

This production credit should be expanded to include electricity produced by "open loop" biomass (including agricultural and forestry residues but excluding municipal solid waste), solar energy, geothermal energy, and landfill gas. The extension and expansion of PTC has recently been garnering strong and consistent support in the U.S. Congress. Investment tax incentives are also needed for smaller-scale renewable energy systems, such as residential photovoltaic panels and solar hot-water heaters, as well as small wind systems used in commercial and farm applications. In these cases, an investment credit in capital or installation expenditures is preferable to a production credit based on electricity generated, due to the relatively high capital cost of these smaller-scale renewable technologies and the fact that the electricity and heat produced is used directly.

Many new energy-efficient technologies have been commercialized in recent years or are nearing commercialization. Tax incentives can help manufacturers justify mass marketing and help buyers and manufacturers offset the relatively high initial capital and installation costs for new technologies. A key element in designing the credits is for only high-efficiency products to be eligible. If eligibility is set too low, there may not be enough energy savings to justify the credits. These tax credits should have limited duration and be reduced in value over time, because once these new technologies become widely available, costs should decline. In this manner, the credits will help innovative technologies get established in the marketplace but will not become permanent subsidies.

Recent federal tax credit legislation to encourage the use of high-efficiency technologies includes incentives for highly efficient clothes washers, refrigerators, and new homes; innovative building technologies such as furnaces, stationary fuel cells, gas-fired pumps, and electric heat-pump water heaters; and investments in commercial buildings that have reduced heating and cooling costs. The incentives currently being proposed in Congress and by the administration will have a relatively modest direct impact on energy use and CO_2 emissions. Savings may only amount to 0.3 quadrillion Btu of energy and 5 million metric tons of carbon emissions per year by 2012. How-

Table 1. U.S. energy consumption and federal expenditures

Fuel source	Primary energy supply 1998 consumption		Direct expenditures and tax expenditures (1999)	
	Value (quads, quadrillion Btu)	Percent	Value (millions of dollars)	Percent
Oil	36.57	40	263	16
Natural gas	21.84	24	1,048	65
Coal	21.62	24	85	5
Oil, gas, and coal combined			205	13
Nuclear	7.16	8	0	–
Renewables	3.48	4	19	1
Total	90.67	100	1620	100

NOTE: The Alternative Fuels Credit accounted for $1,030 of the $1,048 in expenditures for natural gas. Oil, gas, and coal combined includes expenditures that were not allocated to any one of the three individual fuels. Research and development are not included in direct expenditures and tax expenditures. Btu = British thermal units.

SOURCE: Energy Information Administration, *Federal Financial Interventions and Subsidies in Energy Markets 1999: Primary Energy*, U.S. Department of Energy (Washington, D.C., 1999).

ever, if these proposed tax credits help to establish innovative products in the marketplace and reduce the first-cost premium so that the products are viable after the credits are phased out, then the indirect impacts could be many times greater than the direct impacts. It has been estimated that total energy savings could reach 1 quadrillion Btu by 2010 and 2 quadrillion Btu by 2015 if these credits are successfully implemented.[26]

Vehicle Fuel Economy Standards

New vehicles with hybrid gasoline-electric power systems are now produced commercially, and fuel cell–electric vehicles are being produced in prototype quantities. These vehicles combine high-efficiency electric motors with revolutionary power systems to produce a new generation of motor vehicles that are vastly more efficient than today's simple cycle combustion systems. The potential for future hybrid and fuel-cell vehicles to achieve up to 100 miles per gallon (mpg) is believed to be both technically and economically viable in the near future. In the long term, fuel-cell vehicles running directly on hydrogen promise to allow motor vehicle use with very low fuel-cycle emissions.

The improvements in fuel economy that these new vehicles offer will help to slow growth in petroleum demand, reducing our oil import dependency and trade deficit. While the Partnership for a New Generation of Vehicles helped generate some vehicle technology advances, an increase in the Corporate Average Fuel Economy (CAFE) standard, which has been stagnant for 16 years now, is required to provide an incentive for companies to bring these new vehicles to market rapidly.

Recent analyses of the costs and benefits of motor vehicles with higher fuel economy have been conducted by the Union of Concerned Scientists, Massachusetts Institute of Technology, the Office of Technology Assessment, and Oak Ridge National Lab/ACEEE.[27] These studies have generally concluded that

with longer-term technologies, motor vehicle fuel economy can be raised to 45 mpg for cars with a retail price increase of $500 to $1,700 per vehicle, and to 30 mpg for light trucks with a retail price increase of $800 to $1,400 per vehicle.[28] These improvements could be the basis for a new combined fuel economy standard of 40 mpg for both cars and light trucks. The combined standard could be accomplished between 2008 and 2012. The net cost would be negligible once fuel savings are factored in, if the auto industry is given adequate time to retool for this new generation of vehicles. A lower combined standard could be implemented sooner and then raised incrementally each year to achieve the 40-mpg goal by 2012.

Tax credits for hybrid-electric vehicles, battery-electric vehicles, and fuel-cell vehicles are an important part of the puzzle. These funds could, in principle, be raised through a revision of the archaic "gas guzzler" tax, which does not apply to a significant percentage of the light-duty car and truck fleet. The tax penalty and tax credit in combination could be a revenue-neutral "fee-bate" scheme—similar to one recently proposed in California—that would simultaneously reward economical vehicles and penalize uneconomical ones.

Efficiency Standards

A critical strategy for effectively promoting energy efficiency is implementing new standards for buildings, appliances, and equipment. Tax credits do not necessarily remove all the market barriers that prevent clean energy technologies from spreading throughout the marketplace. These barriers include lack of awareness, rush purchases when an existing appliance breaks down, and purchases by builders and landlords who do not pay utility bills.

Significant advances in the efficiency of heating and cooling systems, motors, and appliances have been made in recent years, but more improvements are technologically and economically feasible. A clear federal statement of desired improvements in system efficiency would remove uncertainty about and reduce costs of implementing these changes. Under such a federal mandate, efficiency standards for equipment and appliances could be gradually increased, helping to expand the market share of existing high-efficiency systems.[29]

Extensive transmission and distribution networks waste a significant percentage of electricity generated by traditional power plants. Smaller-scale systems can be located closer to where the energy is used.

Standards remove inefficient products from the market and still leave consumers with a full range of products and features from which to choose. Building, appliance, and equipment efficiency standards have proven to be one of the federal government's most effective energy-saving programs. Analyses by DOE and others indicate that in 2000, appliance and equipment efficiency standards saved 1.2 quadrillion Btu of energy (1.3 percent of U.S. electric use) and reduced consumer energy bills by approximately $9 billion, with energy bill savings far exceeding any increase in product cost. By 2020, standards already enacted will save 4.3 quadrillion Btu per year (3.5 percent of projected U.S. energy use) and reduce peak electric demand by 120,000 MW (more than a 10-percent reduction). ACEEE estimates that by 2020, energy use could be reduced by 1.0 quadrillion Btu by quickly adopting higher standards for equipment that is currently covered under federal law, such as central air conditioners and heat pumps, and by adopting new standards for equipment not covered, such as torchiere (halogen) light fixtures, commercial refrigerators, and appliances that consume power while on standby.[30] Energy bills would decline by approximately $7 billion per year by 2020.[31]

A Renewable Portfolio Standard

The Renewable Portfolio Standard (RPS) is akin to the efficiency standards for vehicles and appliances that have proven successful in the past. A gradually increasing RPS is designed to integrate renewables into the marketplace in the most cost-effective fashion, and it ensures that a growing proportion of electricity sales is provided by renewable energy. An RPS provides the one true means to use market forces most effectively—the market picks the winning and losing technologies.

A number of studies indicate that a national renewable-energy component of 2 percent in 2002, growing to 10 percent in 2010 and 20 percent by 2020, that would include wind, biomass, geothermal, solar, and landfill gas, is broadly good for business and can readily be achieved.[32] States that decide to pursue more aggressive goals could be rewarded through an additional federal incentive program. In the past, federal RPS legislation has been introduced in Congress and it was proposed by the Clinton administration, but it has yet to be re-introduced by either this Congress or the Bush administration.

Including renewables in the United States' power-supply portfolio would protect consumers from fossil fuel price shocks and supply shortages by diversifying the energy options. A properly designed RPS will also create jobs at home and export opportunities abroad. To achieve compliance, a federal RPS should use market dynamics to stimulate innovation through a trading system. National renewable energy credit trading will encourage development of renewables in the regions of the country where they are the most cost-effective, while avoiding expensive long-distance transmission.

The coal, oil, natural gas, and nuclear power industries continue to receive considerable government subsidies, even though they are already well established. Without RPS or a similar mechanism, many renewables will not be able to survive in an increasingly competitive electricity market focused on producing power at the lowest direct cost. And while RPS is designed to deliver renewable that are most ready for the market, additional policies will still be needed to support emerging renewable technologies, like photovoltaics, that have enormous potential to become commercially competitive.

RPS is the surest market-based approach for securing the public benefits of renewables while supplying the greatest amount of clean power at the lowest price. It creates an ongoing incentive to drive down costs by providing a dependable and predictable market. RPS will promote vigorous competition among renewable energy developers and technologies to meet the standard at the lowest cost.

Analysis of the RPS target for 2020 shows renewable energy development in every region of the country, with most coming from wind, biomass, and geothermal sources. In particular, the Plains, Western, and mid-Atlantic states would generate more than 20 percent of their electricity from renewables.[33] Texas has become a leader in developing and implementing a successful RPS that then-Governor Bush signed into law in 1999. The Texas law requires electricity companies to supply 2,000 MW of new renewable resources by 2009, and the state is actually expected to meet this goal by the end of 2002, seven years ahead of schedule. Nine other states have signed an RPS into law: Arizona, Connecticut, Maine, Massachusetts, Nevada, New Jersey, New Mexico, Pennsylvania, and Wisconsin. Minnesota and Iowa have a minimum renewables requirement similar to RPS, and legislation that includes RPS is pending in several other states.

While the participation of 12 states signals a good start, this patchwork of state policies would not be able to drive down the costs of renewable energy technologies and move these technologies fully into the marketplace. Also, state RPS policies have differed substantially from each other thus far. These differences could cause significant market inefficiencies, negating the cost savings that a more comprehensive, streamlined, market-based federal RPS package would provide.

Small-Scale Distributed Energy Generation and Cogeneration

Small-scale distributed electricity generation has several advantages over traditional central-station utility service. Distributed generation reduces energy losses incurred by sending electricity long distances through an extensive transmission and distribution network (often an 8- to 10-percent loss of energy). In addition, generating equipment located close to the end use allows waste heat to be utilized (a process called cogeneration) to meet heating and hot water demands, significantly boosting overall system efficiency.

Distributed generation has faced several barriers in the marketplace, most notably from complicated and expensive utility interconnection requirements. These barriers have led to a push for national safety and power quality standards, now being finalized by the Institute of Electrical and Electronics Engineers (IEEE). Although the adoption of these standards would significantly decrease the economic burden on manufacturers, installers, and customers, the utilities are allowed discretion in adopting or rejecting them.

In designing credits, highest priority should go to renewable or fossil fuel systems that utilize waste heat through combined heat and power (CHP) designs. While a distributed generation system may achieve 35- to 45-percent electrical efficiency, the addition of heat utilization can raise overall efficiency to 80 percent. Industrial CHP potential is estimated to be 88,000 MW, the largest sectors being in the chemicals and paper industries. Commercial CHP potential is estimated to be 75,000 MW, with education, health care, and office building applications making up the most significant percentages.[36]

A National Public Benefits Fund

Electric utilities have historically funded programs to encourage the development of a host of clean energy technologies. Unfortunately, increasing competition and deregulation have led utilities to cut these discretionary expenditures in the last several years. Total utility spending on demand-side management programs fell more than 50 percent from 1993 to 1999. Utilities should be encouraged to invest in the future through rewards (such as tax incentives) for companies that reinvest profits and invigorate the power sector.[37] A national public benefits fund could be financed through a national, competitively neutral wires charge of $0.002 per kWh.

Cost and Benefit Analyses

A range of recent studies are all coming to the same conclusions: that simple but sustained standards and investments in a clean energy economy are not only possible but would also be highly beneficial to future prosperity in the United States.[37] If energy policies proceed as usual, the nation is expected to increase its reliance on coal and natural gas to meet strong growth in electricity use (42 percent by 2020). To meet this demand, it is estimated that 1,300 300-MW power plants would need to be built, with electricity generated by non-hydro renewables only

increasing from 2 percent today to 2.4 percent of total generation in 2020.[38] A set of clean energy polices could meet a much larger share of our future energy needs, with energy efficiency measures projected to almost completely offset the projected growth in electricity use.[39] A clean energy strategy would significantly reduce emissions from utilities. In fact, through a steady shift to clean energy production, power plant carbon dioxide reductions (as proposed in the current legislation before Congress), would not be difficult or expensive to meet.[40]

The United States is becoming increasingly dependent on oil—$120 billion was spent on oil imports last year. Making renewable energy sources a larger part of the energy economy would enable the nation to provide for its own energy needs.

Recent analysis by the Union of Concerned Scientists focused on the costs and environmental impacts of a package of clean energy polices and how the package would affect fossil fuel prices and consumer energy bills. They found that using energy more efficiently and switching from fossil fuels to renewable energy sources will save consumers money by decreasing energy use.[41] A whole-economy analysis carried out by the International Project for Sustainable Energy Paths has also shown that Kyoto-type targets can easily be met, with a net increase of 1 percent in the nation's 2020 GDP, by implementing the right policies.[42]

One of the greatest advantages that energy efficiency and renewable energy sources offer over new power plants, transmission lines, and pipelines is the ability to deploy these technologies very quickly. They can be installed—and benefits can be reaped—immediately.[43] In addition, reductions in CO_2 emissions will have a "clean cascade" effect on the economy because many other pollutants are emitted during fossil fuel combustion.

The renewable and energy-efficient technologies and policies described here have already proven successful and cost-effective at the national and state levels. Supporting them would allow the United States to cost-effectively meet GHG emission targets while providing a sustainable, clean energy future.[44]

We stand at a critical point in the energy, economic, and environmental evolution of the United States. Renewable energy and energy efficiency are now not only affordable, but their expanded use will also open new areas of innovation. Creating opportunities and a fair marketplace for a clean energy economy requires leadership and vision. The tools to implement this evolution are now well known. We must recognize and overcome the current road blocks and create the opportunities needed to put these renewable and energy-efficient measures into effect.

This article is based on testimony provided by D. M. Kammen to the U. S. Senate Commerce, Science and Transportation (July 10, 2001) and U. S. Senate Finance (July 11, 2001) Committees. Antonia V. Herzog and Timothy E. Lipman are postdoctoral researchers and Jennifer L. Edwards is a research assistant at the Renewable and Appropriate Energy Laboratory (RAEL), Energy and Resources Group (ERG), at the University of California at Berkeley. Daniel M. Kammen is a professor of Energy and Society with ERG and a professor of Public Policy with the Goldman School of Public Policy. Address correspondence to D. M. Kammen, 310 Barrows Hall, University of California, Berkeley, CA 94720-3050, or dkammen@socrates.berkeley.edu. Additional material can be found at http://socrates.berkeley.edu/~rael.

NOTES

1. Intergovernmental Panel on Climate Change (IPCC), *Climate Change 2001: The Scientific Basis* (Working Group I of the IPCC, World Meteorological Organization - U.N. Environment Program, Geneva), January 2001; and National Research Council (NRC), *Climate Change Science: An Analysis of Some Key Questions*, Committee on the Science of Climate Change, (National Academy Press, Washington, D.C., 2001). For more information on energy and climate change, see J. P. Holdren, "The Energy-Climate Challenge: Issues for the New U.S. Administration," *Environment*, June 2001, 8–21.

2. D. M. Kammen and R. M. Margolis, "Evidence of Under-Investment in Energy R&D Policy in the United States and the Impact of Federal Policy," *Energy Policy* 27 (1999), 575–84; and R. M. Margolis and D. M. Kammen, "Underinvestment: The Energy Technology and R&D Policy Challenge," *Science* 285 (1999), 690–92.

3. This work appeared in two influential forms that reached dramatically different audiences: A. B. Lovins, "Energy Strategy: The Road Not Taken," *Foreign Affairs* (1976), 65–96; and A. B. Lovins, *Soft Energy Paths: Toward a Durable Peace* (New York: Harper Colophon Books, 1977).

4. A. V. Herzog, T. E. Lipman, and D. M. Kammen, "Renewable Energy Sources," in *Our Fragile World: Challenges and Opportunities for Sustainable Development*, forerunner to the Encyclopedia of Life Support Systems (EOLSS), Volume 1, Section 1 (UNESCO-EOLSS Secretariat, EOLSS Publishers Co. Ltd., 2001).

5. P. Mazza, *Harvesting Clean Energy for Rural Development: Wind,* Climate Solutions Special Report, January 2001.

6. Kammen and Margolis, note 2 above.

7. U.S. Department of Energy (DOE), *Renewable Energy Technology Characterizations*, Topical Report Prepared by DOE Office of Utility Technologies and EPRI, TR-109496, December 1997.

8. B. Haevner and M. Zugel, *Predictably Unpredictable: Volatility in Future Energy Supply and Price From California's Over Dependence on Natural Gas*, CALIPIRG Charitable Trust Research Report, September, 2001).

9. R. L. Ottinger et al., *Environmental Costs of Electricity* (New York: Oceana Publications, Inc., 1991); U.S. Department of Energy (DOE), *Annual Energy Outlook 2000*, DOE/EIA-0383 (00), Energy Information Administration, Washington, D. C., December 2000; and U.S. Department of Energy, 1997.

10. Reuters News Service, "Fuel Cells and New Energies Come of Age Amid Fuel Crisis," 11 September 2000.

11. National Energy Policy, "Reliable, Affordable, and Environmentally Sound Energy for America's Future," Report of the National Energy Policy Development Group, Office of the President, May 2001.

12. Kammen and Margolis, note 2 above.

13. International Institute for Applied Systems Analysis/World Energy Council, *Global Energy Perspectives to 2050 and Beyond* (Laxenburg, Austria, and London, 1995).

14. R. D. Duke and D. M. Kammen, "The Economics of Energy Market Transformation Initiatives," *The Energy Journal* 20 (1999): 15–64.

15. Interlaboratory Working Group, *Scenarios for a Clean Energy Future* (Oak Ridge, Tenn.: Oak Ridge National Laboratory; and Berkeley, Calif.: Lawrence Berkeley National Laboratory), ORNL/CON-476 and LBNL-44029, November 2000.

16. S. Nadel and H. Geller, "Energy Efficiency Polices for a Strong America," American Council for an Energy-Efficient Economy (ACEEE), May 2001 (draft).

17. Clean Energy Partnerships, *A Decade of Success*, Office of Energy Efficiency and Renewable Energy, DOE/EE-0213 (Washington, D.C., 2000).

18. Nadel and Geller, note 16 above.

19. President's Committee of Advisors on Science and Technology (PCAST), *Federal Energy Research and Development for the Challenges of the Twenty-First Century*, Washington, D.C., Energy Research and Development Panel, November 1997.

20. F. Krause, S. DeCanio, and P. Baer, "Cutting Carbon Emissions at a Profit: Opportunities for the U.S." (El Cerrito, Calif.: International Project for Sustainable Energy Paths), May 2001; and A. P. Kinzig and D. M. Kammen, "National Trajectories of Carbon Emissions: Analysis of Proposals to Foster the Transition to Low-Carbon Economies," *Global Environmental Change* 8 (3) (1998): 183–208.

21. Energy Information Administration (EIA), *U.S. Carbon Dioxide Emissions from Energy Sources 2000 Flash Estimate*, based on data from the Monthly Energy Review (May 2001) and the Petroleum Supply Annual 2000, DOE (Washington, D.C.), June 2001.

22. Kammen and Margolis, note 2 above.

23. This does not include revenue outlays for the Alcohol Fuels Excise Tax, which reduces the tax paid on ethanol-blended gasoline. Most ethanol used in the United States is produced from corn, and the GHG emissions impact is uncertain and has been shown to be negligible (M. Delucchi, *A Revised Model of Emissions of Greenhouse Gases from the Use of Transportation Fuels and Electricity*, Institute of Transportation Studies, UCD-ITS-RR-97-22, (Davis, Calif., 1997)); and EIA, *Federal Financial Interventions and Subsidies in*

Energy Markets 1999: Primary Energy, DOE (Washington, D.C., 1999).

24. Established by the Windfall Profit Tax Act of 1980, this tax credit is $3 per barrel of oil equivalent produced, and it phases out when the price of oil rises to $29.50 per barrel (1979 dollars).

25. American Wind Energy Association web site, available at http://www.awea.org, accessed on September 8, 2001.

26. Nadal and Geller, note 16 above.

27. J. Mark, "Greener SUVs: A Blueprint for Cleaner, More Efficient Light Trucks," Union of Concerned Scientists, 1999; M. A. Weiss, J. B. Heywood, E. M. Drake, A. Schafer, and F. F. AuYeung, "On the Road in 2020: A Lifecycle Analysis of New Automobile Technologies," Energy Laboratory, Massachusetts Institute of Technology, MIT EL 00-003 (Cambridge, Mass., October 2000); Office of Technology Assessment, *Advanced Vehicle Technology: Visions of a Super-Efficient Family Car*, U.S. Congress, OTA-ETI-638 (Washington, D.C., September 1995); and D. L. Greene and J. Decicco, "Engineering-Economic Analyses of Automotive Fuel Economy Potential In The United States," *Annual Review of Energy and the Environment*, 25: (2000) 477–536.

28. Greene and Decicco, note 27 above; and Interlaboratory Working Group, note 15 above.

29. S. L. Clemmer, D. Donovan, and A. Nogee, "Clean Energy Blueprint: A Smarter National Energy Policy for Today and the Future, Phase I," Union of Concerned Scientists and Tellus Institute, June 2001.

30. K. B. Rosen and A. K. Meier, *Energy Use of Televisions and Videocassette Recorders in the U.S.*, DOE, LBNL-42393, (Berkeley, Calif.: Lawrence Berkeley National Laboratory), March 1999.

31. Nadal and Geller, note 16 above.

32. Clemmer, Donovan, and Nogee, note 29 above; S. L. Clemmer, A. Nogee, and M. Brower, "A Powerful Opportunity: Making Renewable Electricity the Standard," Union of Concerned Scientists, January 1999; and A. Nogee, S. Clemmer, B. Paulos, and B. Haddad, "Powerful Solutions: 7 Ways to Switch America to Renewable Energy," Union of Concerned Scientists, January 1999.

33. Clemmer, Nogee, and Brower, note 32 above.

34. Distributed generation reflects a new way to manage energy supply and demand. Instead of the old system of large capital-intensive central-station power plants, the improvements in energy efficiency, renewable energy, and small, modular5 "micro-turbines" that burn gas, as well as hydro and other resources, energy supplies could be located closer to users, reducing transmission losses, improving system reliability, and energy security.

35. R. K. Dixon, Office of Power Technologies, U.S. Department of Energy, Second International CHP Symposium, Amsterdam, Netherlands, May 2001.

36. Kammen and Margolis, note 2 above.

37. Interlaboratory Working Group, note 15 above; Krause, DeCanio, and Baer, note 20 above; and Clemmer, Donovan, and Nogee, note 29 above.

38. National Energy Policy, note 11 above.

39. Clemmer, Donovan, and Nogee, note 29 above.

40. Ibid.

41. Ibid.

42. Krause, DeCanio, and Baer, note 20 above.

43. Kinzig and Kammen, note 20 above.

44. P. Baer, et al. "Equity and Greenhouse Gas Responsibility," *Science* 289 (2000), 2287.

Fossil Fuels and Energy Independence

To become self-sufficient in energy resources, the United States needs to combine the conservation strategies learned over the past 30 years with the latest fuel-saving technologies.

B. Samuel Tanenbaum

Following the outbreak of the Arab-Israeli war in 1973, petroleum-producing Arab nations cut back oil production and imposed an embargo that led to an energy crisis in the United States. In response, President Nixon directed Dixy Lee Ray, then-chairman of the Atomic Energy Commission, to "undertake an immediate review of federal and private energy research and development activities" and asked her to submit a national energy plan by December 1 of that year.

Based on the input of hundreds of individuals in workshops, review panels, and government agencies, Ray prepared a comprehensive program, "The Nation's Energy Future." She noted that it was intended to "mobilize the nation's resources toward the attainment of a capacity for energy self-sufficiency by 1980." To achieve that goal, the report recommended five tasks: (1) conserve energy by reducing consumption and adopting processes that use fuel more efficiently; (2) raise domestic production of oil and natural gas; (3) increase the use of coal; (4) greatly expand the use of nuclear power; and (5) promote the use of renewable energy sources.

These recommendations look remarkably similar to suggestions being made by the Bush administration today. Thus it should be useful to review what happened in the 1970s that prevented the country from achieving independence in energy resources and to examine what steps may be taken at this time.

Historic patterns of energy consumption

In 1972, energy consumption in the United States totaled 72 quadrillion Btu (72 followed by 15 zeros)—or, more simply, 72 Quads. [One Btu (British thermal unit) is the energy required to raise the temperature of one pound of water by one degree Fahrenheit.] Of this amount, 45.5 percent was derived from oil, 32.3 percent from natural gas, 17.2 percent from coal, and the remainder from nuclear reactors and renewable sources. In other words, fossil fuels—oil, natural gas, and coal—supplied 95 percent of this nation's energy needs.

At that time, the United States produced all the coal and natural gas it consumed, and it even exported a substantial amount of coal to other countries. The major problem then (as now) was that the country relied heavily on imported oil to supplement domestic production. Oil imports had been growing rapidly, and during the first half of 1973, the import rate exceeded 6 million barrels per day (MBPD), representing about one-third of the nation's oil consumption. [Note that 1 MBPD of oil yields 2.1 Quads of energy per year.]

The 1973 report to the president projected that if no action were taken, the demand for energy would continue to rise at its historic rate, reaching about 100 Quads in 1980 and 200 Quads in 2000. At the same time, oil imports were predicted to grow from 6.5 MBPD in 1970 to about 12 MBPD in 1980 and 24 MBPD in 2000.

To achieve energy self-sufficiency, this country had to eliminate the demand for imported oil. To do so by 1980, Ray's plan laid out three proposals. First, 4.7 MBPD could be saved through conservation measures—such as extra insulation in buildings, more fuel-efficient cars and trucks, and more efficient industrial processes. Second, 1.5 MBPD of oil would be replaced with energy from coal, nuclear, and renewable sources. Third, domestic oil production would need to be raised by 5.8 MBPD.

The report further noted that "self-sufficiency based on fossil fuels can only be temporary. Though large, these resources are finite." For energy independence over the long term, it recommended incentives for the development of renewable energy sources, but the major replacement of fossil fuels was expected to come from rapid growth in nuclear power. Even before the 1973 energy crisis, the Department of Interior projected that the number of large nuclear power plants would grow from 29 in 1973 to 132 in 1980 and 1,200 in 2000.

So, what happened? The energy crisis of 1973 was short-lived, but it led to a two-year drop in total energy consumption by about 5 percent. Thereafter, energy usage rose again, until the Iranian revolution of 1978–79 disrupted that country's oil production. Those circumstances stimulated additional conservation efforts in the United States, so that energy usage fell from 81 Quads in 1979 to 78 Quads in 1980, reaching a low of 73 Quads in 1982 and 1983.

Fossil fuels provided 95 percent of all the energy consumed in the United States in 1972, and they now supply about 85 percent of the energy used.

Since then, energy consumption increased, but at a slower pace than expected. In 2000, the total usage was 99 Quads—less than half the amount predicted in 1973—reflecting the effectiveness of conservation measures.

The efforts to replace some oil with other energy sources also appear to have been successful. Between 1973 and 2000, the fraction of energy derived from oil dropped from 46 to 39 percent, while the fraction obtained from coal increased from 17 to 23 percent. These changes represent striking reversals of the trends of earlier decades.

On the other hand, the goal to achieve energy independence was not met for several reasons. In particular, despite efforts to encourage exploration in the United States and completion of the 800-mile trans-Alaska pipeline in 1977, domestic oil production fell from its peak value of 10.9 MBPD in 1970 to 9.8 MBPD in 1980 and 8.1 MBPD in 2000. Between 1973 and 1980, overall oil consumption decreased slightly, but oil imports rose from 6.2 to 6.4 MBPD, before dropping to a low of 4.3 MBPD in 1985, thanks to conservation efforts. Thereafter, the situation gradually worsened, as U.S. oil production declined and consumption increased. Consequently, by 2000, imports reached 10.7 MBPD.

Furthermore, the growth of nuclear power had been greatly overestimated. Instead of the 1,200 large nuclear power plants projected for the year 2000, only 103 plants are currently operating in the United States. Although the last U.S. nuclear power plant came on line in 1996, no new plants were ordered after the accident at Three Mile Island (TMI) in Pennsylvania in March, 1979.

No one was killed by the mishap at TMI, but it was followed by the much more serious incident in Chernobyl, Ukraine, in April, 1986. In the latter case, 31 people died immediately, hundreds were hospitalized, and many more were expected to die of cancer from exposure to the giant cloud of radiation released.

While nuclear power plants can be made safer, the main hurdle is their high cost. In the 1980s, the cost of completing a nuclear plant with a capacity of one gigawatt (one billion watts) reached $5–6 billion—more than ten times the price tag for a conventional power plant with equivalent capacity. In addition, the spent, radioactive fuel has been stored at reactor sites, pending agreement on a long-term storage area, and the cost of decommissioning an old plant has grown to be far higher than originally anticipated. Thus, the nuclear power industry blossomed for only about a decade before power companies returned to purchasing fossil fuel plants for electricity.

Fossil fuel reserves

Although the energy policies adopted in the 1970s had a greater effect than most people realize, energy production was (and will continue to be) affected by the availability of fossil fuels. These fuels provided 95 percent of all the energy consumed in the United States in 1972, and they now supply about 85 percent of the energy used. The remainder is obtained from nuclear reactors (8 percent) and renewable sources (7 percent). It is therefore important to examine the reserves of oil, natural gas, and coal, and to look for potential substitutes.

Oil reserves. Following the discovery of huge oil deposits in Alaska in 1970, proven reserves of oil in the United States have steadily declined,

from about 210 Quads in 1973 to about 120 Quads today. During this period, domestic oil companies discovered more than 300 Quads, but they produced over 400 Quads, leading to the decline in reserves.

Domestic reserves of natural gas have fallen from about 300 Quads in 1973 to about 170 Quads today.

By contrast, major new discoveries of oil continue to be made overseas. Over the past 30 years, proven oil reserves worldwide have increased from about 3,800 Quads to about 5,900 Quads. Because the worldwide demand for oil is roughly 150 Quads per year, oil is still plentiful on a global scale, and the cost of imported oil remains low.

Natural gas reserves. As a fuel, natural gas offers many advantages. It is clean-burning, inexpensive, and conveniently delivered by pipeline to consumers. The use of natural gas in the United States soared in the decades following the Second World War, as a nationwide network of pipelines was constructed.

Thus far, this country has been able to meet its needs for natural gas through domestic production and modest imports from Canada. Since 1973, U.S. wells have produced over 600 Quads of natural gas. Even so, as with oil, new discoveries have not kept pace with production. Thus, domestic reserves of natural gas have fallen from about 300 Quads in 1973 to about 170 Quads today.

On the other hand, worldwide reserves of natural gas have doubled from about 2,500 Quads in 1973 to about 5,000 Quads today. Given that the global demand is only about 100 Quads per year, natural gas, like oil, is plentiful worldwide. Consequently, the United States may become increasingly reliant on imports of natural gas in the years ahead.

One problem is that, unlike oil, natural gas cannot be shipped easily in tankers. Before shipment, the gas must be compressed and liquefied; refrigerated ships must be used to keep it cool during transit; and special facilities are needed to receive the liquefied fuel and permit it to expand safely for shipment in a normal pipeline. It is therefore much more expensive to import natural gas than oil from overseas. Thus the huge quantities of natural gas available in Russia and Middle Eastern countries cannot be imported as readily as oil.

The United States has plentiful reserves of coal, but coal mining can adversely affect the surrounding land and water unless special protective measures are taken.

Coal reserves. The United States has roughly 25 percent of the world's coal reserves, corresponding to about 7,000 Quads. Domestic annual consumption of coal is only 22 Quads, so there is no likelihood of a coal shortage anytime soon. Coal is also the cheapest fossil fuel. In recent years, the price of coal used in power plants has been about one dollar per million Btu (MBtu), while that of natural gas has been over two dollars per MBtu. Oil has been even more expensive.

Despite the cost advantage, traditional markets for coal in the United States eroded with time. In the 1950s, coal-fired steam locomotives were replaced by more efficient diesel engines, and coal-based heating systems were replaced by cleaner natural gas heaters wherever pipelines made the fuel available.

Today, about 90 percent of the coal sold in the United States is used to produce electricity. New coal-fired plants continue to be built, although tighter requirements for pollution-control equipment have added to the expense and complications of building them. Moreover, coal mining has to be conducted with improved water-management and land-restoration practices to minimize damage to the environment.

Alternative fuels. If coal is heated in hydrogen, it can produce either an oil substitute known as *syncrude* or a gaseous material known as *coal gas*, which has about half the energy content of conventional natural gas. These products are still too costly to be competitive with available fuels, but they may become important during shortages of oil and natural gas.

Another potential fuel is *shale oil*. Shale is a sedimentary rock rich in the organic compound kerogen. Parts of the western United States have rich deposits of shale oil—30 gallons or more of oil per ton of shale, amounting to several times more than the total domestic reserves of conventional oil. To extract shale oil, the rock is mined, crushed, and heated to about 400 1/4°C (750°F). But it takes about six tons of shale to provide as much energy as one ton of coal, and it's not yet possible to produce large amounts of shale oil in a way that protects the environment or competes in price with oil pumped from a well.

Today's technologies, if widely adopted, could reduce the use of fossil fuels in the United States by up to 50 percent.

An asphaltlike oil, called *bitumen*, can be obtained from tar sands. There are sizable deposits of these sands in the United States, but the world's largest deposits are in Canada. Bitumen can be extracted simply by mixing the mined sand with hot water or steam. Again, the major drawback is that the oil cannot be produced at a price that is competitive with conventional oil.

Reducing our consumption of fossil fuels

The annual worldwide consumption of fossil fuels today represents about one percent of known reserves. This is about the same figure as in 1972, because known reserves continue to increase worldwide. Interpretation of these amounts, however, is complicated by the fact that the newly discovered reserves are mainly in remote locations, including at great depths below the surface of the land and sea. Thus, tapping these supplies would require expensive extraction techniques, which would be justified only when the prices of oil and natural gas become sufficiently high.

In any case, there are several reasons why we need to dramatically reduce our consumption of fossil fuels, replacing them with other energy sources. For a country such as the United States, which is heavily dependent on imported oil and is gradually increasing its imports of natural gas, the near-term benefit would be the achievement of energy self-sufficiency. The long-term benefit is that we would save on fuels that are finite in availability. In addition, we would lessen the production of carbon dioxide, which is produced by burning fossil fuels and contributes to global warming—a phenomenon that has been predicted to cause serious problems in the future.

How, then, can we reduce our dependence on fossil fuels? As long as prices for these fuels remain relatively low, most consumers do not see the need to use energy efficiently or to switch to renewable sources. In the last 25 years, however, the United States has demonstrated that sensible steps to conserve energy can be remarkably effective when mandated, or when adequate incentives are provided. Consider the following examples:

- New standards for insulation and efficiency in heating and air-conditioning systems played a leading role in reducing the use of natural gas by about 20 percent in the early 1980s.

- At the same time, requirements for cars to have better gasoline mileage reduced oil consumption by a similar proportion.

- Tax credits and California's requirement that electric utilities buy power from wind and solar generators at a reasonable rate led to a significant use of these alternative energy sources there.

The Toyota Prius, a gasoline-electric hybrid vehicle, has a mileage rating of about 50 miles per gallon of gasoline. If similar engines were to be installed in most new cars and trucks over the next few years, they would dramatically reduce the fuel consumption.

Today's technologies, if widely adopted, could reduce the use of fossil fuels in the United States by up to 50 percent. For instance, hybrid gasoline-electric cars introduced recently by Toyota and Honda have already become quite popular and show that gasoline mileage can be improved by 50–100 percent without sacrificing performance or comfort. If such engines were required of all new cars and trucks within the next five years, the United States could probably save about one-third of its current consumption of 27 Quads per year for transportation, reducing oil imports by more than 4 MBPD.

New, large wind generators can produce electricity at a price that is competitive with any other means of generating power. If used extensively, they could save enormous amounts of fossil fuel. Alternatively, electrical power can be produced with combined-cycle, gas-turbine generators or cogeneration systems that use natural gas twice as efficiently as conventional steam-turbine generating plants [see "Gaining Power by Thinking Small," THE WORLD & I, March 2002, p. 152].

By replacing electric water heaters, stoves, ovens, and clothes dryers with gas units, up to two-thirds of the energy used for these purposes can be saved. Similarly, replacing old air-conditioning systems, refrigerators, heating systems, and pool pumps with efficient new models can reduce energy use by as much as 50 percent. Fossil fuels can be further saved by using solar hot-water heaters and clothes lines to dry clothes. Moreover, about 75 percent of the energy used in incandescent light bulbs can be saved by replacing them with compact fluorescent bulbs or efficient LED (light emitting diode) lamps.

In warm climates, air-conditioning loads can be significantly reduced by making use of reflective roof surfaces, window coatings, attic exhaust fans, and shade trees. In cold climates, passive solar heating technology can be employed.

Fossil fuel consumption can also be reduced by a number of other recent technologies, but they have not yet gained broad acceptance for reasons of cost or reliability. They include photovoltaic solar panels, which produce electricity directly from sunlight; fuel cells, in which hydrogen or methanol is used to produce electricity at high efficiency, reducing air pollution; and systems that extract energy from waste materials, geothermal sources, or ocean tides. It is also possible that advances in nuclear technology may lead to more extensive use of nuclear energy, by processes of either fission (which splits atomic nuclei) or fusion (which brings atomic nuclei together).

In the long run, it seems likely that solar energy will be used directly to meet most of the energy needs of our world. The amount of sunlight reaching the continental United States averages over 500,000 Btu per square foot per year. A circular area with a 50-mile radius thus receives about 110 Quads per year—enough to meet the energy demands of the entire nation. But until we learn how to harness that energy efficiently and cost-effectively, the above-mentioned technologies are available to reduce this country's energy demand by half, without changing our life-styles. By adopting these approaches, we can almost certainly achieve energy independence.

B. Samuel Tanenbaum is professor of engineering at Harvey Mudd College in Claremont, California.

UNIT 4

Biosphere: Endangered Species

Unit Selections

16. **What Is Nature Worth?**, Edward O. Wilson
17. **Where Wildlife Rules**, Frank Graham Jr.
18. **Invasive Species: Pathogens of Globalization**, Christopher Bright
19. **On the Termination of Species**, W. Wayt Gibbs

Key Points to Consider

- Are there ways to assess the value or worth of living organisms other than those from whom we derive direct benefits (our domesticated plant and animals species)? What are the relationships between economic assessments of the biosphere and moral or value judgments on the preservation of species?

- Explain some of the important philosophical and environmental differences between the development of the National Wildlife Refuge system in the United States and other federal conservation systems. Can you assess the success of the NWRS as opposed to the success of national parks and national forests?

- Why is the spread of invasive species through the global trading network so difficult to control? What suggestions would you make to remedy the problem?

- Why is there a disjunction between the rate of species extinction and the action of policy makers to do something about it? Are there ways to calculate the rate of species loss that will aid policy makers in coming to grips with the problem?

 **Links: www.dushkin.com/online/**
These sites are annotated in the World Wide Web pages.

Endangered Species
http://www.endangeredspecie.com/

Friends of the Earth
http://www.foe.co.uk/index.html

Smithsonian Institution Web Site
http://www.si.edu

World Wildlife Federation (WWF)
http://www.wwf.org

Tragically, the modern conservation movement began too late to save many species of plants and animals from extinction. In fact, even after concern for the biosphere developed among resource managers, their effectiveness in halting the decline of herds and flocks, packs and schools, or groves and grasslands has been limited by the ruthlessness and efficiency of the competition. Wild plants and animals compete directly with human beings and their domesticated livestock and crop plants for living space and for other resources such as sunlight, air, water, and soil. As the historical record of this competition in North America and other areas attests, since the seventeenth century human settlement has been responsible—either directly or indirectly—for the demise of many plant and wildlife species. It should be noted that extinction is a natural process—part of the evolutionary cycle—and not always created by human activity; but human actions have the capacity to accelerate a natural process that might otherwise take millennia.

In the opening article of this unit, one of the world's best-known writers on biological issues asks the central question that will control the future of the biosphere. In "What Is Nature Worth?" Edward O. Wilson notes that there are powerful economic reasons for implementing plans to preserve the world's natural biological diversity. Wilson notes that present losses of Earth's plants and animals from human impact are progressing at rates from 1,000 to 10,000 times those of the average over the last half-billion years. If for no other reason than the fact that many of those lost species have incalculable economic value, humankind needs to fully understand just what it is doing to other inhabitants of Earth. But Wilson notes that there are moral arguments as well as economic ones for slowing the rate of extinction.

The second selection in the unit addresses some of the same issues as Wilson but places them in the context of a program of preserving biodiversity that already exists—the National Wildlife Refuge System in the United States. In "Where Wildlife Rules," *Audubon* field editor Frank Graham Jr. describes the NWRS as

"a towering achievement and a testament to our national commitment to conservation." Those are pretty strong words from a source that is generally fairly critical of the conservation policies of the federal government and suggests that, from an environmentalist standpoint, the refuge system must be working. More importantly, the United States' wildlife refuge system (which preserves plants as well as animals) provides the world with a model to preserve biodiversity in areas beyond the borders of the U.S. The third article in the unit also deals with human/environment interaction that begins in smaller spaces but expands to the global scale. Environmental researcher Christopher Bright describes one of the least anticipated results of the growing global economy: the spread of "invasive species." In "Invasive Species: Pathogens of Globalization," Bright notes that thousands of alien species are traveling aboard ships, planes, and railroad cars from one continent to another, often carried in commodities themselves. This consequence of the worldwide global trading network poses enormous hazards for native species in affected areas, and confronting the problem may be as important a challenge to environmental quality as reducing global carbon emissions. As recent events that Bright could not have predicted have shown, when pathogens from one part of the world may be used intentionally as weapons, the challenge becomes even greater.

The unit's final selection deals with the essential issue regarding the biosphere: species extinction. In "On the Termination of Species," senior *Scientific American* writer W. Wayt Gibbs tells us that the scientific evidence of present mass extinction of species is being challenged by skeptics and ignored by politicians. He suggests that part of the problem is the difficulty of defining the dimension of the extinction problem. It is also important that people understand why species loss is important and how it can be dealt with. Part of the solution is figuring out ways to make biodiversity pay in economic as well as ecological terms.

What Is Nature Worth?

There's a powerful economic argument for preserving our living natural environment: The biosphere promotes the long-term material prosperity and health of the human race to a degree that is almost incalculable. But moral reasons, too, should compel us to take responsibility for the natural world.

by Edward O. Wilson

In the early 19th century, the coastal plain of the southern United States was much the same as in countless millenniums past. From Florida and Virginia west to the Big Thicket of Texas, primeval stands of cypress and flatland hardwoods wound around the corridors of longleaf pine through which the early Spanish explorers had found their way into the continental interior. The signature bird of this wilderness, a dweller of the deep bottomland woods, was the ivory-billed woodpecker, *Campephilus principalis*. Its large size, exceeding a crow's, its flashing white primaries, visible at rest, and its loud nasal call—*kent!… kent!… kent!*—likened by John James Audubon to the false high note of a clarinet, made the ivory-bill both conspicuous and instantly recognizable. Mated pairs worked together up and down the boles and through the canopies of high trees, clinging to vertical surfaces with splayed claws while hammering their powerful, off-white beaks through dead wood into the burrows of beetle larvae and other insect prey. The hesitant beat of their strikes—*tick tick… tick tick tick… tick tick*—heralded their approach from a distance in the dark woods. They came to the observer like spirits out of an unfathomed wilderness core.

Alexander Wilson, early American naturalist and friend of Audubon, assigned the ivorybill noble rank. Its manners, he wrote in *American Ornithology* (1808–14), "have a dignity in them superior to the common herd of woodpeckers. Trees, shrubbery, orchards, rails, fence posts, and old prostrate logs are all alike interesting to those, in their humble and indefatigable search for prey; but the royal hunter before us scorns the humility of such situations, and seeks the most towering trees of the forest, seeming particularly attached to those prodigious cypress swamps whose crowded giant sons stretch their bare and blasted or moss-hung arms midway to the sky."

A century later, almost all of the virgin bottomland forest had been replaced by farms, towns, and second-growth woodlots. Shorn of its habitat, the ivory-bill declined precipitously in numbers. By the 1930s, it was down to scattered pairs in the few remaining primeval swamps of South Carolina, Florida, and Louisiana. In the 1940s, the only verifiable sightings were in the Singer Tract of northern Louisiana. Subsequently, only rumors of sightings persisted, and even these faded with each passing year.

The final descent of the ivorybill was closely watched by Roger Tory Peterson, whose classic *A Field Guide to the Birds* had fired my own interest in birds when I was a teenager. In 1995, the year before he died, I met Peterson, one of my heroes, for the first and only time. I asked him a question common in conversations among American naturalists: What of the ivory-billed woodpecker? He gave the answer I expected: "Gone."

I thought, surely not gone *everywhere*, not *globally*! Naturalists are among the most hopeful of people. They require the equivalent of an autopsy report, cremation, and three witnesses before they write a species off, and even then they would hunt for it in séances if they thought there were any chance of at least a virtual image. Maybe, they speculate, there are a few ivorybills in some inaccessible cove, or deep inside a forgotten swamp, known only to a few close-mouthed cognoscenti. In fact, several individuals of a small Cuban race of ivorybills were discovered during the 1960s in an isolated pine forest of Oriente Province. Their current status is unknown. In 1996, the Red List of the World Conservation Union reported the species to be everywhere extinct, including Cuba. I have heard of no further sightings, but evidently no one at this writing knows for sure.

Why should we care about *Campephilus principalis*? It is, after all, only one of 10,000 bird species in the world. Let me give a simple and, I hope, decisive answer: because we knew this particular species, and knew it well. For reasons difficult to understand and express, it became part of our culture, part of the rich mental world of Alexander Wilson and all those afterward who cared about it. There is no way to make a full and final valuation of the ivorybill or any other species in the natural world. The measures we use increase in number and magnitude with no predictable limit. They rise from scattered, unconnected facts and elusive emotions that break through the surface of the subconscious mind, occasionally to be captured by words, though never adequately.

We, *Homo sapiens*, have arrived and marked our territory well. Winners of the Darwinian lottery, bulge-headed paragons of organic evolution, industrious bipedal apes with opposable thumbs, we are chipping away the ivorybills and other miracles around us. As habitats shrink, species decline wholesale in range and abundance. They slide down the Red List ratchet, and the vast majority depart without special notice. Over the past half-billion years, the planet lost perhaps one species per million species each year, including everything from mammals to plants. Today, the annual rate of extinction is 1,000 to 10,000 times faster. If nothing more is done, one-fifth of all the plant and animal species now on earth could be gone or on the road to extinction by 2030. Being distracted and self-absorbed, as is our nature, we have not yet fully understood what we are doing. But future generations, with endless time to reflect, will understand it all, and in painful detail. As awareness grows, so will their sense of loss. There will be thousands of ivory-billed woodpeckers to think about in the centuries and millenniums to come.

Is there any way now to measure even approximately what is being lost? Any attempt is almost certain to produce an underestimate, but let me start anyway with macroeconomics. In 1997, an international team of economists and environmental scientists put a dollar amount on all the ecosystems services provided to humanity free of charge by the living natural environment. Drawing from multiple databases, they estimated the contribution to be $33 trillion or more each year. This amount is nearly twice the 1997 combined gross national product (GNP) of all the countries in the world—$18 trillion. *Ecosystems services* are defined as the flow of materials, energy, and information from the biosphere that support human existence. They include the regulation of the atmosphere and climate; the purification and retention of fresh water; the formation and enrichment of the soil; nutrient cycling; the detoxification and recirculation of waste; the pollination of crops; and the production of lumber, fodder, and biomass fuel.

IF HUMANITY WERE TO TRY TO REPLACE THE FREE SERVICES OF THE NATURAL ECONOMY WITH SUBSTITUTES OF ITS OWN MANUFACTURE, THE GLOBAL GNP WOULD HAVE TO BE INCREASED BY AT LEAST $33 TRILLION.

The 1997 megaestimate can be expressed in another, even more cogent, manner. If humanity were to try to replace the free services of the natural economy with substitutes of its own manufacture, the global GNP would have to be increased by at least $33 trillion. The exercise, however, cannot be performed except as a thought experiment. To supplant natural ecosystems entirely, even mostly, is an economic—and even physical—impossibility, and we would certainly die if we tried. The reason, ecological economists explain, is that the *marginal value*, defined as the rate of change in the value of ecosystems services relative to the rate of decline in the availability of these services, rises sharply with every increment in the decline. If taken too far, the rise will outpace human capacity to sustain the needed services by combined natural and artificial means. Hence, a much greater dependence on artificial means—in other words, environmental prostheses—puts at risk not just the biosphere but humanity itself.

Most environmental scientists believe that the shift has already been taken too far, lending credit to the folk injunction "Don't mess with Mother Nature." The lady is our mother all right, and a mighty dispensational force as well. After evolving on her own for more than three billion years, she gave birth to us a mere million years ago, the blink of an eye in evolutionary time. Ancient and vulnerable, she will not tolerate the undisciplined appetite of her gargantuan infant much longer.

Abundant signs of the biosphere's limited resilience exist all around. The oceanic fish catch now yields $2.5 billion to the U.S. economy and $82 billion worldwide. But it will not grow further, simply because the amount of ocean is fixed and the number of organisms it can generate is static. As a result, all of the world's 17 oceanic fisheries are at or below sustainable yield. During the 1990s, the annual global catch leveled off around 30 million tons. Pressed by ever-growing global demand, it can be expected eventually to drop. Already, fisheries of the western North Atlantic, the Black Sea, and portions of the Caribbean have largely collapsed. Aquaculture, or the farming of fish, crustaceans, and mollusks, takes up part of the slack, but at rising environmental cost. This "fin-and-shell revolution" necessitates the conversion of valuable wetland habitats, which are nurseries for marine life. To feed the captive populations, fodder must be diverted from crop production. Thus, aquaculture competes with other human activities for productive land while reducing natural habitat. What was once free for the taking must now be manufactured. The ultimate result will be an upward inflationary pressure across wide swaths of the world's coastal and inland economies.

Another case in point: Forested watersheds capture rainwater and purify it before returning it by gradual runoffs to the lakes and sea, all for free. They can be replaced only at great cost. For generations, New York City thrived on exceptionally clean water from the Catskill Mountains. The watershed inhabitants were proud that their bottled water was once sold throughout the Northeast. As their population grew, however, they converted more and more of the watershed forest into farms, homes, and resorts. Gradually, the sewage and agricultural runoff adulterated the water, until it fell below Environmental Protection Agency standards. Officials in New York City now faced a choice: They could build a filtration plant to replace the Catskill watershed, at a $6 billion to $8 billion capital cost, followed by $300 million annual running costs, or they could restore the watershed to somewhere near its original purification capacity for $1 billion, with subsequently very low maintenance costs. The decision was easy, even for those born and bred in an urban environment. In 1997, the city raised an environmental bond issue and set out to purchase forested land and to subsidize the upgrading of septic tanks in the Catskills. There is no reason the people of New York City and the Catskills cannot enjoy the double gift from nature in perpetuity of clean water at low cost and a beautiful recreational area at no cost.

There is even a bonus in the deal. In the course of providing natural water management, the Catskill forest region also secures flood control at very little expense. The same benefit is available to the city of Atlanta. When 20 percent of the trees in the metropolitan area were removed during its rapid development, the result was an annual increase in stormwater runoff of 4.4 billion cubic feet. If enough containment facilities were built to capture this volume, the cost would be at least $2 billion. In contrast, trees replanted along streets and in yards, and parking areas are a great deal cheaper than concrete drains and revetments. Their maintenance cost is near zero, and, not least, they are more pleasing to the eye.

In conserving nature, whether for practical or aesthetic reasons, diversity matters. The following rule is now widely accepted by ecologists: The more numerous the species that inhabit an ecosystem, such as a forest or lake, the more productive and stable is the ecosystem. By "production," the scientists mean the amount of plant and animal tissue created in a given unit of time. By "stability," they mean one or the other, or both, of two things: first, how narrowly the summed abundances of all species vary through time; and, second, how quickly the ecosystem recovers from fire, drought, and other stresses that perturb it. Human beings understandably wish to live in the midst of diverse, productive, and stable ecosystems. Who, if

given a choice, would build a home in a wheat field instead of a parkland?

Ecosystems are kept stable in part by the insurance principle of biodiversity: If a species disappears from a community, its niche will be more quickly and effectively filled by another species if there are many candidates for the role instead of few. Example: A ground fire sweeps through a pine forest, killing many of the understory plants and animals. If the forest is biodiverse, it recovers its original composition and production of plants and animals more quickly. The larger pines escape with some scorching of their lower bark and continue to grow and cast shade as before. A few kinds of shrubs and herbaceous plants also hang on and resume regeneration immediately. In some pine forests subject to frequent fires, the heat of the fire itself triggers the germination of dormant seeds genetically adapted to respond to heat, speeding the regrowth of forest vegetation still more.

WHO, IF GIVEN A CHOICE, WOULD BUILD A HOME IN A WHEAT FIELD INSTEAD OF A PARKLAND?

A second example of the insurance principle: When we scan a lake, our macroscopic eye sees only relatively big organisms, such as eelgrass, pondweeds, fishes, water birds, dragonflies, whirligig beetles, and other things big enough to splash and go bump in the night. But all around them, in vastly greater numbers and variety, are invisible bacteria, protistans, planktonic single-celled algae, aquatic fungi, and other microorganisms. These seething myriads are the true foundation of the lake's ecosystem and the hidden agents of its stability. They decompose the bodies of the larger organisms. They form large reservoirs of carbon and nitrogen, release carbon dioxide, and thereby damp fluctuations in the organic cycles and energy flows in the rest of the aquatic ecosystem. They hold the lake close to a chemical equilibrium, and, to a point, they pull it back

from extreme perturbations caused by silting and pollution.

In the dynamism of healthy ecosystems, there are minor players and major players. Among the major players are the ecosystems engineers, which add new parts to the habitat and open the door to guilds of organisms specialized to use them. Biodiversity engenders more biodiversity, and the overall abundance of plants, animals, and microorganisms increases to a corresponding degree.

By constructing dams, beavers create ponds, bogs, and flooded meadows. These environments shelter species of plants and animals that are rare or absent in free-running streams. The submerged masses of decaying wood forming the dams draw still more species, which occupy and feed on them.

Elephants trample and tear up shrubs and small trees, opening glades within forests. The result is a mosaic of habitats that, overall, contains larger numbers of resident species.

Florida gopher tortoises dig 30-foot-long tunnels that diversify the texture of the soil, altering the composition of its microorganisms. Their retreats are also shared by snakes, frogs, and ants specialized to live in the burrows.

Euchondrus snails of Israel's Negev Desert grind down soft rocks to feed on the lichens growing inside. By converting rock to soil and releasing the nutrients photosynthesized by the lichens, the snails multiply niches for other species.

TO EVALUATE INDIVIDUAL SPECIES SOLELY BY THEIR KNOWN PRACTICAL VALUE AT THE PRESENT TIME IS BUSINESS ACCOUNTING IN THE SERVICE OF BARBARISM.

Overall, a large number of independent observations from differing kinds of ecosystems point to the same conclusion: The greater the number of species that live together, the more stable and productive the ecosystems these species compose. On the other hand, mathematical models that attempt to describe the interactions of species in ecosystems show that the apparent opposite also occurs: High levels of diversity can reduce the stability of individual species. Under certain conditions, including random colonization of the ecosystem by large numbers of species that interact strongly with one another, the separate but interlocking fluctuations in species populations can become more volatile, thus making extinction more likely. Similarly, given appropriate species traits, it is mathematically possible for increased diversity to lead to decreased production.

When observation and theory collide, scientists turn to carefully designed experiments for resolution. Their motivation is especially strong in the case of biological systems, which are typically far too complex to be grasped by observation and theory alone. The best procedure, as in the rest of science, is first to simplify the system, then to hold it more or less constant while varying the important parameters one or two at a time to see what happens. In the 1990s a team of British ecologists, in an attempt to approach these ideal conditions, devised the *ecotron*, a growth chamber in which artificially simple ecosystems can be assembled as desired, species by species. Using multiple ecotrons, they found that productivity, measured by the increase of plant bulk, rose with an increase in species numbers. Simultaneously, ecologists monitoring patches of Minnesota grassland—outdoor equivalents of ecotrons—during a period of drought found that patches richer in species diversity underwent less decline in productivity and recovered more quickly than patches with less diversity.

These pioneering experiments appeared to uphold the conclusion drawn earlier from natural history, at least with reference to production. Put more precisely, ecosystems tested thus far do not possess the qualities and starting conditions allowed by theory that can reduce production and produce instability as a result of large species numbers.

But—how can we be sure, the critics asked (pressing on in the best tradition of science), that the increase in production in particular is truly the result of just an increase in the number of species?

Maybe the effect is due to some other factor that just happens to be correlated with species numbers. Perhaps the result is a statistical artifact. For example, the larger the number of plant species present in a habitat, the more likely it is that at least one kind among them will be extremely productive. If that occurs, the increase in the yield of plant tissue—and in the number of the animals feeding on it—is only a matter of luck of the draw, and not the result of some pure property of biodiversity itself. At its base, the distinction made by this alternative hypothesis is semantic. The increased likelihood of acquiring an outstandingly productive species can be viewed as just one means by which the enrichment of biodiversity boosts productivity. (If you draw on a pool of 1,000 candidates for a basketball team, you are more likely to get a star than if you draw on a pool of 100 candidates.)

Still, it is important to know whether other consequences of biodiversity enrichment play an important role. In particular, do species interact in a manner that increases the growth of either one or both? This is the process called *overyielding*. In the mid-1990s, a massive study was undertaken to test the effect of biodiversity on productivity that paid special attention to the presence or absence of overyielding. Multiple projects of BIODEPTH, as the project came to be called, were conducted during a two-year period by 34 researchers in eight European countries. This time, the results were more persuasive. They showed once again that productivity does increase with biodiversity. Many of the experimental runs also revealed the existence of overyielding.

Over millions of years, nature's ecosystems engineers have been especially effective in the promotion of overyielding. They have coevolved with other species that exploit the niches they build. The result is a harmony within ecosystems. The constituent species, by spreading out into multiple niches, seize and cycle more materials and energy than is possible in similar ecosystems. *Homo sapiens* is an ecosystems engineer too, but a bad one. Not having coevolved with the majority of life forms we now encounter around the world, we elimi-

nate far more niches than we create. We drive species and ecosystems into extinction at a far higher rate than existed before, and everywhere diminish productivity and stability.

I will grant at once that economic and production values at the ecosystem level do not alone justify saving every species in an ecosystem, especially those so rare as to be endangered. The loss of the ivory-billed woodpecker has had no discernible effect on American prosperity. A rare flower or moss could vanish from the Catskill forest without diminishing the region's filtration capacity. But so what? To evaluate individual species solely by their known practical value at the present time is business accounting in the service of barbarism. In 1973, the economist Colin W. Clark made this point persuasively in the case of the blue whale, *Balaenopterus musculus*. A hundred feet in length and 150 tons in weight at maturity, the species is the largest animal that ever lived on land or sea. It is also among the easiest to hunt and kill. More than 300,000 blue whales were harvested during the 20th century, with a peak haul of 29,649 in the 1930–31 season. By the early 1970s, the population had plummeted to several hundred individuals. The Japanese were especially eager to continue the hunt, even at the risk of total extinction. So Clark asked, What practice would yield the whalers and humanity the most money: Cease hunting and let the blue whales recover in numbers, then harvest them sustainably forever, or kill the rest off as quickly as possible and invest the profits in growth stocks? The disconcerting answer for annual discount rates over 21 percent: Kill them all and invest the money.

Now, let us ask, what is wrong with that argument?

Clark's implicit answer is simple. The dollars-and-cents value of a dead blue whale was based only on the measures relevant to the existing market—that is, on the going price per unit weight of whale oil and meat. There are many other values, destined to grow along with our knowledge of living *Balaenopterus musculus* and as science, medicine, and aesthetics grow and strengthen, in dimensions and magnitudes still unforeseen. What was the value of the blue whale in A.D. 1000? Close to zero. What will be its value in A.D. 3000? Essentially limitless—to say nothing of the measure of gratitude the generation then alive will feel to those who in their wisdom saved the whale from extinction.

No one can guess the full future value of any kind of animal, plant, or microorganism. Its potential is spread across a spectrum of known and as yet unimagined human needs. Even the species themselves are largely unknown. Fewer than two million are in the scientific register, with a formal Latinized name, while an estimated five to 100 million—or more—await discovery. Of the species known, fewer than one percent have been studied beyond the sketchy anatomical descriptions used to identify them.

> OF THE SPECIES KNOWN, FEWER THAN ONE PERCENT HAVE BEEN STUDIED BEYOND THE SKETCHY ANATOMICAL DESCRIPTIONS USED TO IDENTIFY THEM.

Agriculture is one of the vital industries most likely to be upgraded by attention to the remaining wild species. The world's food supply hangs by a slender thread of biodiversity. Ninety percent is provided by slightly more than 100 plant species out of a quarter-million known to exist. Twenty species carry most of the load, of which only three—wheat, maize, and rice—stand between humanity and starvation. For the most part, the premier 20 are those that happened to be present in the regions where agriculture was independently invented some 10,000 years ago, namely the Mediterranean perimeter and southwestern Asia; Central Asia; the Horn of Africa; the rice belt of tropical Asia; and the uplands of Mexico, Central America, and Andean South America. Yet some 30,000 species of wild plants, most occurring outside these regions, have edible parts consumed at one time or other by hunter-gatherers. Of these species, at least 10,000 can be adapted as domestic crops. A few, including the three species of New World amaranths, the carrotlike arracacha of the Andes, and the winged bean of tropical Asia, are immediately available for commercial development.

In a more general sense, all the quarter-million plant species—in fact, all species of organisms—are potential donors of genes that can be transferred by genetic engineering into crop species in order to improve their performance. With the insertion of the right snippets of DNA, new strains can be created that are, variously, cold resistant, pest resistant, perennial, fast growing, highly nutritious, multipurpose, sparing in their consumption of water, and more easily sowed and harvested. And compared with traditional breeding techniques, genetic engineering is all but instantaneous.

The method, a spinoff of the revolution in molecular genetics, was developed in the 1970s. During the 1980s and 1990s, before the world quite realized what was happening, it came of age. A gene from the bacterium *Bacillus thuringiensis*, for example, was inserted into the chromosomes of corn, cotton, and potato plants, allowing them to manufacture a toxin that kills insect pests. No need to spray insecticides; the engineered plants now perform this task on their own. Other transgenes, as they are called, were inserted from bacteria into soybean and canola plants to make them resistant to chemical weed killers. Agricultural fields can now be cheaply cleared of weeds with no harm to the crops growing there. The most important advance of all, achieved in the 1990s, was the creation of golden rice. This new strain is laced with bacterial and daffodil genes that allow it to manufacture beta-carotene, a precursor of vitamin A. Because rice, the principal food of three billion people, is deficient in vitamin A, the addition of beta-carotene is no mean humanitarian feat. About the same time, the almost endless potential of genetic engineering was confirmed by two circus tricks of the trade: A bacterial gene was implanted into a monkey, and a jellyfish bioluminescence gene into a plant.

But not everyone was dazzled by genetic engineering, and inevitably it stirred opposition. For many, human existence was being transformed in a fundamental and insidious way. With little warning, genetically modified organisms (GMOs) had entered our lives and were all around us, changing incomprehensibly the order of nature and society. A protest movement against the new industry began in the mid-1990s and exploded in 1999, just in time to rank as a millennial event with apocalyptic overtones. The European Union banned transgenic crops, the Prince of Wales compared the methodology to playing God, and radical activists called for a global embargo of all GMOs. "Frankenfoods," "superweeds," and "Farmageddon" entered the vocabulary: GMOs were, according to one British newspaper, the "mad forces of genetic darkness." Some prominent environmental scientists found technical and ethical reasons for concern.

As I write, public opinion and official policy toward genetic engineering have come to vary greatly from one country to the next. France and Britain are vehemently opposed. China is strongly favorable, and Brazil, India, Japan, and the United States cautiously so. In the United States particularly, the public awoke to the issue only after the transgenie (so to speak) was out of the bottle. From 1996 to 1999, the amount of U.S. farmland devoted to genetically modified crops had rocketed from 3.8 million to 70.9 million acres. As the century ended, more than half the soybeans and cotton grown, and nearly a third (28 percent) of the corn, were engineered.

> WITH LITTLE WARNING, GENETICALLY MODIFIED ORGANISMS HAD ENTERED OUR LIVES AND WERE ALL AROUND US, CHANGING INCOMPREHENSIBLY THE ORDER OF NATURE AND SOCIETY.

There are, actually, several sound reasons for anxiety over genetic engineering, which I will now summarize and evaluate.

Many people, not just philosophers and theologians, are troubled by the ethics of transgenic evolution. They grant the benefits but are unsettled by the reconstruction of organisms in bits and pieces. Of course, human beings have been creating new strains of plants and animals since agriculture began, but never at the sweep and pace inaugurated by genetic engineering. And during the era of traditional plant breeding, hybridization was used to mix genes almost always among varieties of the same species or closely similar species. Now it is used across entire kingdoms, from bacteria and viruses to plants and animals. How far the process should be allowed to continue remains an open ethical issue.

The effects on human health of each new transgenic food are hard to predict, and certainly never free of risk. However, the products can be tested just like any other new food products on the market, then certified and labeled. There is no reason at this time to assume that their effects will differ in any fundamental way. Yet scientists generally agree that a high level of alertness is essential, and for the following reason: All genes, whether original to the organism or donated to it by an exotic species, have multiple effects. Primary effects, such as the manufacture of a pesticide, are the ones sought. But destructive secondary effects, including allergenic or carcinogenic activity, are also at least a remote possibility.

Transgenes can escape from the modified crops into wild relatives of the crop where the two grow close together. Hybridization has always occurred widely in agriculture, even before the advent of genetic engineering. It has been recorded at one or another time and place in 12 of the 13 most important crops used worldwide. However, the hybrids have not overwhelmed their wild parents. I know of no case in which a hybrid strain outcompetes wild strains of the same or closely related species in the natural environment. Nor has any hybrid turned into a superweed, in the same class as the worst wild nonhybrid weeds that afflict the planet. As a rule, domesticated species and strains are less competitive than their wild counterparts in both natural and human-modified environments, Of course, transgenes could change the picture. It is simply too early to tell.

Genetically modified crops can diminish biological diversity in other ways. In a now famous example, the bacterial toxin used to protect corn is carried in pollen by wind currents for distances of 60 meters or more from the cultivated fields. Then, landing on milkweed plants, the toxin is capable of killing the caterpillars of monarch butterflies feeding there. In another twist, when cultivated fields are cleared of weeds with chemical sprays against which the crops are protected by transgenes, the food supply of birds is reduced and their local populations decline. These environmental secondary effects have not been well studied in the field. How severe they will become as genetic engineering spreads remains to be seen.

Many people, having become aware of the potential threats of genetic engineering in their food supply, understandably believe that yet another bit of their freedom has been taken from them by faceless corporations (who can even name, say, three of the key players?) using technology beyond their control or even understanding. They also fear that an industrialized agriculture dependent on high technology can by one random error go terribly wrong. At the heart of the anxiety is a sense of helplessness. In the realm of public opinion, genetic engineering is to agriculture as nuclear engineering is to energy.

The problem before us is how to feed billions of new mouths over the next several decades and save the rest of life at the same time—without being trapped in a Faustian bargain that threatens freedom and security. No one knows the exact solution to this dilemma. Most scientists and economists who have studied both sides of it agree that the benefits outweigh the risks. The benefits must come from an evergreen revolution that has as its goal to lift food production well above the level attained by the green revolution of the 1960s, using technology and regulatory policy more advanced, and even safer, than that now in existence.

Genetic engineering will almost certainly play an important role in the evergreen revolution. Energized by recognition of both its promise and its risk, most countries have begun to fashion policies to regulate the marketing of transgenic crops. The ultimate driving force in this rapidly evolving process is international trade. More than 130 countries took an important first step in 2000 to address the issue by tentatively agreeing to the Cartagena Protocol on Biosafety, which provides the right to block imports of transgenic products. The protocol also sets up a joint "biosafety clearing house" to publish information on national policy. About the same time, the U.S. National Academy of Sciences, joined by the science academies of five other countries (Brazil, China, India, Mexico, and the United Kingdom) and the Third World Academy of Sciences, endorsed the development of transgenic crops. They made recommendations for risk assessment and licensing agreements and stressed the needs of the developing countries in future research programs and capital investment.

Medicine is another domain that stands to gain enormously from the world's store of biodiversity, with or without the impetus of genetic engineering. Pharmaceuticals in current use are already drawn heavily from wild species. In the United States, about a quarter of all prescriptions dispensed by pharmacies are substances extracted from plants. Another 13 percent originate from microorganisms, and three percent more from animals—making a total of about 40 percent derived from wild species. What's even more impressive is that nine of the 10 leading prescription drugs originally came from organisms. The commercial value of the relatively small number of natural products is substantial. The over-the-counter cost of drugs from plants alone was estimated in 1998 to be $20 billion in the United States and $84 billion worldwide.

But only a tiny fraction of biodiversity has been utilized in medicine, despite its obvious potential. The narrowness of the base is illustrated by the dominance of ascomycete fungi in the control of bacterial diseases. Although only about 30,000 species of ascomycetes—two percent of the total known species of organisms—have been studied, they have yielded 85 percent of the antibiotics in current use. The underutilization of biodiversity is still greater than these figures alone might suggest—because probably fewer than 10 percent of the world's ascomycete species have even been discovered and given scientific names. The flowering plants have been similarly scanted. Although it is likely that more than 80 percent of the species have received scientific names, only some three percent of this fraction have been assayed for alkaloids, the class of natural products that have proved to be among the most potent curative agents for cancer and many other diseases.

There is an evolutionary logic in the pharmacological bounty of wild species. Throughout the history of life, all kinds of organisms have evolved chemicals needed to control cancer in their own bodies, kill parasites, and fight off predators. Mutations and natural selection, which equip this armamentarium, are processes of endless trial and error. Hundreds of millions of species, evolving by the life and death of astronomical numbers of organisms across geological stretches of time, have yielded the present-day winners of the mutation-and-selection lottery. We have learned to consult them while assembling a large part of our own pharmacopoeia. Thus, antibiotics, fungicides, antimalarial drugs, anesthetics, analgesics, blood thinners, blood-clotting agents, agents that prevent clotting, cardiac stimulants and regulators, immunosuppressive agents, hormone mimics, hormone inhibitors, anticancer drugs, fever suppressants, inflammation controls, contraceptives, diuretics and antidiuretics, antidepressants, muscle relaxants, rubefacients, anticongestants, sedatives,

and abortifacients are now at our disposal, compliments of wild biodiversity.

Revolutionary new drugs have rarely resulted from the pure insights of molecular and cellular biology, even though these sciences have grown very sophisticated and address the causes of disease at the most fundamental level. Rather, the pathway of discovery has usually been the reverse: The presence of the drug is first detected in whole organisms, and the nature of its activity subsequently tracked down to the molecular and cellular levels. Then the basic research begins.

THE PROBLEM BEFORE US IS HOW TO FEED BILLIONS OF NEW MOUTHS OVER THE NEXT SEVERAL DECADES AND SAVE THE REST OF LIFE AT THE SAME TIME.

The first hint of a new pharmaceutical may lie among the hundreds of remedies of Chinese traditional medicine. It may be spotted in the drug-laced rituals of an Amazonian shaman. It may come from a chance observation by a laboratory scientist unaware of its potential importance for medicine. More commonly nowadays, the clue is deliberately sought by the random screening of plant and animal tissues. If a positive response is obtained—say, a suppression of bacteria or cancer cells—the molecules responsible can be isolated and tested on a larger scale, using controlled experiments with animals and then (cautiously!) human volunteers. If the tests are successful, and the atomic structure of the molecule is also in hand, the substance can be synthesized in the laboratory, then commercially, usually at lower cost than by extraction from harvested raw materials. In the final step, the natural chemical compounds provide the prototype from which new classes of organic chemicals can be synthesized, adding or taking away atoms and double bonds here and there. A few of the novel substances may prove more efficient than the natural prototype. And of equal importance to the

pharmaceutical companies, these analogues can be patented.

Serendipity is the hallmark of pharmacological research. A chance discovery can lead not only to a successful drug but to advances in fundamental science, which in time yield other successful drugs. Routine screening, for example, revealed that an obscure fungus growing in the mountainous interior of Norway produces a powerful suppressor of the human immune system. When the molecule was isolated from the fungal tissue and identified, it proved to be a complex molecule of a kind never before encountered by organic chemists. Nor could its effect be explained by the contemporary principles of molecular and cellular biology. But its relevance to medicine was immediately obvious, because when organs are transplanted from one person to another, the immune system of the host must be prevented from rejecting the alien tissue. The new agent, named cyclosporin, became an essential part of the organ transplant industry. It also served to open new lines of research on the molecular events of the immune response itself.

The surprising events that sometimes lead from natural history to medical breakthrough would make excellent science fiction—if only they were untrue. The protagonists of one such plot are the poison dart frogs of Central and South America, which belong to the genera *Dendrobates* and *Phyllobates* in the family Dendrobatidae. Tiny, able to perch on a human fingernail, they are favored as terrarium animals for their beautiful colors: The 40 known species are covered by various patterns of orange, red, yellow, green, or blue, usually on a black background. In their natural habitat, dendrobatids hop about slowly and are relatively unfazed by the approach of potential predators. For the trained naturalist their lethargy triggers an alarm, in observance of the following rule of animal behavior: If a small and otherwise unknown animal encountered in the wild is strikingly beautiful, it is probably poisonous, and if it is not only beautiful but also easy to catch, it is probably deadly. And so it is with dendrobatid frogs, which, it turns out, secrete a powerful toxin from glands on their backs. The potency varies according to species. A single individual of one (perfectly named) Colombian species, *Phyllobates horribilis*, for example, carries enough of the substance to kill 10 men. Indians of two tribes living in the Andean Pacific slope forests of western Colombia, the Emberá Chocó and the Noanamá Chocó, rub the tips of their blowgun darts over the backs of the frogs, very carefully, then release the little creatures unharmed so they can make more poison.

COLLECTING SAMPLES OF VALUABLE SPECIES FROM RICH ECOSYSTEMS AND CULTIVATING THEM IN BULK ELSEWHERE IS NOT ONLY PROFITABLE BUT THE MOST SUSTAINABLE OF ALL.

In the 1970s a chemist, John W. Daly, and a herpetologist, Charles W. Myers, gathered material from a similar Ecuadorian frog, *Epipedobates tricolor*, for a closer look at the dendrobatid toxin. In the laboratory, Daly found that very small amounts administered to mice worked as an opiumlike painkiller, yet otherwise lacked the properties of typical opiates. Would the substance also prove nonaddictive? If so, it might be turned into the ideal anesthetic. From a cocktail of compounds taken from the backs of the frogs, Daly and his fellow chemists isolated and characterized the toxin itself, a molecule resembling nicotine, which they named epibatidine. This natural product proved 200 times more effective in the suppression of pain than opium, but was also too toxic, unfortunately, for practical use. The next step was to redesign the molecule. Chemists at Abbott Laboratories synthesized not only epibatidine but hundreds of novel molecules resembling it. When tested clinically, one of the products, code-named ABT-594, was found to combine the desired properties: It depressed pain like epibatidine, including pain from nerve damage of a kind usually impervious to opiates, and it was nonaddictive.

ABT-594 had two additional advantages: It promoted alertness instead of sleepiness, and it had no side effects on respiration or digestion.

The full story of the poison dart frogs also carries a warning about the conservation of tropical forests. The destruction of much of the habitat in which populations of *Epipedobates* live almost prevented the discovery of epibatidine and its synthetic analogues. By the time Daly and Myers set out to collect enough toxin for chemical analysis, after their initial visit to Ecuador, one of the two prime rainforest sites occupied by the frogs had been cleared and replaced with banana plantations. At the second site, which fortunately was still intact, they found enough frogs to harvest just one milligram of the poison. From that tiny sample, chemists were able, with skill and luck, to identify epibatidine and launch a major new initiative in pharmaceutical research.

It is no exaggeration to say that the search for natural medicinals is a race between science and extinction, and will become critically so as more forests fall and coral reefs bleach out and disintegrate. Another adventure dramatizing this point began in 1987, when the botanist John Burley collected samples of plants from a swamp forest near Lundu in the Malaysian state of Sarawak, on the northwestern corner of the island of Borneo. His expedition was one of many launched by the National Cancer Institute (NCI) to search for new natural substances to add to the fight against cancer and AIDS. Following routine procedure, the team collected a kilogram of fruit, leaves, and twigs from each kind of plant they encountered. Part was sent to the NCI laboratory for assay, and part was deposited in the Harvard University Herbarium for future identification and botanical research.

One such sample came from a small tree at Lundu about 25 feet high. It was given the voucher code label Burley-and-Lee 351. Back at the NCI laboratories, an extract made from it was tested routinely against human cancer cells grown in culture. Like the majority of such preparations, it had no effect. Then it was run through screens designed to test its potency against the AIDS virus.

The NCI scientists were startled to observe that Burley-and-Lee 351 gave, in their words, "100 percent protection against the cytopathic effects of HIV-I infection," having "essentially halted HIV-I replication." In other words, while the substance the sample contained could not cure AIDS, it could stop cold the development of disease symptoms in HIV-positive patients.

The Burley-and-Lee 351 tree was determined to belong to a species of *Calophyllum*, a group of species belonging to the mangosteen family, or Guttiferae. Collectors were dispatched to Lundu a second time to obtain more material from the same tree, with the aim of isolating and chemically identifying the HIV inhibitor. The tree was gone, probably cut down by local people for fuel or building materials. The collectors returned home with samples from other *Calophyllum* trees taken in the same swamp forest, but their extracts were ineffective against the virus.

Peter Stevens, then at Harvard University, and the world authority on *Calophyllum*, stepped in to solve the problem. The original tree, he found, belonged to a rare strain named *Calopsyllum lanigerum*, variety *austrocoriaceum*. The trees sampled on the second trip were another species, which explained their inactivity. No more specimens of *austrocoriaceum* could be found at Lundu. The search for the magic strain widened, and finally a few more specimens were located in the Singapore Botanic Garden. Thus supplied with enough raw material, chemists and microbiologists were able to identify the anti-HIV substance as (+)-calanolide A. Soon afterward the molecule was synthesized, and the synthetic proved as effective as the raw extract. Additional research revealed calanolide to be a powerful inhibitor of reverse transcriptase, an enzyme needed by the HIV virus to replicate itself within the human host cell. Studies are now underway to determine the suitability of calanolide for market distribution.

The exploration of wild biodiversity in the search for useful resources is called *bioprospecting*. Propelled by venture capital, it has in the past 10 years grown into a respectable industry within

131

a global market hungry for new pharmaceuticals. It is also a means for discovering new food sources, fibers, petroleum substitutes, and other products. Sometimes bioprospectors screen many species of organisms in search of chemicals with particular qualities, such as antisepsis or cancer suppression. On other occasions bioprospecting is opportunistic, focusing on one of a few species that show signs of yielding a valuable resource. Ultimately, entire ecosystems will be prospected as a whole, and all of the species assayed for most or all of the products they can yield.

The extraction of wealth from an ecosystem can be destructive or benign. Dynamiting coral reefs and clearcutting forests yield fast profits but are unsustainable. Fishing coral reefs lightly and gathering wild fruit and resins in otherwise undisturbed forest are sustainable. Collecting samples of valuable species from rich ecosystems and cultivating them in bulk elsewhere, in biologically less favored areas, is not only profitable but the most sustainable of all.

Bioprospecting with minimal disturbance is the way of the future. Its promise can be envisioned with the following matrix for a hypothetical forest: To the left, make a list of the thousands of plant, animal, and microbial species, as many as you can, recognizing that the vast majority have not yet been examined, and many still lack even a scientific name. Along the top, prepare a horizontal row of the hundreds of functions imaginable for all the products of these species combined. The matrix itself is the combination of the two dimensions. The spaces filled within the matrix are the potential applications, whose nature remains almost wholly unknown.

The richness of biodiversity's bounty is reflected in the products already extracted by native peoples of the tropical forests, using local knowledge and low technology of a kind transmitted solely by demonstration and oral teaching. Here, for example, is a small selection of the most common medicinal plants used by tribes of the upper Amazon, whose knowledge has evolved from their combined experience with the more than

50,000 species of flowering plants native to the region: motelo sanango, *Abuta grandifolia* (snakebite, fever); dye plant, *Arrabidaea chica* (anemia, conjunctivitis); monkey ladder, *Bauhinia guianensis* (amoebic dysentery); Spanish needles, *Bidens alba* (mouth sores, toothache); firewood tree, or capirona, species of *Calycophyllum* and *Capirona* (diabetes, fungal infection); wormseed, *Chenopodium ambrosioides* (worm infection); caimito, *Chrysophyllum cainito* (mouth sores, fungal infection); toad vine, *Cissus sicyoides* (tumors); renaquilla, *Clusia rosea* (rheumatism, bone fractures); calabash, *Crescentia cujete* (toothache); milk tree, *Couma macrocarpa* (amoebic dysentery, skin inflammation); dragon's blood, *Croton lechleri* (hemorrhaging); fer-de-lance plant, *Dracontium loretense* (snakebite); swamp immortelle, *Erythrina fusca* (infections, malaria); wild mango, *Grias neuberthii* (tumors, dysentery); wild senna, *Senna reticulata* (bacterial infection).

Only a few of the thousands of such traditional medicinals used in tropical forests around the world have been tested by Western clinical methods. Even so, the most widely used already have commercial value rivaling that of farming and ranching. In 1992 a pair of economic botanists, Michael Balick and Robert Mendelsohn, demonstrated that single harvests of wild-grown medicinals from two tropical forest plots in Belize were worth $726 and $3,327 per hectare (2.5 acres) respectively, with labor costs thrown in. By comparison, other researchers estimated per hectare yield from tropical forest converted to farmland at $228 in nearby Guatemala and $339 in Brazil. The most productive Brazilian plantations of tropical pine could yield $3,184 per hectare from a single harvest.

In short, medicinal products from otherwise undisturbed tropical forests can be locally profitable, on condition that markets are developed and the extraction rate is kept low enough to be sustainable. And when plant and animal food products, fibers, carbon credit trades, and ecotourism are added to the mix, the commercial value of sustainable use can be boosted far higher.

Examples of the new economy in practice are growing in number. In the Petén region of Guatemala, about 6,000 families live comfortably by sustainable extraction of rainforest products. Their combined annual income is $4 million to $6 million, more than could be made by converting the forest into farms and cattle ranches. Ecotourism remains a promising but largely untapped additional resource.

Nature's pharmacopoeia has not gone unnoticed by industry strategists. They are well aware that even a single new molecule has the potential to recoup a large capital investment in bioprospecting and product development. The single greatest success to date was achieved with extremophile bacteria living in the boiling-hot thermal springs of Yellowstone National Park. In 1983 Cetus Corporation used one of the organisms, *Thermus aquaticus*, to produce a heat-resistant enzyme needed for DNA synthesis. The manufacturing process, called *polymerase chain reaction* (PCR), is today the foundation of rapid genetic mapping, a stanchion of the new molecular biology and medical genetics. By enabling microscopic amounts of DNA to be multiplied and typed, PCR also plays a key role in crime detection and forensic medicine. Cetus's patents on PCR technology, which have been upheld by the courts, are immensely profitable, with annual earnings now in excess of $200 million—and growing.

Bioprospecting can serve both mainstream economics and conservation when put on a firm contractual basis. In 1991, Merck signed an agreement with Costa Rica's National Institute of Biodiversity (INBio) to assist the search for new pharmaceuticals in Costa Rica's rainforests and other natural habitats. The first deposit was $1 million dispensed over two years, with two similar consecutive grants to follow. During the first period, the field collectors concentrated on plants, in the second on insects, and in the third on microorganisms. Merck is now working through the immense library of materials it gathered during the field program and testing and refining chemical extracts made from them.

Also in 1991, Syntex signed a contract with Chinese science academies to receive up to 10,000 plant extracts a year for pharmaceutical assays. In 1998, Diversa Corporation signed on with Yellowstone National Park to continue bioprospecting the hot springs for biochemicals from thermophilic microbes. Diversa pays the park $20,000 yearly to collect the organisms for study, as well as a fraction of the profits generated by commercial development. Funds returning to Yellowstone will be used to promote conservation of the unique microbes and their habitat, as well as basic scientific research and public education.

Still other agreements have been signed between NPS Pharmaceuticals and the government of Madagascar, between Pfizer and the New York Botanical Garden, and between the international company GlaxoSmithKline and a Brazilian pharmaceutical company, with part of the profits pledged to the support of Brazilian science.

Perhaps it is enough to argue that the preservation of the living world is necessary to our long-term material prosperity and health. But there is another, and in some ways deeper, reason not to let the natural world slip away. It has to do with the defining qualities and self-image of the human species. Suppose, for the sake of argument, that new species can one day be engineered and stable ecosystems built from them. With that distant prospect in mind, should we go ahead and, for short-term gain, allow the original species and ecosystems to be lost? Yes? Erase Earth's living history? Then also burn the art galleries, make cordwood of the musical instruments, pulp the musical scores, erase Shakespeare, Beethoven, and Goethe, and the Beatles too, because all these—or at least fairly good substitutes—can be re-created.

The issue, like all great decisions, is moral. Science and technology are what we can do; morality is what we agree we should or should not do. The ethic from which moral decisions spring is a norm or standard of behavior in support of a value, and value in turn depends on purpose. Purpose, whether personal or global, whether urged by conscience or graven in sacred script, expresses the image we hold of ourselves and our society. A conservation ethic is that which aims to pass on to future generations the best part of the nonhuman world. To know this world is to gain a proprietary attachment to it. To know it well is to love and take responsibility for it.

EDWARD O. WILSON *is Pellegrino University Research Professor and Honorary Curator in Entomology at Harvard University's Museum of Comparative Zoology. His books include* Sociobiology: The New Synthesis (*1975*), Consilience: The Unity of Knowledge (*1998*), *and two Pulitzer Prize winners,* On Human Nature (*1978*) *and* The Ants (*1990, with Burt Holldobler*). *This essay is taken from his latest book,* The Future of Life (*2002*). *Published by arrangement with Alfred A. Knopf, a division of Random House, Inc. Copyright:* © *2002 By Edward O. Wilson.*

From *The Wilson Quarterly*, Winter 2002, pp. 20-39. © 2002 by Edward O. Wilson.

NATIONAL WILDLIFE REFUGE CENTENNIAL

WHERE WILDLIFE
RULES

It started on a tiny island off the coast of Florida. Today, 100 years and 95 million acres later, the refuge system stands as a towering achievement and a testament to our national commitment to conservation.

BY FRANK GRAHAM JR.

THE PETIT MANAN NATIONAL WILDLIFE REFUGE HAS played an ever-present role in my life since the U.S. Fish and Wildlife Service helped restore its seabird colony in the mid-1980s. At night, if the wind is right and the fog rolls in, I hear the throaty blast of the foghorn on Petit Island, a couple of miles off the Maine coast. On late summer days I watch common terns lead their young from the island to rock ledges within 200 yards of our house to feed their chicks on tiny fish snatched from shallow water at the head of Narraguagus Bay. And all through the nonfrigid months, I comb the refuge's mainland segments as a volunteer, not for signs of the showy birdlife but for cryptic spiders sheltered in marshes, woods, and shoreline rocks.

Petit Manan is a 10-acre granite pedestal overlaid by a mat of turf and rank plants and marked dramatically on the sea by a 123-foot stone lighthouse. That it still hosts one of the premier seabird colonies in the Gulf of Maine is a tribute to the efficacy of wildlife protection and management. In 1901, when plume hunters were decimating tern colonies on the Maine coast, William Dutcher (four years later he would become the first president of the new National Association of Audubon Societies) hired the keepers of various lighthouses to protect their islands' nesting birds. On Petit Manan the tern colony rebounded as the warden system took effect.

The local terns again went into eclipse in the 1970s, this time after the Coast Guard automated the lighthouses and removed the keepers. Now the threat was great black-backed and herring gulls, which took over the island, driving out the terns. After the

Coast Guard ceded management of the birds to the Fish and Wildlife Service in 1974, the island became a refuge on which biologists were eventually able to organize a management program that sent the gulls packing. Cooperation has been vital here, as specialists from the National Audubon Society's Seabird Restoration Program, led by Steve Kress, worked with the U.S. Fish and Wildlife Service on a number of Maine islands—particularly in fostering new colonies of Atlantic puffins.

Although the terns have had their ups and downs, today Petit Manan Island presents a noisy, turbulent scene. The air above it is pierced by the cries of thousands of common, Arctic, and endangered roseate terns, while the birds' abundant droppings spread a pervasive acrid odor across the refuge. Atlantic puffins, attracted by the tumult, have now joined laughing gulls, Leach's storm-petrels, and other species that have been given new leases on life by attentive refuge management.

PETIT MANAN ISLAND IS BUT ONE IN A COMPLEX OF 42 offshore refuges that its managing agency, the Fish and Wildlife Service, likes to say is "strung along the Maine coast like a strand of pearls," providing invaluable habitat for seabirds, wading birds, and bald eagles. Yet this complex is a small part of the most extensive system of lands protected for wildlife anywhere on our planet. It reaches down the Atlantic coast to the Caribbean and across the Gulf of Mexico and the continental United States to embrace hundreds of islands along the west coast from the Mexican border to beyond the Arctic Circle in Alaska. From there the system sweeps across the Pacific to provide an umbrella of protection for the remotest U.S. possessions in that vast ocean.

The National Wildlife Refuge System protects 700 species of birds and an equal number of other vertebrate animals; 282 of the species are either threatened or endangered.

The National Wildlife Refuge System is unique in that it was created specifically to protect the nation's wild animals and plants. It carries out its mandate on 540 refuges covering 95 million acres of water, shoreline, desert, rock, marshland, forest, prairie, tundra, and swamp. It protects 700 species of birds and an equal number of other vertebrate animals; 282 of the species are either threatened or endangered.

Consider the diversity in the following.

• **Okefenokee National Wildlife Refuge** (NWR), in southern Georgia and northeastern Florida, where nearly 400,000 acres of peat bogs, lakes, cypress swamps, and pinelands nurture carnivorous plants, 230 species of birds—including the endangered red-cockaded woodpecker and the American wood stork—15,000 alligators, and the greatest concentration of black bears in the Southeast.

• **Ash Meadows NWR**, in southern Nevada, a desert oasis of springs and seeps that is home to two dozen species of obscure and endangered plants and animals, including the Devil's Hole pupfish and an endangered herbaceous plant called the Amargosa niterwort.

• **Quivira NWR**, in central Kansas, attracts clouds of shorebirds, including black-necked stilts, American avocets, and two dozen or more additional shorebird species to its salt flats and shallow, wind-whipped water. Some half-million or so geese and ducks stop over at Quivira on their way to their wintering grounds. Such refuges help Kansas reach a state list of more than 460 bird species, among the richest in the United States.

• **Bitter Lake NWR**, near Roswell, New Mexico, where UFOs are akin to imaginary toads in real gardens, but North America's largest species of dragonfly and smallest damselfly provide genuine sightings for visitors.

The other public lands exist chiefly for people—for recreation or exploitation. The refuges are unique in that they are set aside by the federal government chiefly for wildlife.

Yet the system remains one of America's best-kept secrets. Its 95 million acres (and consequently its constituency and $368 million annual budget) are dwarfed by the holdings of the other big natural resource agencies—the National Park Service and Bureau of Land Management manage 350 million acres between them; the U.S. Forest Service, 200 million acres. There is at least one national wildlife refuge within an hour's drive of every major city, and each year more than 35 million people visit them. Still, national parks attract nearly 10 times that number. Therein lies an explanation: The other public lands exist chiefly for people—for recreation or exploitation. The refuges are unique in that they are set aside chiefly for wildlife. Their reason for being was perhaps best explained by the Fish and Wildlife Service's most illustrious employee, Rachel Carson, who served as both an aquatic biologist and the chief editor of the agency's publications between 1936 and 1952.

"Wild creatures, like men, must have a place to live. As civilization creates cities, builds highways, and drains marshes, it takes away, little by little, the land that is suitable for wildlife."
—Rachel Carson

"Wild creatures, like men, must have a place to live," Carson wrote. "As civilization creates cities, builds highways, and drains marshes, it takes away, little by little, the land that is suitable for wildlife. And as their space for living dwindles, the wildlife populations themselves decline. Refuges resist this trend by saving some areas from encroachment, and by preserving in them, or restoring where necessary, the conditions that wild things need in order to live."

The protection and management of wildlife populations has been the highest priority for the refuges from the start. One can trace a direct line in time and space from the granite base of Petit Manan Island south along the Atlantic coastline to an even smaller site on Florida's Indian River called Pelican Island. They are coevals in the uncertain venture of preserving America's birds. About the same time Audubon's William Dutcher was paying lighthouse keepers to guard the seabirds on Maine islands, President Theodore Roosevelt chose Pelican Island, an unlikely 5.5-acre lump of shells and mud (now eroded to a bit more than 2 acres) as the foundation of what would become the National Wildlife Refuge System.

Roosevelt's passion for the natural world is unequaled among U.S. presidents. "The conservation of our natural resources and their proper use constitute the fundamental problem which underlies almost every other problem of our national life," he once told Congress. He listed birds and bison, as well as timber and minerals, as natural treasures. Among Roosevelt's many naturalist friends was Frank M. Chapman, an ornithologist at New York City's American Museum of Natural

History, a prolific writer, and the founder of *Bird-Lore* (the magazine that became *Audubon*).

"The conservation of our natural resources and their proper use constitute the fundamental problem which underlies almost every other problem of our national life," Theodore Roosevelt once told Congress.

Chapman once described Pelican Island during nesting season as "by far the most fascinating place it has ever been my fortune to see in the world of birds." It was an estimate colored perhaps by the fact that he and his bride, Fanny, had spent part of their honeymoon on the island preparing pelican skins for the museum collection. Plume hunters had destroyed colonies of wading birds on the island in earlier years, and Chapman was aware of the vulnerability of this colony's 3,000 pelicans. It was a dangerous time for bird protectors in Florida: Audubon warden Guy Bradley was to be murdered in the Everglades in 1905. Chapman urged members of various Audubon societies to buy the island from the government and protect it forever. In 1903, when the plan bogged down in red tape, he and Dutcher took the matter to Roosevelt. The sympathetic president promptly declared Pelican Island "a preserve and breeding ground for native birds" and with this stroke created the first federal wildlife refuge.

Before Roosevelt left office in 1909, he had set aside more than 50 refuges on federal property, including 6 in Alaska. During the system's early years, before Congress appropriated money to monitor the refuges, the Audubon societies paid some of the wardens' salaries or supplied them with patrol boats. The practice continued whenever certain refuges were shortchanged by the government, and persisted into the 1920s here and there in Florida and along the Louisiana coast.

Roosevelt's foresight becomes even more apparent with time. During his administration, plume hunters from Japan had ransacked their way across the Pacific and had reached atolls on the fringes of the Hawaiian Islands many hundreds of miles from Honolulu. Specks of land dusted in a faint arc toward nowhere, those atolls are more remote from any continent than all other islands on earth. Yet all lie open to man's tireless penchant for predatory ventures. As marine biologist Carl Safina writes in the *Eye of the Albatross*, "More than ninety percent of the birds that have gone extinct over the last few centuries have been island nesters, including both sea and land birds." In 1909, again by executive order, Roosevelt set aside most of those outlying atolls in what is now the Pacific Remote Islands National Wildlife Refuge Complex.

Today the islands are home to at least 18 species of seabirds, Laysan and black-footed albatrosses prominent among them. Last year Chandler Robbins, biologist and bird bander extraor-

dinaire for the Fish and Wildlife Service, visited sites on Midway Atoll and adjacent refuge lands where he had studied albatrosses 36 years earlier. "I recaptured 200 previously banded albatrosses, one of which was a Laysan albatross I had banded as a five-plus-year-old nesting bird on my first trip there in 1956," he wrote recently. "At age 51, now wearing its fifth successive band, it was raising a newly hatched chick. This sets a new record as the oldest wild bird in the files of the Bird Banding Lab."

As early as 1843, John James Audubon had predicted the extinction of the American bison. In 1905 Theodore Roosevelt, sharing that concern, set aside a "game preserve" for bison in an Oklahoma forest reserve (now part of Wichita Mountains NWR). But by 1908 the bison's plight was dire enough to touch the heartstrings of even legislators. So Congress authorized the purchase of some 13,000 acres on the Flathead Indian Reservation in western Montana to preserve a remnant herd. It was the first federal appropriation to buy land specifically for wildlife protection. Today the 18,763-acre National Bison Range (part of the refuge system) shelters not only the species it was acquired to protect but other animals, too, including pronghorns, elk, deer, and bighorn sheep.

Under the stewardship of the Fish and Wildlife Service the refuge system has expanded into every U.S. state. The agency's mission has been complicated over time as more special interests want a piece of the action. By law, wildlife protection holds priority over other refuge uses. Yet refuges permit cattle grazing, which damages grasslands, polluted water seeps in from nearby developments, and snowmobiles and other unsuitable vehicles destroy wildlife habitat.

Today hunting takes place on more than half of the 540 refuges. How does the agency justify hunting in the light of its mission to protect wildlife? With two reasoned arguments: that it manages wildlife populations, not individual animals (thus putting itself at odds with animal-rights groups); and that the scarcity of predators on many refuges allows populations of prey species to "explode," leading to habitat destruction and, inevitably, to disease and starvation.

Through dedicated refuge staffs, however, protection and management on these lands is strengthened now by wildlife research and population surveys. Refuge biologists try to understand the forces (water shortages, predation, etc.) that reduce waterfowl breeding in the prairie pothole country, or how they can increase vegetative cover for seabirds on nesting islands. From Mille Lacs NWR, two rocky islets comprising one-half acre in Minnesota's Mille Lacs—where the Fish and Wildlife Service struggles to protect a vulnerable colony of common terns against the depredations of ring-billed gulls—to the awesome complexity of the 19.8-million-acre Arctic NWR in Alaska—where the oil industry wants to drill the "American Serengeti"—the system generally works.

LIKE THEIR COLLEAGUES AT MANY OTHER UNITS IN THE system, the staff at Petit Manan has studied its birdlife in great detail but is only beginning in-depth surveys of the full range of species diversity. Freelance biologists and college interns help inventory the plants, reptiles, and amphibians on the island and its adjacent mainland units. A group of professionals and amateurs, organized in the Friends of the Petit Manan National Wildlife Refuge, do volunteer work. I have found my niche assisting Daniel T. Jennings, a retired U.S. Forest Service research entomologist, in collecting spiders. To date we have bagged 178 species in 17 different families, one or two of the species apparently new to science.

Why spiders? These remarkable animals, which perhaps evolved silken webs to counteract the development of flight in their insect prey, are among the planet's dominant terrestrial predators. They make up in numbers for their small size: araneologists (spider experts) have identified more than 37,000 species and believe there may be five or six times that many out there. In other words, spiders are significant as both predators and prey in almost all land-based ecosystems.

It's sweaty work, slogging through a freshwater marsh to sweep crab spiders from grasses with a tough canvas net, or dig-ging pitfall traps with a small trowel among the forest's tangled roots to snare a new species of wolf spider for the refuge list. But there's the thrill of a successful chase, intensified by a sharpened awareness that all around me is a community of drag-onflies, wood frogs, black-bellied plovers, red foxes, and the myriad other creatures whose well-being justifies the existence of this countrywide refuge system.

All over the world species drift into extinction even before biologists have a chance to record them. The cumulative treasures on our refuges have yet to be fully documented, but the precedent set 100 years ago at Pelican Island ensures that most of them will hang around to be counted.

Field editor FRANK GRAHAM JR. *lives in Maine, just five miles from the Petit Manan National Wildlife Refuge.*

Invasive Species: Pathogens of Globalization

by Christopher Bright

Worrld trade has become the primary driver of one of the most dangerous and least visible forms of environmental decline: Thousands of foreign, invasive species are hitch-hiking through the global trading network aboard ships, planes, and railroad cars, while hundreds of others are traveling as commodities. The impact of these bioinvasions can now be seen on every landmass, in nearly all coastal waters (which comprise the most biologically productive parts of the oceans), and probably in most major rivers and lakes. This "biological pollution" is degrading ecosystems, threatening public health, and costing billions of dollars annually. Confronting the problem may now be as critical an environmental challenge as reducing global carbon emissions.

Despite such dangers, policies aimed at stopping the spread of invasive "exotic" species have so far been largely ineffective. Not only do they run up against far more powerful policies and interests that in one way or another encourage invasion, but the national and international mechanisms needed to control the spread of non-native species are still relatively undeveloped. Unlike chemical pollution, for instance, bioinvasion is not yet a working category of environmental decline within the legal culture of most countries and international institutions.

In part, this conceptual blindness can be explained by the fact that even badly invaded landscapes can still look healthy. It is also a consequence of the ancient and widespread practice of introducing exotic species for some tangible benefit: A bigger fish makes for better fishing, a faster-growing tree means more wood. It can be difficult to think of these activities as a form of ecological corrosion—even if the fish or the tree ends up demolishing the original natural community.

The increasing integration of the world's economies is rapidly making a bad situation even worse. The continual expansion of world trade—in ways that are not shaped by any real understanding of their environmental effects—is causing a degree of ecological mixing that appears to have no evolutionary precedent. Under more or less natural conditions, the arrival of an entirely new organism was a rare event in most times and places. Today it can happen any time a ship comes into port or an airplane lands. The real problem, in other words, does not lie with the exotic species themselves, but with the economic system that is continually showering them over the Earth's surface. Bioinvasion has become a kind of globalization disease.

They Came, They Bred, They Conquered

Bioinvasion occurs when a species finds its way into an ecosystem where it did not evolve. Most of the time when this happens, conditions are not suitable for the new arrival, and it enjoys only a brief career. But in a small percentage of cases, the exotic finds everything it needs—and nothing capable of controlling it. At the very least, the invading organism is liable to suppress some native species by consuming resources that they would have used instead. At worst, the invader may rewrite some basic ecosystem "rules"—checks and balances that have developed between native species, usually over many millennia.

Although it is not always easy to discern the full extent of havoc that invasive species can wreak upon an ecosystem, the resulting financial damage is becoming increasingly difficult to ignore. Worldwide, the losses to agriculture might be anywhere from $55 billion to nearly $248 billion annually. Researchers at Cornell University recently concluded that bioinvasion might be costing the United States alone as much as $123 billion per year. In South and Central America, the growth of specialty export crops—upscale vegetables and fruits—has spurred the spread of whiteflies, which are capable of transmitting at least 60 plant viruses. The spread of these viruses has forced the abandonment of more than 1 million hectares of cropland in South America. In the wetlands of northern Nigeria, an exotic cattail is strangling rice paddies, ruining fish habitats, and slowly choking off the Hadejia-Nguru river system. In southern India, a tropical American shrub, the bush morning glory, is causing similar chaos throughout the basin of the Cauvery, one of the region's biggest rivers. In the late 1980s, the accidental release into the Black Sea of *Mnemiopsis leidyi*—a comb jelly native to the east coast of the Americas—provoked the collapse of the already highly stressed Black Sea fisheries, with estimated financial losses as high as $350 million.

Controlling invasive species is difficult enough, but the bigger problem is preventing the machinery of the world trading system from releasing them in the first place.

Controlling such exotics in the field is difficult enough, but the bigger problem is preventing the machinery of the world trading system from releasing them in the first place. That task is becoming steadily more formidable as the trading system continues to grow. Since 1950, world trade has expanded sixfold in terms of value. More important in terms of potential invasions is the vast increase in the volume of goods traded. Look, for instance, at the ship, the primary mechanism of trade—80 percent of the world's goods travel by ship for at least part of their journey from manufacturer to consumer. From 1970 to 1996, the volume of seaborne trade nearly doubled.

Ships, of course, have always carried species from place to place. In the days of sail, shipworms bored into the wooden hulls, while barnacles and seaweeds attached themselves to the sides. A small menagerie of other creatures usually took up residence within these "fouling communities." Today, special paints and rapid crossing times have greatly reduced hull fouling, but each of the 28,700 ships in the world's major merchant fleets represents a honeycomb of potential habitats for all sorts of life, both terrestrial and aquatic.

The most important of these habitats lies deep within a modern ship's plumbing, in the ballast tanks. The ballast tanks of a really big ship—say, a supertanker—may contain more than 200,000 cubic meters of water—equivalent to 2,000 Olympic-sized swimming pools. When those tanks are filled, any little creatures in the nearby water or sediment may suddenly become inadvertent passengers. A few days or weeks later, when the tanks are discharged at journey's end, they may

become residents of a coastal community on the other side of the world. Every year, these artificial ballast currents move some 10 billion cubic meters of water from port to port. Every day, some 3,000 to 10,000 different species are thought to be riding the ballast currents. The result is a creeping homogenization of estuary and bay life. The same creatures come to dominate one coastline after another, eroding the biological diversity of the planet's coastal zones—and jeopardizing their ecological stability.

Some pathways of invasion extend far beyond ships. Another prime mechanism of trade is the container: the metal box that has revolutionized the transportation of just about every good not shipped in bulk. The container's effect on invasion ecology has been just as profound. For centuries, shipborne exotics were largely confined to port areas—but no longer. Containers move from ship to harbor crane to the flatbed of a truck or railroad car and then on to wherever a road or railroad leads. As a result, all sorts of stowaways that creep aboard containers often wind up far inland. Take the Asian tiger mosquito, for example, which can carry dengue fever, yellow fever, and encephalitis. The huge global trade in containers of used tires—which are, under the right conditions, an ideal mosquito habitat—has dispersed this species from Asia and the Indo-Pacific into Australia, Brazil, the eastern United States, Mozambique, New Zealand, Nigeria, and southern Europe. Even packing material within containers can be a conduit for exotic species. Untreated wood pallets, for example, are to forest pests what tires are to mosquitoes. One creature currently moving along this pathway is the Asian long-horn beetle, a wood-boring insect from China with a lethal appetite for deciduous trees. It has turned up at more than 30 locations around the United States and has also been detected in Great Britain. The only known way to eradicate it is to cut every tree suspected of harboring it, chip all the wood, and burn all the chips.

As other conduits for global trade expand, so does the potential for new invasions. Air cargo service, for example, is building a global network of virtual canals that have great potential for transporting tiny, short-lived creatures such as microbes and insects. In 1989, only three airports received more than 1 million tons of cargo; by 1996, there were 13 such airports. Virtually everywhere you look, the newly constructed infrastructure of the global economy is forming the groundwork for an ever-greater volume of biological pollution.

The Global Supermarket

Bioinvasion cannot simply be attributed to trade in general, since not all trade is "biologically dirty." The natural resource industries—especially agriculture, aquaculture, and forestry—are causing a disproportionate share of the problem. Certain trends within each of these industries are liable to exacerbate the invasion pressure. The migration of crop pests can be attributed, in part, to a global agricultural system that has become increasingly uniform and integrated. (In China, for example, there were about 10,000 varieties of wheat being grown at mid-century; by

1970 there were only about 1,000.) Any new pest—or any new form of an old pest—that emerges in one field may eventually wind up in another.

The key reason that South America has suffered so badly from white-flies, for instance, is because a pesticide-resistant biotype of that fly emerged in California in the 1980s and rapidly became one of the world's most virulent crop pests. The fly's career illustrates a common dynamic: A pest can enter the system, disperse throughout it, and then develop new strains that reinvade other parts of the system. The displacement of traditional developing-world crop varieties by commercial, homogenous varieties that require more pesticide, and the increasing development of pesticide resistance among all the major pest categories—insects, weeds, and fungi—are likely to boost this trend.

Similar problems pertain to aquaculture—the farming and exporting of fish, shellfish, and shrimp. Partly because of the progressive depletion of the world's most productive fishing grounds, aquaculture is a booming business. Farmed fish production exceeded 23 million tons by 1996, more than triple the volume just 12 years before. Developing countries in particular see aquaculture as a way of increasing protein supply.

But many aquaculture "crops" have proved very invasive. In much of the developing world, it is still common to release exotic fish directly into natural waterways. It is hardly surprising, then, that some of the most popular aquaculture fish have become true cosmopolitans. The Mozambique tilapia, for example, is now established in virtually every tropical and subtropical country. Many of these introductions—not just tilapia, but bass, carp, trout, and other types of fish—are implicated in the decline of native species. The constant flow of new introductions catalogued with such enthusiasm in the industry's publications are a virtual guarantee that tropical freshwater ecosystems are unraveling beneath the surface.

Aquaculture is also a spectacularly efficient conduit of disease. Perhaps the most virulent set of wildlife epidemics circling the Earth today involves shrimp production in the developing world. Unlike fish, shrimp are not a subsistence crop: They are an extremely lucrative export business that has led to the bulldozing of many tropical coasts to make way for shrimp ponds. One of the biggest current developments is an Indonesian operation that may eventually cover 200,000 hectares. A horde of shrimp pathogens—everything from viruses to protozoa—is chasing these operations, knocking out ponds, and occasionally ruining entire national shrimp industries: in Taiwan in 1987, in China in 1993, and in India in 1994. Shrimp farming has become, in effect, a form of "managed invasion." Since shrimp are important components of both marine and freshwater ecosystems worldwide, it is anybody's guess at this point what impact shrimpborne pathogens will ultimately have.

Managed invasion is an increasingly common procedure in another big biopolluting industry: forestry. Industrial roundwood production (basically, the cutting of logs for uses other than fuel) currently hovers at around 1.5 billion cubic meters annually, which is more than twice the level of the 1950s. An increasing amount of wood and wood pulp is coming out of tree plantations (not inherently a bad idea, given the rate at which

the world is losing natural forests). In North America and Europe, plantation forestry generally uses native species, so the gradation from natural forest to plantation is not usually as stark as it is in developing countries, where exotics are the rule in industrial-plantation development.

For the most part, these developing-country plantations bear about as much resemblance to natural forests as corn fields do to undisturbed prairies. And like corn fields, they are maintained with heavy doses of pesticides and subjected to a level of disturbance—in particular, the use of heavy equipment to harvest the trees—that tends to degrade soil. Some plantation trees have launched careers as king-sized weeds. At least 19 species of exotic pine, for example, have invaded various regions in the Southern Hemisphere, where they have displaced native vegetation and, in some areas, apparently lowered the water tables by "drinking" more water than the native vegetation would consume. Even where the trees have not proved invasive, the exotic plantations themselves are displacing natural forest and traditional forest peoples. This type of tree plantation is almost entirely designed to feed wood to the industrialized world, where 77 percent of industrial roundwood is consumed. As with shrimp production, local ecological health is being sacrificed for foreign currency.

There is another, more poignant motive for the introduction of large numbers of exotic trees into the developing world. In many countries severely affected by forest loss, reforestation is recognized as an important social imperative. But the goal is often nothing more than increasing tree cover. Little distinction is made between plantation and forest or between foreign and native species. Surayya Khatoon, a botanist at the University of Karachi, observes that "awareness of the dangers associated with invasive species is almost nonexistent in Pakistan, where alien species are being planted on a large scale in so-called afforestation drives."

Even international agreements that focus specifically on ecological problems have generally given bioinvasion short shrift.

The industrial sources of biological pollution are very diverse, but they reflect a common mindset. Whether it is a tree plantation, a shrimp farm, or even a bit of landscaping in the back yard, the Earth has become a sort of "species supermarket"; if a species looks good for whatever it is that you have in mind, pull it off the shelf and take it home. The problem is that many of the traits you want the most—adaptability, rapid growth, and easy reproduction—also tend to make the organism a good candidate for invasion.

Launching a Counter-Attack

Since the processes of invasion are deeply embedded in the globalizing economy, any serious effort to root them out will run the risk of exhausting itself. Most industries and policymakers are striving to open borders, not erect new barriers to trade. Moreover, because bioinvasion is not yet an established policy category, jurisdiction over it is generally badly fragmented—or even absent—on both the national and international levels. Most countries have some relevant legislation—laws intended to discourage the movement of crop pests, for example—but very few have any overall legislative authority for dealing with the problem. (New Zealand is the noteworthy exception: Its Biosecurity Act of 1993 and its Hazardous Substances and New Organisms Act of 1996 do establish such an authority.) Although it is true that there are many treaties that bear on the problem in one way or another—23 at least count—there is no such thing as a bioinvasion treaty.

Even agreements that focus specifically on ecological problems have generally given bioinvasion short shrift. Agenda 21, for example—the blueprint for sustainable development that emerged from the 1992 Earth Summit in Rio de Janeiro—reflects little awareness of the dangers of exotic forestry and aquaculture. Among international agencies, only certain types of invasion seem to get much attention. There are treaties—such as the 1951 International Plant Protection Convention—that limit the movement of agricultural pests, but there is currently no clear international mechanism for dealing with ballast water releases. Obviously, in such a context, you need to pick your fights carefully. They have to be important, winnable, and capable of yielding major opportunities elsewhere. The following three-point agenda offers some hope of slowing invasion over the near term.

The first item: Plug the ballast water pathway. As a technical problem, this objective is probably just on the horizon of feasibility, making it an excellent policy target. Strong national and international action could push technologies ahead rapidly. At present, the most effective technique is ballast water exchange, in which the tanks of a ship are pumped out and refilled in the open sea. (Coastal organisms, pumped into the tanks at the ship's last port of call, usually will not survive in the open ocean; organisms that enter the tanks in mid-ocean probably will not survive in the next port of call.) But it can take several days to exchange the water in all of a ship's ballast tanks, so the procedure may not be feasible for every leg of a journey, and the tanks never empty completely. In bad weather, the process can be too dangerous to perform at all. Consequently, other options will be necessary—filters or even toxins (that may not sound very appealing, but some common water treatment compounds may be environmentally sound). It might even be possible to build port-side ballast water treatment plants. Such a mixture of

technologies already exists as the standard means of controlling chemical pollution.

This objective is drifting into the realm of legal possibility as well. As of July 1 this year, all ships entering U.S. waters must keep a record of their ballast water management. The United States has also issued voluntary guidelines on where those ships can release ballast water. These measures are a loose extension of the regulations that the United States and Canada have imposed on ship traffic in the Great Lakes, where foreign ballast water release is now explicitly forbidden. In California, the State Water Resources Control Board has declared San Francisco Bay "impaired" because it is so badly invaded—a move that may allow authorities to use regulations written for chemical pollution as a way of controlling ballast water. Australia now levies a small tax on all incoming ships to support ballast water research.

Internationally, the problem has acquired a high profile at the UN International Maritime Organization (IMO), which is studying the possibility of developing a ballast management protocol that would have the force of international law. No decision has been made on the legal mechanism for such an agreement, although the most likely possibility is an annex to MARPOL, the International Convention for the Prevention of Pollution from Ships.

Within the shipping industry, the responses to such proposals have been mixed. Although industry officials concede the problem in the abstract, the prospect of specific regulations has tended to provoke unfavorable comment. After an IMO meeting last year on ballast water management, a spokesperson for the International Chamber of Shipping argued that rigorous ballast exchange would cost the industry millions of dollars a year—and that internationally binding regulations should be avoided in favor of local regulation, wherever particular jurisdictions decide to address the problem. Earlier this year in California, a proposed bill that would have essentially prohibited foreign ballast water release in the state's ports provoked outcries from local port representatives, who argued that such regulations might encourage ship traffic to bypass California ports in favor of the Pacific Northwest or Mexico. Of course, any management strategy is bound to cost something, but the important question is: What impact will this additional cost have? It may not have much impact at all. In Canada, for example, the Vancouver Port Authority reported that its ballast water program has had no detectable effect on port revenues.

The second item on the agenda: Fix the World Trade Organization (WTO) Agreement on the Application of Sanitary and Phytosanitary Measures. This agreement, known as the SPS, was part of the diplomatic package that created the WTO in 1994. The SPS is supposed to promote a common set of procedures for evaluating risks of contamination in internationally traded commodities. The contaminants can be chemical (pesticide residues in food) or they can be living things (Asian longhorn beetles in raw wood).

One of the procedures required by the SPS is a risk assessment, which is supposed to be done before any trade-constricting barriers are imposed to prevent a contaminated good from entering a country. If you want to understand the fundamental flaw in this approach as it applies to bioinvasion, all you have to do is recall the famous observation by the eminent biologist E. O. Wilson: "We dwell on a largely unexplored planet." When it comes to the largest categories of living things—insects, fungi, bacteria, and so on—we have managed to name only a tiny fraction of them, let alone figure out what damage they can cause. Consider, for example, the rough, aggregate risk assessments done by the United States Department of Agriculture (USDA) for wood imported into the United States from Chile, Mexico, and New Zealand. The USDA found dozens of "moderate" and "high" risk pests and pathogens that have the potential for doing economic damage on the order of hundreds of millions of dollars at least—and ecological damage that is incalculable. But even with wide-open thoroughfares of invasion such as these, the SPS requirement in its current form is likely to make preemptive action vulnerable to trade complaints before the WTO.

Another SPS requirement intended to insure a consistent application of standards is that a country must not set up barriers against an organism that is already living within its borders unless it has an "official control program" for that species. This approach is unrealistic for both biological and financial reasons. Thousands of exotic species are likely to have invaded most of the world's countries and not even the wealthiest country could possibly afford to fight them all. Yet it certainly is possible to exacerbate a problem by introducing additional infestations of a pest, or by boosting the size of existing infestations, or even by increasing the genetic vigor of a pest population by adding more "breeding stock." The SPS does not like "inconsistencies"—if you are not controlling a pest, you have no right to object to more of it; if you try to block one pathway of invasion, you had better be trying to block all the equivalent pathways. Such an approach may be theoretically neat, but in the practical matter of dealing with exotics, it is a prescription for paralysis.

In the near term, however, any effort to repair the SPS is likely to be difficult. The support of the United States, a key member of the WTO, will be critical for such reforms. And although the United States has demonstrated a heightened awareness of the problem—as evidenced by President Bill Clinton's executive order to create an Invasive Species Council—it is not clear whether that commitment will be reflected in the administration's trade policy. During recent testimony before Congress, the U.S. Trade Representative's special trade negotiator for agricultural issues warned that the United States was becoming impatient with the "increasing use of SPS barriers as the 'trade barrier of choice.'" In the developing world, it is reasonable to assume that any country with a strong export sector in a natural resource industry would not welcome tougher regulations. Some developed countries, however, may be sympathetic to change. The European Union (EU) has sought very strict standards in its disputes with the United States over bans on beef from cattle fed with growth hormones and on genetically altered foods. It is possible that the EU might be willing to entertain a stricter SPS. The same might be true of Japan, which has attempted to secure stricter testing of U.S. fruit imports.

The third item: Build a global invasion database. Currently, the study of bioinvasion is an obscure and rather fractured enterprise. It can be difficult to locate critical information or relevant expertise. The full magnitude of the issue is still not registering on the public radar screen. A global database would consolidate existing information, presumably into some sort of central research institution with a major presence on the World Wide Web. One could "go" to such a place—either physically or through cyberspace—to learn about everything from the National Ballast Water Information Clearinghouse that the U.S. Coast Guard is setting up, to the database on invasive woody plants in the tropics that is being assembled at the University of Wales. The database would also stimulate the production of new media to encourage additional research and synthesis. It is a telling indication of how fragmented this field is that, after more than 40 years of formal study, it is just now getting its first comprehensive journal: *Biological Invasions*.

Better information should have a number of practical effects. The best way to control an invasion—when it cannot be prevented outright—is to go after the exotic as soon as it is detected. An emergency response capability will only work if officials know what to look for and what to do when they find it. But beyond such obvious applications, the database could help bring the big picture into focus. In the struggle with exotics, you can see the free-trade ideal colliding with some hard ecological realities. Put simply: It may never be safe to ship certain goods to certain places—raw wood from Siberia, for instance, to North America. The notion of real, permanent limits to economic activity will for many politicians (and probably some economists) come as a strange and unpalatable idea. But the global economy is badly in need of a large dose of ecological realism. Ecosystems are very diverse and very different from each other. They need to stay that way if they are going to continue to function.

WANT TO KNOW MORE?

Although the scientific literature on bioinvasion is enormous and growing rapidly, most of it is too technical to attract a readership outside the field. For a nontechnical, broad overview of the problem, readers should consult Robert Devine's *Alien Invasion: America's Battle with Non-Native Animals and Plants* (Washington: National Geographic Society, 1998) or Christopher Bright's *Life Out of Bounds: Bioinvasion in a Borderless World* (New York: W.W. Norton & Company, 1998).

If you have a long-term interest in bioinvasion, you will want to get acquainted with the book that founded the field: Charles Elton's *The Ecology of Invasions by Animals and Plants* (London: Methuen, 1958). A historical overview of bioinvasions can be found in Alfred Crosby's book *Ecological Imperialism: The Biological Expansion of Europe, 900–1900* (Cambridge: Cambridge University Press, 1986).

Many studies focus on invasion of particular regions. The focus can be very broad, as in P. S. Ramakrishnan, ed., *Ecology of Biological Invasions in the Tropics*, proceedings of an international workshop held at Nainital, India, (New Delhi: International Scientific Publications, 1989). Generally, however, the coverage is much narrower, as in Daniel Simberloff, Don Schmitz, and Tom Brown, eds., *Strangers in Paradise: Impact and Management of Nonindigenous Species in Florida* (Washington: Island Press, 1997). The other standard research tack has been to look at a particular type of invader. The most accessible results of this exercise are encyclopedic surveys such as Christopher Lever's *Naturalized Mammals of the World* (London: Longman, 1985) and his companion volumes on naturalized birds and fish. In the plant kingdom, the genre is represented by Leroy Holm, et al., *World Weeds: Natural Histories and Distribution* (New York: John Wiley and Sons, 1997).

There are many worthwhile documents available for anyone who is interested not just in the ecology of invasion, but also in its economic, social, and epidemiological implications. Just about every aspect of the problem is discussed in Odd Terje Sandlund, Peter Johan Schei, and Aslaug Viken, eds., *Proceedings of the Norway/UN Conference on Alien Species* (Trondheim: Directorate for Nature Management and Norwegian Institute for Nature Research, 1996). A groundbreaking study of invasion in the United States, with particular emphasis on economic effects, is *Harmful Nonindigenous Species in the United States* (Washington: Office of Technology Assessment, September 1993). An assessment of the ballast water problem is available from the National Research Council's Commission on Ships' Ballast Operations' *Stemming the Tide: Controlling Introductions of Nonindigenous Species by Ships' Ballast Water* (Washington: National Academy Press, 1996). Readers who are interested in exotic tree plantations as a form of "managed invasion" might look through Ricardo Carrere and Larry Lohmann's *Pulping the South: Industrial Tree Plantations and the World Paper Economy* (London: Zed Books, 1996) and the World Rainforest Movement's *Tree Plantations: Impacts and Struggles* (Montevideo: WRM, 1999). Unfortunately, there are no analogous studies of shrimp farms.

For links to relevant Web sites, as well as a comprehensive index of related FOREIGN POLICY articles, access **www.foreignpolicy.com.**

Christopher Bright is a research associate at the Worldwatch Institute in Washington, DC, and author of *Life Out of Bounds: Bioinvasion in a Borderless World (New York: W.W. Norton & Company, 1998).*

On the Termination of Species

Ecologists' warnings of an ongoing mass extinction are being challenged by skeptics and largely ignored by politicians. In part that is because it is surprisingly hard to know the dimensions of the die-off, why it matters and how it can best be stopped.

By W. Wayt Gibbs

HILO, HAWAII—Among the scientists gathered here in August at the annual meeting of the Society for Conversation Biology, the despair was almost palpable. "I'm just glad I'm retiring soon and won't be around to see everything disappear," said P. Dee Boersma, former president of the society, during the opening night's dinner. Other veteran field biologists around the table murmured in sullen agreement.

At the next morning's keynote address, Robert M. May, a University of Oxford zoologist who presides over the Royal Society and until last year served as chief scientific adviser to the British government, did his best to disabuse any remaining optimists of their rosy outlook. According to his latest rough estimate, the extinction rate—the pace at which species vanish—accelerated during the past 100 years to roughly 1,000 times what it was before humans showed up. Various lines of argument, he explained, "suggest a speeding up by a further factor of 10 over the next century or so.... And that puts us squarely on the breaking edge of the sixth great wave of extinction in the history of life on Earth."

From there, May's lecture grew more depressing. Biologists and conservationists alike, he complained, are afflicted with a "total vertebrate chauvinism." Their bias toward mammals, birds and fish—when most of the diversity of life lies elsewhere—undermines scientists' ability to predict reliably the scope and consequences of biodiversity loss. It also raises troubling questions about the high-priority "hotspots" that environmental groups are scrambling to identify and preserve.

"Ultimately we have to ask ourselves why we care" about the planet's portfolio of species and its diminishment, May said. "This central question is a political and social question of values, one in which the voice of conservation scientists has no particular standing." Unfortunately, he concluded, of "the three kinds of argument we

Overview/Extinction Rates

- Eminent ecologists warn that humans are causing a mass extinction event of a severity not seen since the age of dinosaurs came to an end 65 million years ago. But paleontologists and statisticians have called such comparisons into doubt.
- It is hard to know how fast species are disappearing. Models based on the speed of tropical deforestation or on the growth of endangered species lists predict rising extinction rates. But biologists' bias toward plants and vertebrates, which represent a minority of life, undermine these predictions. Because 90 percent of species do not yet have names, let alone censuses, they are impossible to verify.
- In the face of uncertainty about the decline of biodiversity and its economic value, scientists are debating whether rare species should be the focus of conservation. Perhaps, some suggest, we should first try to save relatively pristine—and inexpensive—land where evolution can progress unaffected by human activity.

use to try to persuade politicians that all this is important... none is totally compelling."

Although May paints a truly dreadful picture, his is a common view for a field in which best-sellers carry titles such as *Requiem for Nature*. But is despair justified? *The Skeptical Environmentalist*, the new English translation of a recent book by Danish statistician Bjørn Lomborg, charges that reports of the death of biodiversity have

Mass Extinctions Past—and Present?

TIMELINE OF EXTINCTION marks the five most widespread die-offs in the fossil history of life on Earth.

Placoderm

Phytosaur teeth

END ORDOVICIAN
DURATION: **10 million years (my)**
MARINE GENERA OBSERVED EXTINGUISHED: **60%**
CALCULATED MARINE SPECIES EXTINCT: **85%**
SUSPECTED CAUSE: Dramatic fluctuations in sea level

END PERMIAN
DURATION: **Unknown**
MARINE GENERA OBSERVED EXTINGUISHED: **82%**
CALCULATED MARINE SPECIES EXTINCT: **95%**
SUSPECTED CAUSES: Dramatic fluctuations in climate or sea level; asteroid or comet impacts; severe volcanic activity

END CRETACEOUS
DURATION: **<1 my**
MARINE GENERA OBSERVED EXTINGUISHED: **47%**
CALCULATED MARINE SPECIES EXTINCT: **76%**
SUSPECTED CAUSES: Impact; severe volcanism

LATE DEVONIAN
DURATION: **<3 my**
MARINE GENERA OBSERVED EXTINGUISHED: **57%**
CALCULATED MARINE SPECIES EXTINCT: **83%**
SUSPECTED CAUSES: Impact; global cooling; loss of oxygen in oceans

END TRIASSIC
DURATION: **3 to 4 my**
MARINE GENERA OBSERVED EXTINGUISHED: **53%**
CALCULATED MARINE SPECIES EXTINCT: **80%**
SUSPECTED CAUSES: Severe volcanism; global warming

Trilobite

Rugose coral

Mosasaur

Millions of years ago

570	510	439	409	363	290	248	210	146	65	1.64
Cambrian	Ordovician	Silurian	Devonian	Carboniferous	Permian	Triassic	Jurassic	Cretaceous	Tertiary Quaternary	

With more than 1,100 species (*eight at right*) suspected to have disappeared in the past 500 years, ecologists fear a sixth mass extinction event is imminent. The die-offs so far, however, would probably not signal anything unusual to future paleontologists looking back at our time.

SPECIES (*Scientific name*)	LAST SEEN, LOCATION	EXTINCTION CAUSES
Deepwater ciscoe (*Coregonus johannae*)	1952, Lakes Huron and Michigan	Overfishing, hybridization
Pupfish (*Cyprinodon ceciliae*)	1988, Ojo de Agua La Presa, Mexico	Loss of food supply
Dobson's fruit bat (*Dobsonia chapmani*)	1970s, Cebu Islands, Philippines	Forest destruction, overhunting
Caribbean monk seal (*Monachus tropicalis*)	1950s, Caribbean Sea	Overhunting, harassment
Guam flycatcher (*Myiagra freycineti*)	1983, Guam	Predation by introduced brown tree snakes
Kaua'i 'O'o (*Moho braccatus*)	1987, Island of Kaua'i, Hawaii	Disease, rat predation
Xerces Blue Butterfly (*Glaucopsyche xerces*)	1941, San Francisco Peninsula	Land conversion
Tobias' Caddis Fly (*Hydropsyche tobiasi*)	1950s, Rhine River, Germany	Industrial and urban pollution

SOURCES: *Committee on Recently Extinct Organisms; BirdLife International; Xerces Society; World Wildlife Fund*
LESTER V. BERGMAN *Corbis* (trilobite); JAMES L. AMOS *Corbis* (*Placoderm*); RICHARD PASELK *Humboldt State University/Natural History Museum* (*Rugose coral and Phytosaur teeth*); MIKE EVERHART *Oceans of Kansas Paleontology* (*Mosasaur*)

been greatly exaggerated. In the face of such external skepticism, internal uncertainty and public apathy, some scientists are questioning the conservation movement's overriding emphasis on preserving rare species and the threatened hotspots in which they are concentrated. Perhaps, they suggest, we should focus instead on saving something equally at risk but even more valuable: evolution itself.

Doom...

MAY'S CLAIM that humans appear to be causing a cataclysm of extinctions more severe than any since the one that erased the dinosaurs 65 million years ago may shock those who haven't followed the biodiversity issue. But it prompted no gasps from the conservation biologists. They have heard variations of this dire forecast since at least 1979, when Normal Myers guessed in *The Sinking*

Ark that 40,000 species lose their last member each year and that one million would be extinct by 2000. In the 1980s Thomas Lovejoy similarly predicted that 15 to 20 percent would die off by 2000; Paul Ehrlich figured half would be gone by now. "I'm reasonably certain that [the elimination of one fifth of species] didn't happen," says Kirk O. Winemiller, a fish biologist at Texas A&M University who just finished a review of the scientific literature on extinction rates.

More recent projections factor in a slightly slower demise because some doomed species have hung on longer than anticipated. Indeed, a few have even returned from the grave. "It was discovered only this summer that the Bavarian vole, continental Eurasia's one and only presumed extinct mammal [since 1500], is in fact still with us," says Ross D. E. MacPhee, curator of mammalogy at the American Museum of Natural History (AMNH) in New York City.

The Portfolio of Life

How severe is the extinction crisis? That depends in large part on how many species there are altogether. The greater the number, the more species will die out every year from natural causes and the more new ones will naturally appear. But although the general outlines of the tree of life are clear, scientists are unsure how many twigs lie at the end of each branch. When it comes to bacteria, viruses, protists and archaea (a whole kingdom of single-celled life-forms discovered just a few decades ago), microbiologists have only vague notions of how many branches there are.

Birds, fish, mammals and plants are the exceptions. Sizing up the global workforce of about 5,000 professional taxonomists, zoologist Robert M. May of the University of Oxford noted that about equal numbers study vertebrates, plants and invertebrates. "You may wish to think this record reflects some

judicious appreciation of what's important," he says. "My view of that is: absolute garbage. Whether you are interested in how ecosystems evolved, their current functioning or how they are likely to respond to climate change, you're going to learn a lot more by looking at soil microorganisms than at charismatic vertebrates."

For every group except birds, says Peter Hammond of the National History Museum in London, new species are now being discovered faster than ever before, thanks to several new international projects. An All Taxa Biodiversity Inventory under way in Great Smoky Mountains National Park in North Carolina and Tennessee has discovered 115 species—80 percent of them insects or arachnids—in its first 18 months of work. Last year 40 scientists formed the All Species Project, a society devoted to the (probably quixotic) goal of cataloguing every living species, microbes included, within 25 years.

Other projects, such as the Global Biodiversity Information Facility and Species2000, are building Internet databases that will codify species records that are now scattered among the world's museums and universities. If biodiversity is defined in strictly pragmatic terms as the variety of life-forms we know about, it is growing prodigiously.

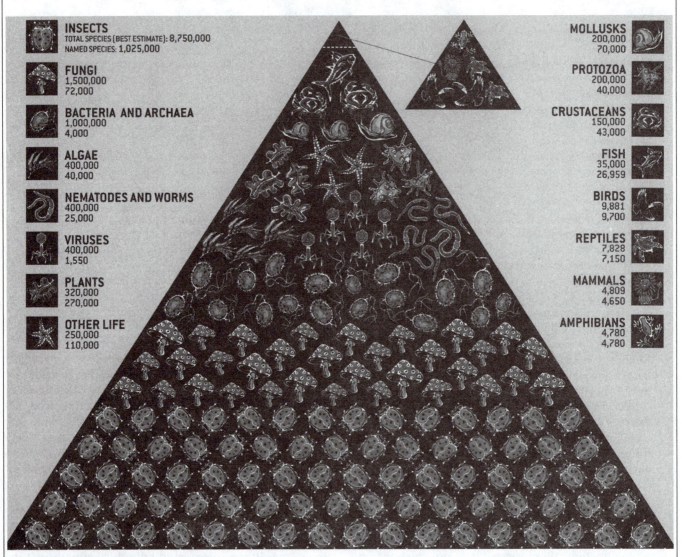

INSECTS
TOTAL SPECIES (BEST ESTIMATE): 8,750,000
NAMED SPECIES: 1,025,000

FUNGI
1,500,000
72,000

BACTERIA AND ARCHAEA
1,000,000
4,000

ALGAE
400,000
40,000

NEMATODES AND WORMS
400,000
25,000

VIRUSES
400,000
1,550

PLANTS
320,000
270,000

OTHER LIFE
250,000
110,000

MOLLUSKS
200,000
70,000

PROTOZOA
200,000
40,000

CRUSTACEANS
150,000
43,000

FISH
35,000
26,959

BIRDS
9,881
9,700

REPTILES
7,828
7,150

MAMMALS
4,809
4,650

AMPHIBIANS
4,780
4,780

PYRAMID OF DIVERSITY

TO A FIRST APPROXIMATION, all multicellular species are insects. Biologists know the least about the true diversity and ecological importance of the very groups that are most common.

SOURCES: Encyclopedia of Biodiversity, edited by S. A. Levin; "Biodiversity Hotspots for Conservation Priorities," by N. Myers et al. in *Nature*, Vol. 403, pages 853–858, February 24, 2000; William Eschemeyer (fish species); Marc Van Regenmortel (virus species); IUCN Red List 2000

Still, in the 1999 edition of his often-quoted book *The Diversity of Life*, Harvard University biologist E. O. Wilson cites current estimates that between 1 and 10 percent of species are extinguished every decade, at least 27,000 a year. Michael J. Novacek, AMNH's provost of science, wrote in a review article this spring that "figures approaching 30 percent extermination of all species by the mid-21st century are not unrealistic." And in a 1998 survey of biologists, 70 percent said they believed that a mass extinction is in progress; a third of them expected to lose 20 to 50 percent of the world's species within 30 years.

"Although these assertions of massive extinctions of species have been repeated everywhere you look, they do not equate with the available evidence," Lomborg argues in *The Skeptical Environmentalist*. A professor of statistics and political science at the University of Århus, he alleges that environmentalists have ignored recent evidence that tropical deforestation is not taking the toll that was feared. "No well-investigated group of animals shows a pattern of loss that is consistent with greatly heightened extinction rates," MacPhee concurs. The best models, Lomborg suggests, project an extinction rate of 0.15 percent of species per decade, "not a catastrophe but a problem—one of many that mankind still needs to solve."

… or Gloom?

"IT'S A TOUGH question to put numbers on," Wilson allows. May agrees but says "that isn't an argument for not asking the question" of whether a mass extinction event is upon us.

> "If you are looking for hard evidence of tens or hundreds or thousands of species disappearing each year, you aren't going to find it."
>
> —KIRK O. WINEMILLER, TEXAS A&M

To answer that question, we need to know three things: the natural (or "background") extinction rate, the current rate and whether the pace of extinction is steady or changing. The first step, Wilson explains, is to work out the mean life span of a species from the fossil record. "The background extinction rate is then the inverse of that. If species are born at random and all live exactly one million years—and it varies, but it's on that order—then that means one species in a million naturally goes extinct each year," he says.

In a 1995 article that is still cited in almost every scientific paper on this subject (even in Lomborg's book), May used a similar method to compute the background rate. He relied on estimates that put the mean species life span at five million to 10 million years, however; he thus came up with a rate that is five to 10 times lower than Wilson's.

But according to paleontologist David M. Raup (then at the University of Chicago), who published some of the figures May and Wilson relied on, their calculations are seriously flawed by three false assumptions.

One is that species of plants, mammals, insects, marine invertebrates and other groups all exist for about the same time. In fact, the typical survival time appears to vary among groups by a factor of 10 or more, with mammal species among the least durable. Second, they assume that all organisms have an equal chance of making it into the fossil record. But paleontologists estimate that fewer than 4 percent of all species that ever lived are preserved as fossils. "And the species we do see are the widespread, very successful ones," Raup says. "The weak species confined to some hilltop or island all went extinct before they could be fossilized," adds John Alroy of the University of California at Santa Barbara.

The third problem is that May and Wilson use an average life span when they should use a median. Because "the vast majority of species are short-lived," Raup says, "the average is distorted by the very few that have very long life spans." All three oversimplifications underestimate the background rate—and make the current picture scarier in comparison.

Earlier this year U.C.S.B. biomathematician Helen M. Regan and several of her colleagues published the first attempt ever to correct for the strong biases and uncertainties in the data. They looked exclusively at mammals, the best-studied group. They estimated how many of the mammals now living, and how many of those recently extinguished, would show up as fossils. They also factored in the uncertainty for each number rather than relying on best guesses. In the end they concluded that "the current rate of mammalian extinction lies between 17 and 377 times the background extinction rate." The best estimate, they wrote, is a 36- to 78-fold increase.

Regan's method is still imperfect. Comparing the past 400 years with the previous 65 million unavoidably assumes, she says, "that the current extinction rate will be sustained over millions of years." Alroy recently came up with a way to measure the speed of extinctions that doesn't suffer from such assumptions. Over the past 200 years, he figures, the rate of loss among mammal species has been some 120 times higher than natural.

A Grim Guessing Game

ATTEMPTS TO FIGURE out the current extinction rate are fraught with even more uncertainties. The international conservation organization IUCN keeps "Red Lists" of organisms suspected to be extinct in the wild. But MacPhee complains that "the IUCN methodology for recognizing extinction is not sufficiently rigorous to be reliable." He and other extinction experts have formed the Committee on Recently Extinct Organisms, which combed the Red Lists to identify those species that were clearly unique and that had not been found despite a reasonable search. They certified 60 of the 87 mammals listed by IUCN as extinct but claim that only 33 of the 92 freshwater fish presumed extinct by IUCN are definitely gone forever.

For every species falsely presumed absent, however, there may be hundreds or thousands that vanish unknown to science. "We are uncertain to a factor of 10 about how many species we share the planet with," May points out. "My guess would be roughly seven million, but credible guesses range from five to 15 million," excluding microorganisms.

Taxonomists have named approximately 1.8 million species, but biologists know almost nothing about most of them, especially the insects, nematodes and crustaceans that dominate the animal kingdom. Some 40 percent of the 400,000 known beetle species have each been recorded at just one location—and with no idea of individual species' range, scientists have no way to confirm its extinction. Even invertebrates known to be extinct often go unrecorded: when the passenger pigeon was eliminated in 1914, it took two species of parasitic lice with it. They still do not appear on IUCN's list.

"It is extremely difficult to observe an extinction; it's like seeing an airplane crash," Wilson says. Not that scientists aren't trying. Articles on the "biotic holocaust," as Myers calls it, usually figure that the vast majority of extinctions have been in the tropical Americas. Freshwater fishes are especially vulnerable, with more than a quarter listed as threatened. "I work in Venezuela, which has substantially more freshwater fishes than all of North America. After 30 years of work, we've done a reasonable job of cataloguing fish diversity there," observes Winemiller of Texas A&M, "yet we can't point to one documented case of extinction."

A similar pattern emerges for other groups of organisms, he claims. "If you are looking for hard evidence of tens or hundreds or thousands of species disappearing each year, you aren't going to find it. That could be because the database is woefully inadequate," he acknowledges. "But one shouldn't dismiss the possibility that it's not going to be the disaster everyone fears."

The Logic of Loss

THE DISASTER SCENARIOS are based on several independent lines of evidence that seem to point to fast and rising extinction rates. The most widely accepted is the species-area relation. "Generally speaking, as the area of habitat falls, the number of species living in its drops proportionally by the third root to the sixth root," explains Wilson, who first deduced this equation more than 30 years ago. "A middle value is the fourth root, which means that when you eliminate 90 percent of the habitat, the number of species falls by half."

"From that rough first estimate and the rate of the destruction of the tropical forest, which is about 1 percent a

Extinction *Filters*

SURVIVAL OF THE FITTEST takes on a new meaning when humans develop a region. Among four Mediterranean climate regions, those developed more recently have lost larger fractions of their vascular plant species in modern times. Once the species least compatible with agriculture are filtered out by "artificial selection," extinction rates seem to fall.

REGION [in order of development]	EXTINCT [per 1,000]	THREATENED [percent]
Mediterranean	1.3	14.7
South African Cape	3.0	15.2
California	4.0	10.2
Western Australia	6.6	17.5

SOURCE: "Extinctions in Mediterranean Areas." Werner Greuter in Extinction Rates. Edited by J. H. Lawton and R. H. May. Oxford University Press, 1995

year," Wilson continues, "we can predict that about one quarter of 1 percent of species either become extinct immediately or are doomed to much earlier extinction." From a pool of roughly 10 million species, we should thus expect about 25,000 to evaporate annually.

Lomborg challenges that view on three grounds, however. Species-area relations were worked out by comparing the number of species on islands and do not necessarily apply to fragmented habitats on the mainland. "More than half of Costa Rica's native bird species occur in largely deforested countryside habitats, together with similar fractions of mammals and butterflies," Stanford University biologist Gretchen Daily noted recently in *Nature*. Although they may not thrive, a large fraction of forest species may survive on farmland and in woodlots—for how long, no one yet knows.

That would help explain Lomborg's second observation, which is that in both the eastern U.S. and Puerto Rico, clearance of more than 98 percent of the primary forests did not wipe out half of the bird species in them. Four centuries of logging "resulted in the extinction of only one forest bird" out of 200 in the U.S. and seven out of 60 native species in Puerto Rico, he asserts.

Such criticisms misunderstand the species-area theory, according to Stuart L. Pimm of Columbia University. "Habitat destruction acts like a cookie cutter stamping out poorly mixed dough," he wrote last year in *Nature*. "Species found only within the stamped-out area are themselves stamped out. Those found more widely are not."

Of the 200 bird types in the forests of the eastern U.S., Pimm states, all but 28 also lived elsewhere. Moreover, the forest was cleared gradually, and gradually it regrew as farmland was abandoned. So even at the low point, around 1872, woodland covered half the extent of the original forest. The species-area theory predicts that a 50

Why Biodiversity Doesn't [Yet] Pay

FOZ DO IGUAÇU, BRAZIL—At the International Congress of Entomologists last summer, Ebbe Nielsen, director of the Australian National Insect Collection in Canberra, reflected on the reasons why, despite the 1992 Convention on Biological Diversity signed here in Brazil by 178 countries, so little has happened since to secure the world's threatened species. "You and I can say extinction rates are too high and we have to stop it, but to convince the politicians we have to have convincing reasons," he said. "In developing countries, the economic pressures are so high, people use whatever they can find today to survive until tomorrow. As long as that's the case, there will be no support for biodiversity at all."

Not, that is, unless it can be made more profitable to leave a forest standing or a wetland wet than it is to convert the land to farm, pasture or parking lot. Unfortunately, time has not been kind to the several arguments environmentalists have made to assign economic value to each one of perhaps 10 million species.

A Hedge against Disease and Famine

"Narrowly utilitarian arguments say: The incredible genetic diversity contained in the population and species diversity that we are heirs to is ultimately the raw stuff of tomorrow's biotechnological revolution," observes Robert May of Oxford. "It is the source of new drugs." Or new foods, adds E. O. Wilson of Harvard, should something happen to the 30 crops that supply 90 percent of the calories to the human diet, or to the 14 animal species that make up 90 percent of our livestock.

"Some people who say that may even believe it," May continues. "I don't. Give us 20 or 30 years and we will design new drugs from the molecule up, as we are already beginning to do."

Hopes were raised 10 years ago by reports that Merck has paid $1.14 million to InBio, a Costa Rican conservation group, for novel chemicals extracted from rain-forest species. The contract would return royalties to InBio if any of the leads became drugs. But none have, and Merck terminated the agreement in 1999. Shaman Pharmaceuticals, founded in 1989 to commercialize traditional medicinal plants, got as far as late-stage clinical trials but then went bankrupt. And given, as Wilson himself notes in *The Diversity of Life*, that more than 90 percent of the known varieties of the basic food plants are on deposit in seed banks, national parks are hardly the cheapest form of insurance against crop failures.

Ecosystem Services

"Potentially the strongest argument," May says, "is a broadly utilitarian one: ecological systems deliver services we're only just beginning to think of trying to estimate. We do not understand how much you can simplify these sys-

tems and yet still have them function. As Aldo Leopold once said, the first rule of intelligent tinkering is to keep all the pieces."

The trouble with this argument, explains Columbia University economist Geoffrey Heal, is that "it does not make sense to ask about the value of replacing a life-support system." Economics can only assign values to things for which there are markets, he says. If all oil were to vanish, for example, we could switch to alternative fuels that cost $50 a barrel. But that does not determine the price of oil.

And although recent experiments suggest that removing a large fraction of species from a small area lowers its biomass and ability to soak up carbon dioxide, scientists cannot say yet whether the principle applies to whole ecosystems. "It may be that a grievously simplified world—the world of the cult movie *Blade Runner*—can be so run that we can survive in it," May concedes.

A Duty of Stewardship

Because science knows so little of the millions of species out there, let alone what complex roles each one plays in the ecosystems it inhabits, it may never be possible for economics to come to the aid of endangered species. A moral argument may thus be the best last hope—certainly it is appeals to leaders' sense of stewardship that have accomplished the most so far. But is it hazardous for scientists to make it?

They do, of course, in various forms. To Wilson, "a species is a masterpiece of evolution, a million-year-old entity encoded by five billion genetic letters, exquisitely adapted to the niche it inhabits." For that reason, conservation biologist David Ehrenfeld proposed in *The Arrogance of Humanism*, "long-standing existence in Nature is deemed to carry with it the unimpeachable right to continued existence."

Winning public recognition of such a right will take much education and persuasion. According to a poll last year, fewer than one quarter of Americans recognized the term "biological diversity." Three quarters expressed concern about species and habitat loss, but that is down from 87 percent in 1996. And May observes that the concept of biodiversity stewardship "is a developed-world luxury. If we were in abject poverty trying to put food in the mouth of the fifth child, the argument would have less resonance."

But if scientists "proselytize on behalf of biodiversity"—as Wilson, Lovejoy, Ehrlich and many others have done—they should realize that "such work carries perils," advises David Takacs of California State University at Monterey Bay. "Advocacy threatens to undermine the perception of value neutrality and objectivity that leads laypersons to listen to scientists in the first place." And yet if those who know rare species best and love them most cannot speak openly on their behalf, who will?

percent reduction should knock out 16 percent of the endemic species: in this case, four birds. And four species did go extinct. Lomborg discounts one of those four that may have been a subspecies and two others that perhaps succumbed to unrelated insults.

But even if the species-area equation holds, Lomborg responds, official statistics suggest that deforestation has been slowing and is now well below 1 percent a year. The U.N. Food and Agriculture Organization recently estimated that from 1990 to 2000 the world's forest cover dropped at an average annual rate of 0.2 percent (11.5 million hectares felled, minus 2.5 million hectares of new growth).

Annual forest loss was around half a percent in most of the tropics, however, and that is where the great majority of rare and threatened species live. So although "forecasters may get these figures wrong now and then, perhaps colored by a desire to sound the alarm, this is just a matter of timescale," replies Carlos A. Peres, a Brazilian ecologist at the University of East Anglia in England.

An Uncertain Future

ECOLOGISTS HAVE TRIED other means to project future extinction rates. May and his co-workers watched how vertebrate species moved through the threat categories in IUCN's database over a four-year period (two years for plants), projected those very small numbers far into the future and concluded that extinction rates will rise 12- to 55-fold over the next 300 years. Georgina M. Mace, director of science at the Zoological Society of London, came to a similar conclusion by combining models that plot survival odds for a few very well known species. Entomologist Nigel E. Stork of the Natural History Museum in London noted that a British bird is 10 times more likely than a British bug to be endangered. He then extrapolated such ratios to the rest of the world to predict 100,000 to 500,000 insect extinctions by 2300. Lomborg favors this latter model, from which he concludes that "the rate for all animals will remain below 0.208 percent per decade and probably be below 0.7 percent per 50 years."

It takes a heroic act of courage for any scientist to erect such long and broad projections on such a thin and lopsided base of data. Especially when, according to May, the data on endangered species "may tell us more about the vagaries of sampling efforts, of taxonomists' interests and of data entry than about the real changes in species' status."

Biologists have some good theoretical reasons to fear that even if mass extinction hasn't begun yet, collapse is imminent. At the conference in Hilo, Kevin Higgins of the University of Oregon presented a computer model that tracks artificial organisms in a population, simulating their genetic mutation rates, reproductive behavior and ecological interactions. He found that "in small populations, mutations tend to be mild enough that natural selection doesn't filter them out. That dramatically shortens the time to extinction." So as habitats shrink and populations are wiped out—at a rate of perhaps 16 million a year, Daily has estimated—"this could be a time bomb, an extinction event occurring under the surface," Higgins warns. But proving that that bomb is ticking in the wild will not be easy.

And what will happen to fig trees, the most widespread plant genus in the tropics, if it loses the single parasitic wasp variety that pollinates every one of its 900 species? Or to the 79 percent of canopy-level trees in the Samoan rain forests if hunters kill off the flying foxes on which they depend? Part of the reason so many conservationists are so fearful is that they expect the arches of entire ecosystems to fall once a few "keystone" species are removed.

Others distrust that metaphor. "Several recent studies seem to show that there is some redundancy in ecosystems," says Melodie A. McGeoch of the University of Pretoria in South Africa, although she cautions that what is redundant today may not be redundant tomorrow. "It really doesn't make sense to think the majority of species would go down with marginally higher pressures than if humans weren't on the scene," MacPhee adds. "Evolution should make them resilient."

If natural selection doesn't do so, artificial selection might, according to work by Werner Greuter of the Free University of Berlin, Thomas M. Brooks of Conservation International and others. Greuter compared the rate of recent plant extinctions in four ecologically similar regions and discovered that the longest-settled, most disturbed area—the Mediterranean—had the lowest rate. Plant extinction rates were higher in California and South Africa, and they were highest in Western Australia. The solution to this apparent paradox, they propose, is that species that cannot coexist with human land use tend to die out soon after agriculture begins. Those that are left are better equipped to dodge the darts we throw at them. Human-induced extinctions may thus fall over time.

> "It turns out to be a lot easier to persuade a corporate CEO or a billionaire of the importance of the issue than it is to convince the American public."
> —EDWARD O. WILSON, HARVARD UNIVERSITY

If true, that has several implications. Millennia ago our ancestors may have killed off many more species than we care to think about in Europe, Asia and other long-settled regions. On the other hand, we may have more time than we fear to prevent future catastrophes in areas where humans have been part of the ecosystem for a while—and less time than we hope to avoid them in what little wilderness remains pristine.

"The question is how to deal with uncertainty, because there really is no way to make that uncertainty go away," Winemiller argues. "We think the situation is extremely serious; we just don't think the species extinction issue is the peg that conservation movement should hang its hat on. Otherwise, if it turns out to be wrong, where does that leave us?"

Long-Term Savings

IT COULD LEAVE conservationists with less of a sense of urgency and with a handful of weak political and economic arguments [*see box "Why Biodiversity Doesn't (Yet) Pay"*]. It might also force them to realize that "many of the species in trouble today are in fact already members of the doomed, living dead," as David S. Woodruff wrote in the *Proceedings of the National Academy of Sciences* this past May. "Triage" is a dirty word to many environmentalists. "Unless we say no species loss is acceptable, then we have no line in the sand to defend, and we will be pushed back and back as losses build," Brooks argued at the Hilo meetings. But losses are inevitable, Wilson says, until the human population stops growing.

"I call that the bottleneck," Wilson elaborates, "because we have to pass through that scramble for remaining resources in order to get to an era, perhaps sometime in the 22nd century, of declining population. Our goal is to carry as much of the biodiversity through as possible." Biologists are divided, however, on whether the few charismatic species now recognized as endangered should determine what gets pulled through the bottleneck.

"The argument that when you protect birds and mammals, the other things come with them just doesn't stand up to close examination," May says. A smarter goal is "to try to conserve the greatest amount of evolutionary history." Far more valuable than a panda or rhino, he suggests, are relic life-forms such as the tuatara, a large iguanalike reptile that lives only on islets off the coast of New Zealand. Just two species of tuatara remain from a group that branched off from the main stem of the reptilian evolutionary tree so long ago that this couple make up a genus, an order and almost a subclass all by themselves.

But Woodruff, who is an ecologist at the University of California at San Diego, invokes an even broader principle. "Some of us advocate a shift from saving things, the products of evolution, to saving the underlying process, evolution itself," he writes. "This process will ultimately provide us with the most cost-effective solution to the general problem of conserving nature."

There are still a few large areas where natural selection alone determines which species succeed and which fail. "Why not save functioning ecosystems that haven't been despoiled yet?" Winemiller asks. "Places like the Guyana shield region of South America contain far more species than some of the so-called hotspots." To do so would mean purchasing tracts large enough to accommodate entire ecosystems as they roll north and south in response to the shifting climate. It would also mean prohibiting all human uses of land. It may not be impossible: utterly undeveloped wilderness is relatively cheap, and the population of potential buyers has recently exploded.

"It turns out to be a lot easier to persuade a corporate CEO or a billionaire of the importance of the issue than it is to convince the American public," Wilson says. "With a Ted Turner or a Gordon Moore or a Craig McCaw involved, you can accomplish almost as much as a government of a developed country would with a fairly generous appropriation."

"Maybe even more," agrees Richard E. Rice, chief economist for Conservation International. With money from Moore, McCaw, Turner and other donors, CI has outcompeted logging companies for forested land in Suriname and Guyana. In Bolivia, Rice reports, "we conserved an area the size of Rhode Island for half the price of a house in my neighborhood," and the Nature Conservancy was able to have a swath of rain forest as big as Yellowstone National Park set aside for a mere $1.5 million. In late July, Peru issued to an environmental group the country's first "conservation concession"—essentially a renewable lease for the right to *not* develop the land—for 130,000 hectares of forest. Peru has now opened some 60 million hectares of its public forests to such concessions, Rice says. And efforts are under way to negotiate similar deals in Guatemala and Cameroon.

"Even without massive support in public opinion or really effective government policy in the U.S., things are turning upward," Wilson says, with a look of cautious optimism on his face. Perhaps it is a bit early to despair after all.

MORE TO EXPLORE

Extinction Rates. Edited by John H. Lawton and Robert M. May. Oxford University Press, 1995.

The Currency and Tempo of Extinction. Helen M. Regan et al. in the *American Naturalist*, Vol. 157, No. 1, pages 1–10; January 2001.

Encyclopedia of Biodiversity. Edited by Simon Asher Levin. Academic Press, 2001.

The Skeptical Environmentalist. Bj&&0248rn Lomborg. Cambridge University Press, 2001.

W. Wayt Gibbs is senior writer.

UNIT 5

Resources: Land and Water

Unit Selections

Key Points to Consider

• Why is the number of farmers in the world decreasing? What kinds of social, economic, and cultural impacts will be produced by decreasing the number of family farmers and increasing the amount of land farmed by corporate farmers?

• What are the conflicting uses of water that prompt such environmental problems as those recently experienced in the Klamath River basin in Oregon? Are there ways to resolve conflicts over the use of water that have not yet been implemented?

• Explain the relationship between agricultural production and irrigation. Are irrigation systems inefficient and wasteful of water? If so, why, and what could be done to eliminate such inefficiencies?

• To what extent do we depend upon groundwater for all purposes? Explain the two central problems of groundwater dependence: pollution and overwithdrawal. How can dependence upon groundwater be balanced with the maintenance of the quality and quantity of the groundwater resource?

• How has overfishing contributed to a decline in the food supply from the oceans? Has oceanic pollution contributed to this decline?

 **Links: www.dushkin.com/online/**
These sites are annotated in the World Wide Web pages.

Agriculture Production Statistics
http://www.wri.org/statistics/fao-prd.html

Global Climate Change
http://www.puc.state.oh.us/consumer/gcc/index.html

National Oceanic and Atmospheric Administration (NOAA)
http://www.noaa.gov

National Operational Hydrologic Remote Sensing Center (NOHRSC)
http://www.nohrsc.nws.gov

Virtual Seminar in Global Political Economy/Global Cities & Social Movements
http://csf.colorado.edu/gpe/gpe95b/resources.html

Terrestrial Sciences
http://www.cgd.ucar.edu/tss/

The worldwide situations regarding reduction of biodiversity, scarce energy resources, and pollution of the environment have received the greatest amount of attention among members of the environmentalist community. But there are a number of other resource issues that demonstrate the interrelated nature of all human activities and the environments in which they occur. One such issue is the declining quality of agricultural land. In the developing world, excessive rural populations have forced the overuse of lands and sparked a shift into marginal areas, and the total availability of new farmland is decreasing at an alarming rate of 2 percent per year. In the developed world, intensive mechanized agriculture has resulted in such a loss of topsoil that some agricultural experts are predicting a decline in food production. Other natural resources, such as minerals and timber, are declining in quantity and quality as well; in some cases they are no longer usable at present levels of technology. The overuse of groundwater reserves has resulted in potential shortages beside which the energy crisis pales in significance. And the very productivity of Earth's environmental systems—their ability to support human and other life—is being threatened by processes that derive at least in part from energy overuse and inefficiency and from pollution. Many environmentalists believe that both the public and private sectors, including individuals, are continuing to act in a totally irresponsible manner with regard to the natural resources upon which we all depend.

Uppermost in the minds of many who think of the environment in terms of an integrated whole, as evidenced by many of the selections in this unit, is the concept of the threshold or critical limit of human interference with natural systems of land, water, and air. This concept suggests that the environmental systems we occupy have been pushed to the brink of tolerance in terms of stability and that destabilization of environmental sys-

tems has consequences that can only be hinted at, rather than predicted. Although the broader issue of system change and instability, along with the lesser issues such as the quantity of agricultural land, the quality of iron ore deposits, the sustained yield of forests, or the availability of freshwater seems to be quite diverse, all are closely tied to a pair of concepts—that of resource marginality and of the globalization of the economy that has made marginality a global rather than a regional problem. Many of these ideas are brought together in the lead article of this unit. In "Where Have All the Farmers Gone?" Brian Halweil of the Worldwatch Institute discusses the globalization of industry and trade that is creating a uniform approach to all forms of economic management, including management of agricultural resources. Halweil notes that increasing agribusiness and decreasing numbers of family farmers represent a loss of both biological and cultural diversity, as standardized farming and business practices that may work for some areas but not for others are applied uniformly.

In the second article in this unit, farming also figures importantly—but in entirely different ways. Journalist Bruce Barcott, writing in *Mother Jones,* discusses the conflicts between various water users, from farmers to fishermen, that have erupted in the drought-stricken American West. In "What's a River For?" Barcott focuses his attention on the Klamath River basin of northern California and southern Oregon where a struggle for an increasingly scarce resource has erupted between farmers who need water for irrigation, factories and cities that need water for manufacturing and domestic purposes, ranchers whose livestock depend upon water for their very existence, and fishermen—both recreational and subsistence—who depend upon the river's salmon habitat for nourishment for both body and soul. The Klamath simply could not and cannot support all of these conflicting uses, and in 2002, a decision was made to release irrigation water from the river, resulting in one of the largest fish die-backs in American history.

Use of water for irrigation impacts not only on fish populations, however. We now know that the loss of topsoil to accelerated erosion, the loss of living and farming space to reservoirs, and increasing the salinity in irrigation waters and soil are problems greater than the dryness that prompts irrigation to begin with. In the third and fourth articles in the unit, the struggle between different stakeholders in the land and water use wars also takes center stage. In "A Human Thirst," UN consultant Don Hinrichsen reports that human uses now consume more than half the world's freshwater, often leaving other animal species to go thirsty. As water is withdrawn from rivers and other sources to feed the demands of agriculture, industry, and over six billion humans, the aquatic ecosystems, the plants and animals they support, and the very surface of the land itself suffers.

The next article in the unit moves the discussion from primarily surface waters to those that lie below the surface—the groundwater resource that is estimated to be more than thirty times greater in volume than surface water. In "Our Perilous Dependence on Groundwater," geologist Marc J. Defant discusses the issues of the aquifer, an underground storage system for water upon which farmers, industrialists, and homeowners alike depend. While cities and factories have long tapped into aquifers via wells, widespread agricultural use did not begin until the invention of pivot-irrigation systems that are simply huge sprinkler systems that use electrical pumps to bring water from underground storage to the surface. Aquifers face two major problems that will limit their future use and create major problems for those accustomed to this "free" resource: (1) they are becoming increasingly polluted as a result of toxic chemical residues, largely from agricultural chemicals such as fertilizers, insecticides, and herbicides; (2) they are becoming depleted as the rate of water withdrawal far exceeds the rate of recharge (the replacement of water in the aquifer from precipitation and seepage from surface waters) in most areas.

In the final article in this unit, "Oceans Are on the Critical List," researcher Anne Platt McGinn conends that the primary threats to the health of the world's oceanic ecosystems are human-induced and include overharvesting of fish, pollution of both costal and deep-water zones, introduction of alien species and the consequent threat to oceanic biodiversity, and climate change, which also poses threats to biodiversity. Efforts to protect the oceans, McGinn claims, lag far behind what is needed.

There are two possible solutions to all these problems posed by the use of increasingly marginal and scarce resources and by the continuing pollution of the global atmosphere. One is to halt the basic cause of the problems—increasing population and consumption. The other is to provide incentives and techniques for the conservation and management of existing resources and for the discovery of alternative resources to eliminate the demand for more marginal resources and the use of heavily polluting ones.

Where Have All the Farmers Gone?

The globalization of industry and trade is bringing more and more uniformity to the management of the world's land, and a spreading threat to the diversity of crops, ecosystems, and cultures. As Big Ag takes over, farmers who have a stake in their land—and who often are the most knowledgeable stewards of the land—are being forced into servitude or driven out.

by Brian Halweil

Since 1992, the U.S. Army Corps of Engineers has been developing plans to expand the network of locks and dams along the Mississippi River. The Mississippi is the primary conduit for shipping American soybeans into global commerce—about 35,000 tons a day. The Corps' plan would mean hauling in up to 1.2 million metric tons of concrete to lengthen ten of the locks from 180 meters to 360 meters each, as well as to bolster several major wing dams which narrow the river to keep the soybean barges moving and the sediment from settling. This construction would supplement the existing dredges which are already sucking 85 million cubic meters of sand and mud from the river's bank and bottom each year. Several different levels of "upgrade" for the river have been considered, but the most ambitious of them would purportedly reduce the cost of shipping soybeans by 4 to 8 cents per bushel. Some independent analysts think this is a pipe dream.

Around the same time the Mississippi plan was announced, the five governments of South America's La Plata Basin—Bolivia, Brazil, Paraguay, Argentina, and Uruguay—announced plans to dredge 13 million cubic meters of sand, mud, and rock from 233 sites along the Paraguay-Paraná River. That would be enough to fill a convoy of dump trucks 10,000 miles long. Here, the plan is to straighten natural river meanders in at least seven places, build dozens of locks, and construct a major port in the heart of the Pantanal—the world's largest wetland. The Paraguay-Paraná flows through the center of Brazil's burgeoning soybean heartland—second only to the

United States in production and exports. According to statements from the Brazilian State of Mato Grasso, this "Hidrovía" (water highway) will give a further boost to the region's soybean export capacity.

Lobbyists for both these projects argue that expanding the barge capacity of these rivers is necessary in order to improve competitiveness, grab world market share, and rescue farmers (either U.S. or Brazilian, depending on whom the lobbyists are addressing) from their worst financial crisis since the Great Depression. Chris Brescia, president of the Midwest River Coalition 2000, an alliance of commodity shippers that forms the primary lobbying force for the Mississippi plan, says, "The sooner we provide the waterway infrastructure, the sooner our family farmers will benefit." Some of his fellow lobbyists have even argued that these projects are essential to feeding the world (since the barges can then more easily speed the soybeans to the world's hungry masses) and to saving the environment (since the hungry masses will not have to clear rainforest to scratch out their own subsistence).

Probably very few people have had an opportunity to hear both pitches and compare them. But anyone who has may find something amiss with the argument that U.S. farmers will become more competitive versus their Brazilian counterparts, at the same time that Brazilian farmers will, for the same reasons, become more competitive with their U.S. counterparts. A more likely outcome is that farmers of these two nations will be pitted against each other in a costly race to maximize production, resulting in short-cut practices that essentially strip-mine their soil and throw long-term investments in the land to the wind. Farmers in Iowa will have stronger incentives to plow up land along stream banks, triggering faster erosion of topsoil. Their brethren in Brazil will find themselves needing to cut deeper into the savanna, also accelerating erosion. That will increase the flow of soybeans, all right—both north and south. But it will also further depress prices, so that even as the farmers are shipping more, they're getting less income per ton shipped. And in any case, increasing volume can't help the farmers survive in the long run, because sooner or later they will be swallowed by larger, corporate, farms that can make up for the smaller per-ton margins by producing even larger volumes.

So, how can the supporters of these river projects, who profess to be acting in the farmer's best interests, not notice the illogic of this form of competition? One explanation is that from the advocates' (as opposed to the farmers') standpoint, this competition isn't illogical at all—because the lobbyists aren't really representing farmers. They're working for the commodity processing, shipping, and trading firms who want the price of soybeans to fall, because these are the firms that buy the crops from the farmers. In fact, it is the same three agribusiness conglomerates—Archer Daniels Midland (ADM), Cargill, and Bunge—that are the top soybean processors and traders along both rivers.

Welcome to the global economy. The more brutally the U.S. and Brazilian farmers can batter each-other's prices (and standards of living) down, the greater the margin of profit these three giants gain. Meanwhile, another handful of companies controls the markets for genetically modified seeds, fertilizers, and herbicides used by the farmers—charging oligopolistically high prices both north and south of the equator.

In assessing what this proposed digging-up and reconfiguring of two of the world's great river basins really means, keep in mind that these projects will not be the activities of private businesses operating inside their own private property. These are proposed public works, to be undertaken at huge public expense. The motive is neither the plight of the family farmer nor any moral obligation to feed the world, but the opportunity to exploit poorly informed public sentiments about farmers' plights or hungry masses as a means of usurping public policies to benefit private interests. What gets thoroughly Big Muddied, in this usurping process, is that in addition to subjecting farmers to a gladiator-like attrition, these projects will likely bring a cascade of damaging economic, social, and ecological impacts to the very river basins being so expensively remodeled.

What's likely to happen if the lock and dam system along the Mississippi is expanded as proposed? The most obvious effect will be increased barge traffic, which will accelerate a less obvious cascade of events that has been underway for some time, according to Mike Davis of the Minnesota Department of Natural Resources. Much of the Mississippi River ecosystem involves aquatic rooted plants, like bullrush, arrowhead, and wild celery. Increased barge traffic will kick up more sediment, obscuring sunlight and reducing the depth to which plants can survive. Already, since the 1970s, the number of aquatic plant species found in some of the river has been cut from 23 to about half that, with just a handful thriving under the cloudier conditions. "Areas of the river have reached an ecological turning point," warns Davis. "This decline in plant diversity has triggered a drop in the invertebrate communities that live on these plants, as well as a drop in the fish, mollusk, and bird communities that depend on the diversity of insects and plants." On May 18, 2000, the U.S. Fish and Wildlife Service released a study saying that the Corps of Engineers project would threaten the 300 species of migratory birds and 12 species of fish in the Mississippi watershed, and could ultimately push some into extinction. "The least tern, the pallid sturgeon, and other species that evolved with the ebbs and flows, sandbars and depths, of the river are progressively eliminated or forced away as the diversity of the river's natural habitats is removed to maximize the barge habitat," says Davis.

The outlook for the Hidrovía project is similar. Mark Robbins, an ornithologist at the Natural History Museum at the University of Kansas, calls it "a key step in creating a Florida Everglades-like scenario of destruction in the Pantanal, and an American Great Plains-like scenario in the Cerrado in southern Brazil." The Paraguay-Paraná feeds the Pantanal wetlands, one of the most diverse habitats on the planet, with its populations of woodstorks, snailkites, limpkins, jabirus, and more than 650 other species of birds, as well as more than 400 species of fish and hundreds of other less-studied plants, mussels, and marshland organisms. As the river is dredged and the banks are built up to funnel the surrounding wetlands water into the navigation path, bird nesting habitat and fish spawning grounds will be eliminated, damaging the indigenous and other traditional societies that depend on these resources. Increased barge traffic will suppress river species here just as it will on the Mississippi. Meanwhile, herbicide-intensive soybean monocultures—on farms so enormous that they dwarf even the biggest operations in the U.S. Midwest—are rapidly replacing diverse grasslands in the fragile Cerrado. The heavy plowing and periodic absence of ground cover associated with such farming erodes 100 million tons of soil per year. Robbins notes that "compared to the Mississippi, this southern river system and surrounding grassland is several orders of magnitude more diverse and has suffered considerably less, so there is much more at stake."

Supporters of such massive disruption argue that it is justified because it is the most "efficient" way to do business. The perceived efficiency of such farming might be compared to the perceived efficiency of an energy system based on coal. Burning coal looks very efficient if you ignore its long-term impact on air quality and climate stability. Similarly, large farms look more efficient than small farms if you don't count some of their largest costs—the loss of the genetic diversity that underpins agriculture, the pollution caused by agro-chemicals, and the dislocation of rural cultures. The simultaneous demise of small, independent farmers and rise of multinational food giants is troubling not just for those who empathize with dislocated farmers, but for anyone who eats.

An Endangered Species

Nowadays most of us in the industrialized countries don't farm, so we may no longer really understand that way of life. I was born in the apple orchard and dairy country of Dutchess County, New York, but since age five have spent most of my life in New York City—while most of the farms back in Dutchess County have given way to spreading subdivisions. It's also hard for those of us who get our food from supermarket shelves or drive-thru windows to know how dependent we are on the viability of rural communities.

Whether in the industrial world, where farm communities are growing older and emptier, or in developing nations where population growth is pushing the number of farmers continually higher and each generation is inheriting smaller family plots, it is becoming harder and harder to make a living as a farmer. A combination of falling incomes, rising debt, and worsening rural poverty is forcing more people to either abandon farming as their primary activity or to leave the countryside altogether—a bewildering juncture, considering that farmers produce perhaps the only good that the human race cannot do without.

Since 1950, the number of people employed in agriculture has plummeted in all industrial nations, in some regions by more than 80 percent. Look at the numbers, and you might think farmers are being singled out by some kind of virus:

- In Japan, more than half of all farmers are over 65 years old; in the United States, farmers over 65 outnumber those under 35 by three to one. (Upon retirement or death, many will pass the farm on to children who live in the city and have no interest in farming themselves.)
- In New Zealand, officials estimate that up to 6,000 dairy farms will disappear during the next 10 to 15 years—dropping the total number by nearly 40 percent.
- In Poland, 1.8 million farms could disappear as the country is absorbed into the European Union—dropping the total number by 90 percent.
- In Sweden, the number of farms going out of business in the next decade is expected to reach about 50 percent.

- In the Philippines, Oxfam estimates that over the next few years the number of farm households in the corn–producing region of Mindanao could fall by some 500,000—a 50 percent loss.
- In the United States, where the vast majority of people were farmers at the time of the American Revolution, fewer people are now full-time farmers (less than 1 percent of the population) than are full-time prisoners.
- In the U.S. states of Nebraska and Iowa, between a fifth and a third of farmers are expected to be out of business within two years.

Of course, the declining numbers of farmers in industrial nations does not imply a decline in the importance of the farming sector. The world still has to eat (and 80 million more mouths to feed each year than the year before), so smaller numbers of farmers mean larger farms and greater concentration of ownership. Despite a precipitous plunge in the number of people employed in farming in North America, Europe, and East Asia, half the world's people still make their living from the land. In sub-Saharan Africa and South Asia, more than 70 percent do. In these regions, agriculture accounts, on average, for half of total economic activity.

Some might argue that the decline of farmers is harmless, even a blessing, particularly for less developed nations that have not yet experienced the modernization that moves peasants out of backwater rural areas into the more advanced economies of the cities. For most of the past two centuries, the shift toward fewer farmers has generally been assumed to be a kind of progress. The substitution of high-powered diesel tractors for slow-moving women and men with hoes, or of large mechanized industrial farms for clusters of small "old fashioned" farms, is typically seen as the way to a more abundant and affordable food supply. Our urban-centered society has even come to view rural life, especially in the form of small family-owned businesses, as backwards or boring, fit only for people who wear overalls and go to bed early—far from the sophistication and dynamism of the city.

Urban life does offer a wide array of opportunities, attractions, and hopes—some of them falsely created by urban-oriented commercial media—that many farm families decide to pursue willingly. But city life often turns out to be a disappointment, as displaced farmers find themselves lodged in crowded slums, where unemployment and ill-health are the norm and where they are worse off than they were back home. Much evidence suggests that farmers aren't so much being lured to the city as they are being driven off their farms by a variety of structural changes in the way the global food chain operates. Bob Long, a rancher in McPherson County, Nebraska, stated in a recent *New York Times* article that passing the farm onto his son would be nothing less than "child abuse."

As long as cities are under the pressure of population growth (a situation expected to continue at least for the next three or four decades), there will always be pressure for a large share of humanity to subsist in the countryside. Even in highly urbanized North America and Europe, roughly 25 percent of the population—275 million people—still reside in rural areas. Meanwhile, for the 3 billion Africans, Asians, and Latin Americans who remain in the countryside—and who will be there for the foreseeable future—the marginalization of farmers has set up a vicious cycle of low educational achievement, rising infant mortality, and deepening mental distress.

Hired Hands on Their Own Land

In the 18th and 19th centuries, farmers weren't so trapped. Most weren't wealthy, but they generally enjoyed stable incomes and strong community ties. Diversified farms yielded a range of raw and processed goods that the farmer could typically sell in a local market. Production costs tended to be much lower than now, as many of the needed inputs were home-grown: the farmer planted seed that he or she had saved from the previous

year, the farm's cows or pigs provided fertilizer, and the diversity of crops—usually a large range of grains, tubers, vegetables, herbs, flowers, and fruits for home use as well as for sale—effectively functioned as pest control.

Things have changed, especially in the past half-century, according to Iowa State agricultural economist Mike Duffy. "The end of World War II was a watershed period," he says. "The widespread introduction of chemical fertilizers and synthetic pesticides, produced as part of the war effort, set in motion dramatic changes in how we farm—and a dramatic decline in the number of farmers." In the post-war period, along with increasing mechanization, there was an increasing tendency to "outsource" pieces of the work that the farmers had previously done themselves—from producing their own fertilizer to cleaning and packaging their harvest. That outsourcing, which may have seemed like a welcome convenience at the time, eventually boomeranged: at first it enabled the farmer to increase output, and thus profits, but when all the other farmers were doing it too, crop prices began to fall.

Before long, the processing and packaging businesses were adding more "value" to the purchased product than the farmer, and it was those businesses that became the

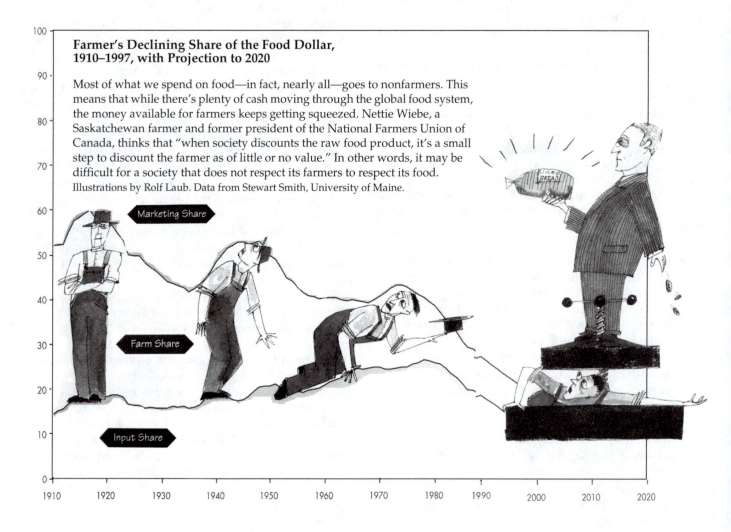

Farmer's Declining Share of the Food Dollar, 1910–1997, with Projection to 2020

Most of what we spend on food—in fact, nearly all—goes to nonfarmers. This means that while there's plenty of cash moving through the global food system, the money available for farmers keeps getting squeezed. Nettie Wiebe, a Saskatchewan farmer and former president of the National Farmers Union of Canada, thinks that "when society discounts the raw food product, it's a small step to discount the farmer as of little or no value." In other words, it may be difficult for a society that does not respect its farmers to respect its food. Illustrations by Rolf Laub. Data from Stewart Smith, University of Maine.

Marketing Share

Farm Share

Input Share

ConAgra: *Vertical Integration, Horizontal Concentration, Global Omnipresence*

Three conglomerates (ConAgra/DuPont, Cargill/Monsanto, and Novartis/ADM) dominate virtually every link in the North American (and increasingly, the global) food chain. Here's a simplified diagram of one conglomerate.

KEY: ⬇ Vertical integration of production links, from seed to supermarket ⬌ Concentration within a link

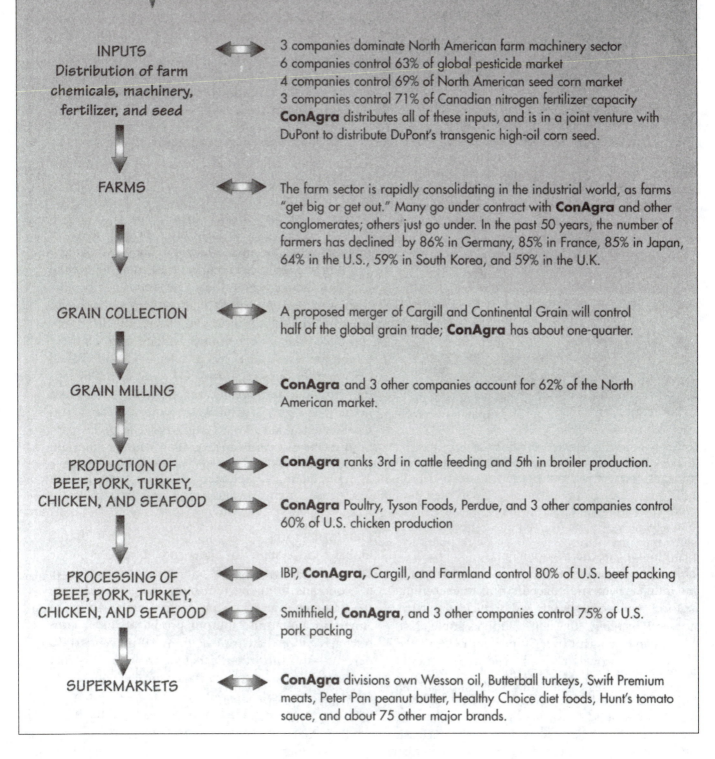

INPUTS
Distribution of farm chemicals, machinery, fertilizer, and seed
⬌
3 companies dominate North American farm machinery sector
6 companies control 63% of global pesticide market
4 companies control 69% of North American seed corn market
3 companies control 71% of Canadian nitrogen fertilizer capacity
ConAgra distributes all of these inputs, and is in a joint venture with DuPont to distribute DuPont's transgenic high-oil corn seed.

FARMS
⬌
The farm sector is rapidly consolidating in the industrial world, as farms "get big or get out." Many go under contract with **ConAgra** and other conglomerates; others just go under. In the past 50 years, the number of farmers has declined by 86% in Germany, 85% in France, 85% in Japan, 64% in the U.S., 59% in South Korea, and 59% in the U.K.

GRAIN COLLECTION
⬌
A proposed merger of Cargill and Continental Grain will control half of the global grain trade; **ConAgra** has about one-quarter.

GRAIN MILLING
⬌
ConAgra and 3 other companies account for 62% of the North American market.

PRODUCTION OF BEEF, PORK, TURKEY, CHICKEN, AND SEAFOOD
⬌
ConAgra ranks 3rd in cattle feeding and 5th in broiler production.
⬌
ConAgra Poultry, Tyson Foods, Perdue, and 3 other companies control 60% of U.S. chicken production

PROCESSING OF BEEF, PORK, TURKEY, CHICKEN, AND SEAFOOD
⬌
IBP, **ConAgra,** Cargill, and Farmland control 80% of U.S. beef packing
⬌
Smithfield, **ConAgra**, and 3 other companies control 75% of U.S. pork packing

SUPERMARKETS
⬌
ConAgra divisions own Wesson oil, Butterball turkeys, Swift Premium meats, Peter Pan peanut butter, Healthy Choice diet foods, Hunt's tomato sauce, and about 75 other major brands.

dominant players in the food industry. Instead of farmers outsourcing to contractors, it became a matter of large food processors buying raw materials from farmers, on the processors' terms. Today, most of the money is in the work the farmer no longer does—or even controls. In the United States, the share of the consumer's food dollar that trickles back to the farmer has plunged from nearly 40 cents in 1910 to just above 7 cents in 1997, while the shares going to input (machinery, agrochemicals, and seeds) and marketing (processing, shipping, brokerage, advertising, and retailing) firms have continued to expand. (See graph "Farmer's Declining Share of the Food Dollar") The typical U.S. wheat farmer, for instance, gets just 6 cents of the dollar spent on a loaf of bread—so when you buy that loaf, you're paying about as much for the wrapper as for the wheat.

Ironically, then, as U.S. farms became more mechanized and more "productive," a self-destructive feedback loop was set in motion: over-supply and declining crop prices cut into farmers' profits, fueling a demand for more technology aimed at making up for shrinking margins by increasing volume still more. Output increased dramatically, but expenses (for tractors, combines, fertilizer, and seed) also ballooned—while the commodity prices stagnated or declined. Even as they were looking more and more modernized, the farmers were becoming less and less the masters of their own domain.

On the typical Iowa farm, the farmer's profit margin has dropped from 35 percent in 1950 to 9 percent today. In order to generate the same income, this farm would need to be roughly four times as large today as in 1950—or the farmer would need to get a night job. And that's precisely what we've seen in most industrialized nations: fewer farmers on bigger tracts of land producing a greater share of the total food supply. The farmer with declining margins buys out his neighbor and expands or risks being cannibalized himself.

There is an alternative to this huge scaling up, which is to buck the trend and bring some of the input-supplying and post-harvest processing—and the related profits—back onto the farm. But more self-sufficient farming would be highly unpopular with the industries that now make lucrative profits from inputs and processing. And since these industries have much more political clout than the farmers do, there is little support for rescuing farmers from their increasingly servile condition—and the idea has been largely forgotten. Farmers continue to get the message that the only way to succeed is to get big.

The traditional explanation for this constant pressure to "get big or get out" has been that it improves the efficiency of the food system—bigger farms replace smaller farms, because the bigger farms operate at lower costs. In some respects, this is quite true. Scaling up may allow a farmer to spread a tractor's cost over greater acreage, for example. Greater size also means greater leverage in purchasing inputs or negotiating loan rates—increasingly important as satellite-guided combines and other equipment make farming more and more capital-intensive. But these economies of scale typically level off. Data for a wide range of crops produced in the United States show that the lowest production costs are generally achieved on farms that are much smaller than the typical farm now is. But large farms can tolerate lower margins, so while they may not *produce* at lower cost, they can afford to *sell* their crops at lower cost, if forced to do so—as indeed they are by the food processors who buy from them. In short, to the extent that a giant farm has a financial benefit over a small one, it's a benefit that goes only to the processor—not to the farmer, the farm community, or the environment.

This shift of the food dollar away from farmers is compounded by intense concentration in every link of the food chain—from seeds and herbicides to farm finance and retailing. In Canada, for example, just three companies control over 70 percent of fertilizer sales, five banks provide the vast majority of agricultural credit, two companies control over 70 percent of beef packing, and five companies dominate food retailing. The merger of Philip Morris and Nabisco will create an empire that collects nearly 10 cents of every dollar a U.S. consumer spends on food, according to a company spokesperson. Such high concentration can be deadly for the bottom line, allowing agribusiness firms to extract higher prices for the products farmers buy from them, while offering lower prices for the crop they buy from the farmers.

An even more worrisome form of concentration, according to Bill Heffernan, a rural sociologist at the University of Missouri, is the emergence of several clusters of firms that—through mergers, takeovers, and alliances with other links in the food chain—now possess "a seamless and fully vertically integrated control of the food system from gene to supermarket shelf." (See diagram "ConAgra") Consider the recent partnership between Monsanto and Cargill, which controls seeds, fertilizers, pesticides, farm finance, grain collection, grain processing, livestock feed processing, livestock production, and slaughtering, as well as some well-known processed food brands. From the standpoint of a company like Cargill, such alliances yield tremendous control over costs and can therefore be extremely profitable.

But suppose you're the farmer. Want to buy seed to grow corn? If Cargill is the only buyer of corn in a hundred mile radius, and Cargill is only buying a particular Monsanto corn variety for its mills or elevators or feedlots, then if you don't plant Monsanto's seed you won't have a market for your corn. Need a loan to buy the seed? Go to Cargill-owned Bank of Ellsworth, but be sure to let them know which seed you'll be buying. Also mention that you'll be buying Cargill's Saskferco brand fertilizer. OK, but once the corn is grown, you don't like the idea of having to sell to Cargill at the prices it dictates? Well, maybe you'll feed the corn to your pigs, then, and sell them to the highest bidder. No problem—Cargill's Excel Corporation buys pigs, too. OK, you're moving to the city, and renouncing the farm life! No more home-made grits for breakfast, you're buying corn flakes. Well, good news: Cargill Foods supplies corn flour to the top cereal makers. You'll notice, though, that all the big brands of corn flakes seem to have pretty much the same hefty price per ounce. After all, they're all made by the agricultural oligopoly.

As these vertical food conglomerates consolidate, Heffernan warns, "there is little room left in the global food system for independent farmers"—the farmers being increasingly left with "take it or leave it" contracts from the remaining conglomerates. In the last two decades, for example, the share of American agricultural output produced under contract has more than tripled, from 10 percent to 35 percent—and this doesn't include the contracts that farmers must sign to plant genetically engineered seed. Such centralized control of the food system, in which farmers are in effect reduced to hired hands on their own land, reminds Heffernan of the Soviet-style state farms, but with the Big Brother role now being played by agribusiness executives. It is also reminiscent of the "company store" which once dominated small American mining or factory towns, except that if you move out of town now, the store is still with you. The company store has gone global.

With the conglomerates who own the food dollar also owning the political clout, it's no surprise that agricultural policies—including subsidies, tax breaks, and environmental legislation at both the national and international levels—do not generally favor the farms. For example, the conglomerates command growing influence over both private and public agricultural research priorities, which might explain why the U.S. Department of Agriculture (USDA), an agency ostensibly beholden to farmers, would help to develop the seed-sterilizing Terminator technology—a biotechnology that offers farmers only greater dependence on seed companies. In some cases the influence is indirect, as manifested in government funding decisions, while in others it is more blatant. When Novartis provided $25 million to fund a research partnership with the plant biology department of the University of California at Berkeley, one of the conditions was that Novartis has the first right of refusal for any patentable inventions. Under those circumstances, of course, the UC officials—mindful of where their funding comes from—have strong incentives to give more attention to technologies like the Terminator seed, which shifts profit away from the farmer, than to technologies that directly benefit the farmer or the public at large.

Even policies that are touted to be in the best interest of farmers, like liberalized trade in agricultural products, are increasingly shaped by non-farmers. Food traders, processors, and distributors, for example, were some of the principal architects of recent revisions to the General Agreement on Trade and Tariffs (GATT)—the World Trade Organization's predecessor—that paved the way for greater trade flows in agricultural commodities. Before these revisions, many countries had mechanisms for assuring that their farmers wouldn't be driven out of their own domestic markets by predatory global traders. The traders, however, were able to do away with those protections.

The ability of agribusiness to slide around the planet, buying at the lowest possible price and selling at the highest, has tended to tighten the squeeze already put in place by economic marginalization, throwing every farmer on the planet into direct competition with every other farmer. A recent UN Food and Agriculture Organization assessment of the experience of 16 developing nations in implementing the latest phase of the GATT concluded that "a common reported concern was with a general trend towards the concentration of farms," a process that tends to further marginalize small producers and exacerbate rural poverty and unemployment. The sad irony, according to Thomas Reardon, of Michigan State University, is that while small farmers in all reaches of the world are increasingly affected by cheap, heavily subsidized imports of foods from outside of their traditional rural markets, they are nonetheless often excluded from opportunities to participate in food exports themselves. To keep down transaction costs and to keep processing standardized, exporters and other downstream players prefer to buy from a few large producers, as opposed to many small producers.

As the global food system becomes increasingly dominated by a handful of vertically integrated, international corporations, the servitude of the farmer points to a broader society-wide servitude that OPEC-like food cartels could impose, through their control over food prices and food quality. Agricultural economists have already noted that the widening gap between retail food prices and farm prices in the 1990s was due almost exclusively to exploitation of market power, and not to extra services provided by processors and retailers. It's questionable whether we should pay as much for a bread wrapper as we do for the nutrients it contains. But beyond this, there's a more fundamental question. Farmers are professionals, with extensive knowledge of their local soils, weather, native plants, sources of fertilizer or mulch, native pollinators, ecology, and community. If we are to have a world where the land is no longer managed by such professionals, but is instead managed by distant corporate bureaucracies interested in extracting maximum output at minimum cost, what kind of food will we have, and at what price?

Agrarian Services

No question, large industrial farms can produce lots of food. Indeed, they're designed to maximize quantity. But when the farmer becomes little more than the lowest-cost producer of raw materials, more than his own welfare will suffer. Though the farm sector has lost power and profit, it is still the one link in the agrifood chain accounting for the largest share of agriculture's public goods—including half the world's jobs, many of its most vital communities, and many of its most diverse landscapes. And in providing many of these goods, small farms clearly have the advantage.

Local economic and social stability: Over half a century ago, William Goldschmidt, an anthropologist working at the USDA, tried to assess how farm structure and size affect the health of rural communities. In California's San Joaquin Valley, a region then considered to be at the cutting edge of agricultural industrialization, he identified two small towns that were alike in all basic economic and geographic dimensions, including value of agricultural production, except in farm size. Comparing the two, he found an inverse correlation between the sizes of the farms and the well-being of the communities they were a part of.

The small-farm community, Dinuba, supported about 20 percent more people, and at a considerably higher level of living—including lower poverty rates, lower levels of economic and social class distinctions, and a lower crime rate—than the large-farm community of Arvin. The majority of Dinuba's residents were independent entrepreneurs, whereas fewer than 20 percent of Arvin's residents were—most of the others being agricultural laborers. Dinuba had twice as many business establishments as Arvin, and did 61 percent more retail business. It had more schools, parks, newspapers, civic organizations, and churches, as well as better physical infrastructure—paved streets, sidewalks, garbage disposal, sewage disposal and other public services. Dinuba also had more institutions for democratic decision making, and a much broader participation by its citizens. Political scientists have long recognized that a broad base of independent entrepreneurs and property owners is one of the keys to a healthy democracy.

The distinctions between Dinuba and Arvin suggest that industrial agriculture may be limited in what it can do for a community. Fewer (and less meaningful) jobs, less local spending, and a hemorrhagic flow of profits to absentee landowners and distant suppliers means that industrial farms can actually be a net drain on the local economy. That hypothesis has been corroborated by Dick Levins, an agricultural economist at the University of Minnesota. Levins studied the economic receipts from Swift County, Iowa, a typical Midwestern corn and soybean community, and found that although total farm sales are near an all-time high, farm income there has been dismally low—and that many of those who were once the financial stalwarts of the community are now deeply in debt. "Most of the U.S. Corn Belt, like Swift County, is a colony, owned and operated by people who don't live there and for the benefit of those who don't live there," says Levin. In fact, most of the land in Swift County is rented, much of it from absentee landlords.

This new calculus of farming may be eliminating the traditional role of small farms in anchoring rural economies—the kind of tradition, for example, that we saw in the emphasis given to the support of small farms by Japan, South Korea, and Taiwan following World War II. That emphasis, which brought radical land reforms and targeted investment in rural areas, is widely cited as having been a major stimulus to the dramatic economic boom those countries enjoyed.

Not surprisingly, when the economic prospects of small farms decline, the social fabric of rural communities begins to tear. In the United States, farming families are more than twice as likely as others to live in poverty. They have less education and lower rates of medical protection, along with higher rates of infant mortality, alcoholism, child abuse, spousal abuse, and mental stress. Across Europe, a similar pattern is evident. And in sub-Saharan Africa, sociologist Deborah Bryceson of the Netherlands-based African Studies Centre has studied the dislocation of small farmers and found that "as de-agrarianization proceeds, signs of social dysfunction associated with urban areas [including petty crime and breakdowns of family ties] are surfacing in villages."

People without meaningful work often become frustrated, but farmers may be a special case. "More so than other occupations, farming represents a way of life and defines who you are," says Mike Rosemann, a psychologist who runs a farmer counseling network in Iowa. "Losing the family farm, or the prospect of losing the family farm, can generate tremendous guilt and anxiety, as if one has failed to protect the heritage that his ancestors worked to hold onto." One measure of the despair has been a worldwide surge in the number of farmers committing suicide. In 1998, over 300 cotton farmers in Andhra Pradesh, India, took their lives by swallowing pesticides that they had gone into debt to purchase but that had nonetheless failed to save their crops. In Britain, farm workers are two-and-a-half times more likely to commit suicide than the rest of the population. In the United States, official statistics say farmers are now five times as likely to commit suicide as to die from farm accidents, which have been traditionally the most frequent cause of unnatural death for them. The true number may be even higher, as suicide hotlines report that they often receive calls from farmers who want to know which sorts of accidents (Falling into the blades of a combine? Getting shot while hunting?) are least likely to be investigated by insurance companies that don't pay claims for suicides.

Whether from despair or from anger, farmers seem increasingly ready to rise up, sometimes violently, against government, wealthy landholders, or agribusiness giants. In recent years we've witnessed the Zapatista revolution in Chiapas, the seizing of white-owned farms by landless blacks in Zimbabwe, and the attacks of European farmers on warehouses storing genetically engineered seed. In the book *Harvest of Rage*, journalist Joel Dyer links the 1995 Oklahoma City bombing that killed nearly 200 people—

In the Developing World, an Even Deeper Farm Crisis

"One would have to multiply the threats facing family farmers in the United States or Europe five, ten, or twenty times to get a sense of the handicaps of peasant farmers in less developed nations," says Deborah Bryceson, a senior research fellow at the African Studies Centre in the Netherlands. Those handicaps include insufficient access to credit and financing, lack of roads and other infrastructure in rural areas, insecure land tenure, and land shortages where population is dense.

Three forces stand out as particularly challenging to these peasant farmers:

Structural adjustment requirements, imposed on indebted nations by international lending institutions, have led to privatization of "public commodity procurement boards" that were responsible for providing public protections for rural economies. "The newly privatized entities are under no obligation to service marginal rural areas," says Rafael Mariano, chairman of a Filipino farmers' union. Under the new rules, state protections against such practices as dumping of cheap imported goods (with which local farmers can't compete) were abandoned at the same time that state provision of health care, education, and other social services was being reduced.

Trade liberalization policies associated with structural adjustment have reduced the ability of nations to protect their agricultural economies even if they want to. For example, the World Trade Organization's Agreement on Agriculture will forbid domestic price support mechanisms and tariffs on imported goods—some of the primary means by which a country can shield its own farmers from overproduction and foreign competition.

The growing emphasis on agricultural grades and standards—the standardizing of crops and products so they can be processed and marketed more "efficiently"—has tended to favor large producers, and to marginalize smaller ones. Food manufacturers and supermarkets have emerged as the dominant entities in the global agri-food chain, and with their focus on brand consistency, ingredient uniformity, and high volume, smaller producers often are unable to deliver—or aren't even invited to bid.

Despite these daunting conditions, many peasant farmers tend to hold on long after it has become clear that they can't compete. One reason, says Peter Rosset of the Institute for Food and Development Policy, is that "even when it gets really bad, they will cling to agriculture because of the fact that it at least offers some degree of food security—that you can feed yourself." But with the pressures now mounting, particularly as export crop production swallows more land, even that fallback is lost.

as well as the rise of radical right and antigovernment militias in the U.S. heartland—to a spreading despair and anger stemming from the ongoing farm crisis. Thomas Homer-Dixon, director of the Project on Environment, Population, and Security at the University of Toronto, regards farmer dislocation, and the resulting rural unemployment and poverty, as one of the major security threats for the coming decades. Such dislocation is responsible for roughly half of the growth of urban populations across the Third World, and such growth often occurs in volatile shantytowns that are already straining to meet the basic needs of their residents. "What was an extremely traumatic transition for Europe and North America from a rural society to an urban one is now proceeding at two to three times that speed in developing nations," says Homer-Dixon. And, these nations have considerably less industrialization to absorb the labor. Such an accelerated transition poses enormous adjustment challenges for India and China, where perhaps a billion and a half people still make their living from the land.

Ecological stability: In the Andean highlands, a single farm may include as many as 30 to 40 distinct varieties of potato (along with numerous other native plants), each having slightly different optimal soil, water, light, and temperature regimes, which the farmer—given enough time—can manage. (In comparison, in the United States, just four closely related varieties account for about 99 percent of all the potatoes produced.) But, according to Karl Zimmerer, a University of Wisconsin sociologist, declining farm incomes in the Andes force more and more growers into migrant labor forces for part of the year, with serious effects on farm ecology. As time becomes constrained, the farmer manages the system more homogenously—cutting back on the number of traditional varieties (a small home garden of favorite culinary varieties may be the last refuge of diversity), and scaling up production of a few commercial varieties. Much of the traditional crop diversity is lost.

Complex farm systems require a highly sophisticated and intimate knowledge of the land—something small-scale, full-time farmers are more able to provide. Two or three different crops that have different root depths, for example, can often be planted on the same piece of land, or crops requiring different drainage can be planted in close proximity on a tract that has variegated topography. But these kinds of cultivation can't be done with heavy tractors moving at high speed. Highly site-specific and management-intensive cultivation demands ingenuity and awareness of local ecology, and can't be achieved by heavy equipment and heavy applications of agrochemicals. That isn't to say that being small is always sufficient to ensure ecologically sound food production, because economic adversity can drive small farms, as well as big ones, to compromise sustainable food production by transmogrifying the craft of land stewardship into the crude labor of commodity production. But a large-scale, highly mechanized farm is simply not equipped to preserve landscape complexity. Instead, its normal modus is to use blunt management tools, like crops that have been genetically engineered to churn out insecticides, which obviate the need to scout the field to see if spraying is necessary at all.

In the U.S. Midwest, as farm size has increased, cropping systems have gotten more simplified. Since 1972, the number of counties with more than 55 percent of their acreage planted in corn and soybeans has nearly tripled, from 97 to 267. As farms scaled up, the great simplicity of managing the corn-soybean rotation—an 800 acre farm, for instance, may require no more than a couple of weeks planting in the spring and a few weeks harvesting in the fall—became its big selling point. The various arms of the agricultural economy in the region, from extension services to grain elevators to seed suppliers, began to solidify around this corn-soybean rotation, reinforcing the farmers' movement away from other crops. Fewer and fewer farmers kept livestock, as beef and hog production became "economical" only in other parts of the country where it was becoming more concentrated. Giving up livestock meant eliminating clover, pasture mixtures, and a key source of fertilizer in the Midwest, while creating tremendous manure concentrations in other places.

But the corn and soybean rotation—one monoculture followed by another—is extremely inefficient or "leaky" in its use of applied fertilizer, since low levels of biodiversity tend to leave a range of vacant niches in the field, including different root depths and different nutrient preferences. Moreover, the Midwest's shift to monoculture has subjected the country to a double hit of nitrogen pollution, since not only does the removal and concentration of livestock tend to dump inordinate amounts of feces in the places (such as Utah and North Carolina) where the livestock operations are now located, but the monocultures that remain in the Midwest have much poorer nitrogen retention than they would if their cropping were more complex. (The addition of just a winter rye crop to the corn-soy rotation has been shown to reduce nitrogen runoff by nearly 50 percent.) And maybe this disaster-in-the-making should really be regarded as a triple hit, because in addition to contaminating Midwestern water supplies, the runoff ends up in the Gulf of Mexico, where the nitrogen feeds massive algae blooms. When the algae die, they are decomposed by bacteria, whose respiration depletes the water's oxygen—suffocating fish, shellfish,

and all other life that doesn't escape. This process periodically leaves 20,000 square kilometers of water off the coast of Louisiana biologically dead. Thus the act of simplifying the ecology of a field in Iowa can contribute to severe pollution in Utah, North Carolina, Louisiana, *and* Iowa.

The world's agricultural biodiversity—the ultimate insurance policy against climate variations, pest outbreaks, and other unforeseen threats to food security—depends largely on the millions of small farmers who use this diversity in their local growing environments. But the marginalization of farmers who have developed or inherited complex farming systems over generations means more than just the loss of specific crop varieties and the knowledge of how they best grow. "We forever lose the best available knowledge and experience of place, including what to do with marginal lands not suited for industrial production," says Steve Gleissman, an agroecologist at the University of California at Santa Cruz. The 12 million hogs produced by Smithfield Foods Inc., the largest hog producer and processor in the world and a pioneer in vertical integration, are nearly genetically identical and raised under identical conditions—regardless of whether they are in a Smithfield feedlot in Virginia or Mexico.

As farmers become increasingly integrated into the agribusiness food chain, they have fewer and fewer controls over the totality of the production process—shifting more and more to the role of "technology applicators," as opposed to managers making informed and independent decisions. Recent USDA surveys of contract poultry farmers in the United States found that in seeking outside advice on their operations, these farmers now turn first to bankers and then to the corporations that hold their contracts. If the contracting corporation is also the same company that is selling the farm its seed and fertilizer, as is often the case, there's a strong likelihood that the company's procedures will be followed. That corporation, as a global enterprise with no compelling local ties, is also less likely to be concerned about the pollution and resource degradation created by those procedures, at least compared with a farmer who is rooted in that community. Grower contracts generally disavow any environmental liability.

And then there is the ecological fallout unique to large-scale, industrial agriculture. Colossal confined animal feeding operations (CAFOs)—those "other places" where livestock are concentrated when they are no longer present on Midwestern soy/corn farms—constitute perhaps the most egregious example of agriculture that has, like a garbage barge in a goldfish pond, overwhelmed the scale at which an ecosystem can cope. CAFOs are increasingly the norm in livestock production, because, like crop monocultures, they allow the production of huge populations of animals which can be slaughtered and marketed at rock-bottom costs. But the disconnection between the livestock and the land used to produce their feed means that such CAFOs generate gargantuan amounts of waste, which the surrounding soil cannot possibly absorb. (One farm in Utah will raise over five million hogs in a year, producing as much waste each day as the city of Los Angeles.) The waste is generally stored in large lagoons, which are prone to leak and even spill over during heavy storms. From North Carolina to South Korea, the overwhelming stench of these lagoons—a combination of hydrogen sulfide, ammonia, and methane gas that smells like rotten eggs—renders miles of surrounding land uninhabitable.

A different form of ecological disruption results from the conditions under which these animals are raised. Because massive numbers of closely confined livestock are highly susceptible to infection, and because a steady diet of antibiotics can modestly boost animal growth, overuse of antibiotics has become the norm in industrial animal production. In recent months, both the Centers for Disease Control and Prevention in the United States and the World Health Organization have identified such industrial feeding operations as principal causes of the growing antibiotic resistance in food-borne bacteria like *salmonella* and *campylobacter*. And as decisionmaking in the food chain grows ever more concentrated—confined behind fewer corporate doors—there may be other food safety issues that you won't even hear about, particularly in the burgeoning field of genetically modified organisms (GMOs). In reaction to growing public concern over GMOs, a coalition that ingenuously calls itself the "Alliance for Better Foods"—actually made up of large food retailers, food processors, biotech companies and corporate-financed farm organizations—has launched a $50 million public "educational" campaign, in addition to giving over $676,000 to U.S. lawmakers and political parties in 1999, to head off the mandatory labeling of such foods.

Perhaps most surprising, to people who have only casually followed the debate about small-farm values versus factory-farm "efficiency," is the fact that a wide body of evidence shows that small farms are actually more productive than large ones—by as much as 200 to 1,000 percent greater output per unit of area. How does this jive with the often-mentioned productivity advantages of large-scale mechanized operations? The answer is simply that those big-farm advantages are always calculated on the basis of how much of *one crop* the land will yield per acre. The greater productivity of a smaller, more complex farm, however, is calculated on the basis of how much food *overall* is produced per acre. The smaller farm can grow several crops utilizing different root depths, plant heights, or nutrients, on the same piece of land simultaneously. It is this "polyculture" that offers the small farm's productivity advantage.

To illustrate the difference between these two kinds of measurement, consider a large Midwestern corn farm. That farm may produce more corn per acre than a small farm in which the corn is grown as part of a polyculture that also includes beans, squash, potato, and "weeds"

that serve as fodder. But in overall output, the polycrop—under close supervision by a knowledgeable farmer—produces much more food overall, whether you measure in weight, volume, bushels, calories, or dollars.

The inverse relationship between farm size and output can be attributed to the more efficient use of land, water, and other agricultural resources that small operations afford, including the efficiencies of intercropping various plants in the same field, planting multiple times during the year, targeting irrigation, and integrating crops and livestock. So in terms of converting inputs into outputs, society would be better off with small-scale farmers. And as population continues to grow in many nations, and the agricultural resources per person continue to shrink, a small farm structure for agriculture may be central to meeting future food needs.

Rebuilding Foodsheds

Look at the range of pressures squeezing farmers, and it's not hard to understand the growing desperation. The situation has become explosive, and if stabilizing the erosion of farm culture and ecology is now critical not just to farmers but to everyone who eats, there's still a very challenging question as to what strategy can work. The agribusiness giants are deeply entrenched now, and scattered protests could have as little effect on them as a mosquito bite on a tractor. The prospects for farmers gaining political strength on their own seem dim, as their numbers—at least in the industrial countries—continue to shrink.

A much greater hope for change may lie in a joining of forces between farmers and the much larger numbers of other segments of society that now see the dangers, to their own particular interests, of continued restructuring of the countryside. There are a couple of prominent models for such coalitions, in the constituencies that have joined forces to fight the Mississippi River Barge Capacity and Hidrovía Barge Capacity projects being pushed forward in the name of global soybean productivity.

The American group has brought together at least the following riverbedfellows:

- National environmental groups, including the Sierra Club and National Audubon Society, which are alarmed at the prospect of a public commons being damaged for the profit of a small commercial interest group;

- Farmers and farmer advocacy organizations, concerned about the inordinate power being wielded by the agribusiness oligopoly;

- Taxpayer groups outraged at the prospect of a corporate welfare payout that will drain more than $1 billion from public coffers;

- Hunters and fishermen worried about the loss of habitat;

- Biologists, ecologists, and birders concerned about the numerous threatened species of birds, fish, amphibians, and plants;
- Local-empowerment groups concerned about the impacts of economic globalization on communities;
- Agricultural economists concerned that the project will further entrench farmers in a dependence on the export of low-cost, bulk commodities, thereby missing valuable opportunities to keep money in the community through local milling, canning, baking, and processing.

A parallel coalition of environmental groups and farmer advocates has formed in the Southern hemisphere to resist the Hidrovía expansion. There too, the river campaign is part of a larger campaign to challenge the hegemony of industrial agriculture. For example, a coalition has formed around the Landless Workers Movement, a grassroots organization in Brazil that helps landless laborers to organize occupations of idle land belonging to wealthy landlords. This coalition includes 57 farm advocacy organizations based in 23 nations. It has also brought together environmental groups in Latin America concerned about the related ventures of logging and cattle ranching favored by large landlords; the mayors of rural towns who appreciate the boost that farmers can give to local economies; and organizations working on social welfare in Brazil's cities, who see land occupation as an alternative to shantytowns.

The Mississippi and Hidrovía projects, huge as they are, still constitute only two of the hundreds of agro-industrial developments being challenged around the world. But the coalitions that have formed around them represent the kind of focused response that seems most likely to slow the juggernaut, in part because the solutions these coalitions propose are not vague or quixotic expressions of idealism, but are site-specific and practical. In the case of the alliance forming around the Mississippi River project, the coalition's work has included questioning the assumptions of the Corps of Engineers analysis, lobbying for stronger antitrust examination of agribusiness monopolies, and calling for modification of existing U.S. farm subsidies, which go disproportionately to large farmers. Environmental groups are working to re-establish a balance between use of the Mississippi as a barge mover and as an intact watershed. Sympathetic agricultural extensionists are promoting alternatives to the standard corn-soybean rotation, including certified organic crop production, which can simultaneously bring down

Past and Future: Connecting the Dots

Given the direction and speed of prevailing trends, how far can the decline in farmers go? The lead editorial in the September 13, 1999 issue of *Feedstuffs*, an agribusiness trade journal, notes that "Based on the best estimates of analysts, economists and other sources interviewed for this publication, American agriculture must now quickly consolidate all farmers and livestock producers into about 50 production systems… each with its own brands," in order to maintain competitiveness. Ostensibly, other nations will have to do the same in order to keep up.

To put that in perspective, consider that in traditional agriculture, each farm is an independent production system. In this map of Ireland's farms circa 1930, each dot represents 100 farms, so the country as a whole had many thousands of independent production systems. But if the *Feedstuffs* prognosis were to come to pass, this map would be reduced to a single dot. And even an identically keyed map of the much larger United States would show the country's agriculture reduced to just one dot.

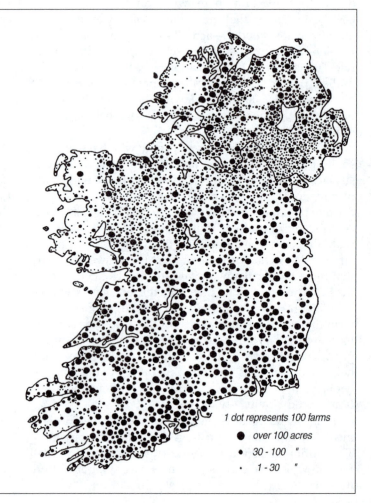

1 dot represents 100 farms

● over 100 acres
• 30 - 100 "
· 1 - 30 "

input costs and garner a premium for the final product, and reduce nitrogen pollution.

The United States and Brazil may have made costly mistakes in giving agribusiness such power to reshape the rivers and land to its own use. But the strategy of interlinked coalitions may be mobilizing in time to save much of the world's agricultural health before it is too late. Dave Brubaker, head of the Spira/GRACE Project on Industrial Animal Production at the Johns Hopkins University School of Public Health, sees these diverse coalitions as "the beginning of a revolution in the way we look at the food system, tying in food production with social welfare, human health, and the environment." Brubaker's project brings together public health officials focused on antibiotic overuse and water contamination resulting from hog waste; farmers and local communities who oppose the spread of new factory farms or want to close down existing ones; and a phalanx of natural allies with related campaigns, including animal rights activists, labor unions, religious groups, consumer rights activists, and environmental groups.

"As the circle of interested parties is drawn wider, the alliance ultimately shortens the distance between farmer and consumer," observes Mark Ritchie, president of the Institute for Agriculture and Trade Policy, a research and advocacy group often at the center of these partnerships. This closer proximity may prove critical to the ultimate sustainability of our food supply, since socially and ecologically sound buying habits are not just the passive *result* of changes in the way food is produced, but can actually be the most powerful *drivers of* these changes. The explosion of farmers' markets, community-supported agriculture, and other direct buying arrangements between farmers and consumers points to the growing numbers of nonfarmers who have already shifted their role in the food chain from that of choosing from the tens of thousands of food brands offered by a few dozen companies to bypassing such brands altogether. And, since many of the additives and processing steps that take up the bulk of the food dollar are simply the inevitable consequence of the ever-increasing time commercial food now spends in global transit and storage, this shortening of distance between grower and consumer will not only benefit the culture and ecology of farm communities. It will also give us access to much fresher, more flavorful, and more nutritious food. Luckily, as any food marketer can tell you, these characteristics aren't a hard sell.

Brian Halweil is a staff researcher at the Worldwatch Institute.

WHAT'S A RIVER FOR?

Thousands of dead salmon, acres of dying crops, pesticide-poisoned birds: How the Klamath River became the first casualty in the West's new water wars

By Bruce Barcott

ON THE MORNING of September 19, 2002, the Yurok fishermen who set their gill nets near the mouth of the Klamath River arrived to find the largest salmon run in years fully under way. The fish had returned from the ocean to the Klamath, on the Northern California coast, to begin their long trip upstream to spawn; there were thousands of them, as far as the eye could see. And they were dying. Fully-grown 30-pounders lay beached on shoreline rocks. Smaller fish floated in midriver eddies. Day after day they kept washing up; by the third day, biologists were estimating that 33,000 fish had been killed in one of the largest salmon die-offs in U.S. history.

The Yurok knew immediately what had happened. For months they, along with state experts and commercial fishermen, had been pleading with the federal government to stop diverting most of the river's water into the potato and alfalfa fields of Oregon's upper Klamath Basin. But the Bureau of Reclamation, the agency in charge of federal irrigation projects, refused to intervene. No one had proved, it argued, that the fish really needed the water.

When the die-off was discovered, federal authorities raced to send a flush of water downriver. But they didn't change their long-term policy—using the Klamath's water to support farmers over fish, fishermen, and some of the most environmentally critical wildlands in the nation.

From Montana to New Mexico, the Bush administration is favoring the water rights of farmers, ranchers, and developers over endangered species, Indian tribes—and the federal government itself.

FOR THE BUSH ADMINISTRATION, and for the advocacy groups that have joined the battle, the fight over the Klamath is more than a regional dispute. It's a bellwether signaling a key shift in the federal government's stance in a new generation of Western water wars. From Montana to New Mexico, conflicts over rivers, wetlands, and irrigation projects are pitting federal water rights against local and state governments and private interests. And in each case, the Bush administration is favoring farmers, ranchers, and developers over the rights of endangered species, Indian tribes, and the federal government itself. In the past year alone, administration officials have backed away from water policies designed to protect fish and birds along the Rio Grande and in California's Central Valley, given up their claim to protect thousands of acres of wetlands from being filled in for subdivisions and shopping malls, and moved toward ceding federal rights to water in several national parks and wildlife refuges.

"The administration sees water rights as property rights that come before other rights, including the right of ecosystems to exist," says Steve Malloch, executive director of the Western Water Alliance, a regional environmental advocacy group. "When faced with the choice between [environmental protection] and what they perceive to be inviolable property rights, they've made it clear what side they're going to come down on."

Nowhere have these issues played out more dramatically than on the 250-mile course the Klamath takes from Oregon's arid high country to the redwood forests of California's northern coast. When the federal government was forced by the courts in 2001 to withhold irrigation water and keep the Klamath flowing, national property-rights groups rallied to support struggling farmers. A year later, when the administration took water from the fish and gave it to farmers, the salmon kill made the river a cause célèbre for environmental groups. At least four separate lawsuits, with plaintiffs ranging from fishermen to property-rights advocates to environmentalists, are challenging government policies on the river. And a National Marine Fisheries Service biologist has filed for whistleblower protection, arguing that, under pressure from "a very high level," his agency changed a key scientific report to justify withholding water from the fish.

Decisions in some of the cases are expected soon, just as another season of trouble gets under way on the Klamath. After a winter of meager snowfalls, another drought is likely. Someone will have to go without water again. And that, in this part of the world, is the one thing no one can afford.

THE KLAMATH IS BORN A CRIPPLE. The river begins in Klamath Falls, Oregon, at the southern outlet of Upper Klamath Lake, a body of water so shallow that a tall man could nearly cross it without wetting his hat. Before the river becomes a river, the U.S. Bureau of Reclamation's "A" Canal diverts half its water into a labyrinthine system of aqueducts that irrigate more than 1,400 farms and ranches. The falls? There are no falls, never have been. The town's founders invented the name to attract homesteaders.

This was what you found when you arrived in Klamath Falls a century ago: swamps everywhere, and birds thick as flies. The Klamath Basin, a high arid plateau the size of Connecticut, didn't get much rain—about as much, on average, as the Texas Panhandle—but what it got it kept, draining the runoff from surrounding mountains into three shallow lakes and a vast system of wetlands. Ten million birds paused to feed here during their migration from Canada to Mexico. Egrets, terns, mallards, pelicans, eagles, tundra swans, and herons browsed amid thickets of 10-foot-tall bulrushes known as tule (too-lee). The Klamath Indians pulled tens of thousands of sucker fish from the lakes every year. This was the Everglades of the American West.

Where some saw a thriving wetland, the U.S. Bureau of Reclamation saw farmland drowning under water. Starting in 1906, the Bureau drained and replumbed nearly the entire basin, building a complex set of canals that shrank Lower Klamath Lake and Tule Lake to one-quarter their original size and sending the water to irrigate thousands of acres of crops and pasture. After the farmers had their share, some of the water would drain into the Tule Lake and Lower Klamath National Wildlife refuges, and some would be pumped back into the Klamath River. Over time, a map of the project came to resemble a guide to the London Underground.

A few weeks after low river levels caused one of the largest salmon kills in U.S. history, a federal biologist said his superiors had changed a key scientific assessment under pressure from "a very high level."

Thanks to irrigation, things hummed along for nearly a century. Homesteaders built farm towns along the Oregon-California border: Merrill, Malin, Tulelake. Teenage girls were crowned queen of the potato festival. Teenage boys hung out at the Tulelake bowling alley. A local historian proclaimed the Klamath Basin Project "one of the most successful government projects in American history."

But over those 100 years, any life that didn't grow in neat harrowed rows began to drain out of the basin.

The Klamath Indians noticed first. The shortnose sucker, a staple of the tribal diet, began growing scarce in Upper Klamath Lake. Farther downstream, the coho salmon that provided sustenance to the Yurok tribe—and to thousands of commercial fishermen along the coast—all but vanished. Environmentalists successfully fought to have both fish protected under the Endangered Species Act. The tribes, whose treaties guaranteed them the right to harvest sucker and salmon in perpetuity, filed lawsuits demanding protection for the fish. By 1995, the Interior Department's regional solicitor—the department's in-house lawyer—issued an opinion reflecting the new legal reality: Tribes, and the endangered species they sought to protect, would have first right to disputed federal waters. Irrigators would have to come second.

The Bureau of Reclamation, an agency created not to protect nature but to overcome it, was slow to implement the new priorities. But by the spring of 2001, it had no choice. A drought had struck the Klamath and a new lawsuit, from commercial fishermen along the coast, required that river levels be kept high enough for the fish. That meant withholding some of the farmers' irrigation water and sending it downriver—at least until the drought broke.

The drought did not break.

"WE WERE LOOKING GOOD AND GOING into 2001," says Dick Carleton. "We had our three tractors and potato equipment nearly paid off. Then we got word that the water wouldn't be coming." The 59-year-old farmer steers his red F-250 truck down a muddy track separating fields of alfalfa stubble. Carleton's farm sits three days' flow from the Klamath headwaters. He and his son Jim, 34, grow alfalfa and potatoes on 1,500 acres in Merrill, Oregon, 15 miles south of Klamath Falls. If you've eaten Campbell's cream of potato soup or snacked on Frito-Lay potato chips you may have tasted their crop. But you might not again. The Carletons are bankrupt.

"It's Chapter 12," Dick explains. "You don't lose everything, but you still owe your debts. Gives us a chance to regroup."

Without irrigation water, the Carletons lost their 2001 potato crop. Their bills piled up. A typical potato farmer can carry a $300,000 loan just on his equipment. Multiply Dick Carleton by 1,400 (the number of local growers who depend on Bureau irrigation) and you end up with a lot of angry farmers.

By the summer of 2001, those farmers were making national news. They paraded their tractors through Klamath Falls and carried signs that said "Klamath Basic Betrayed." National property-rights advocates flocked to the basin, accusing the government of engaging in "rural cleansing." Western members of Congress held hearings in Klamath Falls and vowed to rewrite the Endangered Species Act. The farmers started a vigil at the "A" Canal headgate. When somebody surreptitiously opened the canal—and local police refused to make arrests—federal marshals were called in.

"I wasn't a political person before 2001," says Dick Carleton. "But I spent a lot of time up at the headgate." For the farmers, things grew desperate. Some started selling off equipment for 10 cents on the dollar.

Many farmers lashed out at the Bush administration for locking up their water. But behind the scenes, top officials at the Interior Department, which oversees the Bureau of Reclamation, were scrambling to keep the farmers in business. "The department was in no position to say, 'We're not going to comply with the law,'" recalls Sue Ellen Wooldridge, deputy chief of staff for Interior Secretary Gale Norton. "We tried to see if there was any flex"—any wiggle room in the law that would allow delivery of irrigation water—"and there wasn't."

Officials knew that the only way to change the irrigation plan was a new scientific assessment of how much water the fish needed. In October 2001, Norton asked a committee of the National Research Council to review the available research on the Klamath. Four months later, the committee delivered a draft report: There wasn't enough evidence, it concluded, to know exactly how much water was enough. Environmentalists argue that the document left out key studies of the river; the council says it is still working on a final report.

But if the committee's findings were ambivalent, the Bureau of Reclamation's response was anything but. As long as science hadn't proved that low flows would harm the salmon, the agency announced, farmers would receive their full allocation of water. Wildlife experts—especially those at the National Marine Fisheries Service, which is charged with protecting the salmon—objected, but they were overruled. In a ceremony on March 29, 2002, with national TV cameras in attendance and farmers chanting, "Let the waters flow," Norton herself helped open the Klamath headgate.

The bureau's allocation to the farmers was so generous that by summer, irrigation ditches in Klamath County were overflowing with water, prompting Oregon state officials to complain about flooding. By September, dead salmon were washing up on the Yurok reservation. And the following month, Michael Kelly—the National Marine Fisheries Service's lead biologist in the Klamath case—filed a claim for protection as a whistleblower, charging that his superiors had changed a key scientific assessment, known as a biological opinion, in response to "political pressure." Environmentalists are now challenging the biological opinion—the foundation for the Klamath irrigation plan—in court. Kelly isn't speaking to the press while the claim is pending.

Throughout the controversy, the Interior Department has argued that its policy on the Klamath is driven by science; if the science changes, says Wooldridge, so will the irrigation plan. But the department's stance also reflects Interior Secretary Norton's own long-standing philosophy. The secretary cut her teeth working for the Rocky Mountain Legal Foundation, a property-rights group in Colorado where her boss was Reagan's controversial Interior Secretary James Watt, who once proposed selling off public lands, including some national parks. As Colorado attorney general in the 1990s, Norton frequently challenged the federal government's land and water rights in her state. Among her first actions at Interior was to appoint Colorado attorney Bennett Raley, who has represented cities and farmers in water disputes, as the assistant secretary overseeing the Bureau of Reclamation.

"There's a new wind blowing out of Washington," says Janet Neuman, an environmental law professor at Lewis & Clark College who, in the '90s, served on a federal water commission charged with evaluating Bureau of Reclamation policies. "At one point, the bureau was trying to move away from being perceived as being captive to [private interests]. Now they are pulling back from that."

In Colorado, for example, the administration has announced that it will not insist on its right to keep water flowing in the Gunnison River—a move that would endanger the unique ecosystems of Black Canyon of the Gunnison National Park, but could allow the water to be diverted toward Denver's booming suburbs. And on the Rio Grande, Norton is backing the city of Albuquerque in its attempts to withhold water from the river, even if that means destroying the habitat of an endangered fish called the silvery minnow.

Similar conflicts are likely to erupt throughout the West, experts believe, as drought and booming development exacerbate the pressure on already overtaxed water systems. "The Klamath is looked at as symptomatic," says Neuman. "There but for the grace of God go many, many other basins where there has been decades of over-commitment of the water resources. Endangered species listings and irrigation demands and unmet tribal demands—those circumstances exist in lots of places around the West, and all it takes is a particularly low water year to bring them to a head. It's just a matter of how low, and how soon."

"THERE—IN THAT TREE," says Bob Hunter. "Northern harrier." The marsh hawk perches in a bare cottonwood tree, scowling at a flock of bufflehead ducks bobbing on the marsh. A flock of white swans flaps overhead, their long necks slanted like 737s at takeoff.

Hunter is a lawyer for the Oregon environmental group Water Watch, and today he is guiding me through some of the most bizarre wetlands in the nation's wildlife refuge system. "What you're looking at is the only wildlife refuge in America that grows potatoes and hay," he says as we drive down a gravel levee road. To our left is the water of Tule Lake. To our right is an overwintering alfalfa field. Both are within the refuge's borders.

The peculiar setup goes back to the days of Teddy Roosevelt, who created both the National Wildlife Refuge System and the Bureau of Reclamation. The Klamath Basin became their battlefield. Reclamation engineers wanted it for farmland. Conservationists claimed it as a bird sanctuary. Over the years a compromise emerged that allowed local farmers to lease about one-third of the land in the refuges at Lower Klamath Lake and Tule Lake. To grow crops, they need irrigation water—the "excess" water that would otherwise keep the wetlands wet.

For years, environmentalists have argued for an end to farming in the refuge: "If you take those 32,000 acres of lease land out of the irrigation loop," says Hunter, "that's more than

15 percent of the Klamath Project put back into the river—at no cost to the government." In the '90s, the refuge's manager began doing just that, withholding water from the lease lands in drought years. But the Bush administration scrapped that policy last year.

If you imagine the Pacific flyway as an hourglass stretching from Alaska to South America, the Klamath marshes sit at its waist; 80 percent of all the migratory birds in the West stop here at some point on their journeys. "We don't really know if there's any other place in the United States that has the same significance for wildlife," says Wendell Wood, southern Oregon field representative for the Oregon Natural Resources Council. But since irrigation began a century ago, the number of birds in the basin has dropped by 90 percent, and those that remain—including nearly 1,000 bald eagles—find their breeding grounds drained and polluted. During the summer, many of the wetlands near Tule Lake turn into fields of cracked mud, even as the alfalfa fields next to them sprout a lush crop. At least 46 different insect- and weed-killing chemicals are regularly applied to the fields in the refuge, sometimes mixed in with irrigation water in a process called "chemigation"; the Bureau of Reclamation itself uses several toxic herbicides to keep irrigation canals weed-free. The refuge's longtime manager, Phil Norton, once said that when he first came to Tule Lake, he "couldn't believe they called it a wildlife refuge."

NOT FAR FROM THE REFUGE'S DUCK PONDS, on about 600 acres of what used to be Tule Lake, John Anderson grows alfalfa and mint. His father, Robert, won the land in a 1947 government lottery that offered World War II veterans the chance to be America's last homesteaders. "This was a thriving community back then," recalls Anderson, now 50. "Lot of neighbors, lot of kids around. Prices were better. Lot of folks got out of the business since then. At my church I'm one of the only farmers younger than 75."

Sure enough, a stroll down Tulelake's dusty streets reveals a town whose decline began long before the 2001 water crisis. The bowling alley is shuttered, and the hardware store closes all winter. The only signs of life come from a small grocery store, a county ag extension office, and Tulelake High School.

Years ago, subsidizing water-intensive crops like potatoes in a region that gets only 18 inches of rain each year seemed like an efficient use of federal resources. But, John Anderson says, farmers can tell that the winds have shifted. "Americans have changed their priorities," he says. "Now they want rivers, wetlands, clean water, wildlife. I can understand. But the American people should be willing to pay for it."

Anderson is among a growing number of farmers who support a seemingly simple solution to the water dilemma: What if the government paid some of the farmers to quit irrigating? A buyout of, say, 30 percent of the Klamath Projects' 200,000 acres of irrigated land, at roughly $2,500 an acre, would cost $150 million. The scheme would remove a lot of claims on the river and could leave enough water for both the fish and the remaining farmers. Variations of the plan have been floated by numerous conservation groups and some members of Oregon's congressional delegation.

When it comes to water in the West, it's appropriate to borrow from Faulkner: The past isn't dead here. It's not even past.

But when it comes to water in the West, it's appropriate to borrow from Faulkner: The past isn't dead here. It's not even past. Decades-old contracts and double crosses are recalled as if they happened last week. A century ago, farmers in California's Owens Valley took a water buyout and saw the lifeblood of their valley diverted to Los Angeles. Roman Polanski immortalized the scam in *Chinatown*. "Some of the Klamath pioneers," says Dan Keppen, head of the organization that represents the valley's irrigators, "moved here from the Owens Valley."

Keppen's group, the Klamath Water Users Association, is dead set against a buyout, which it says would simply usher in the end of farming in the basin. Last year the group successfully lobbied to block a $175 million congressional aid package for the Klamath because some of the money could have been used to buy out farms. National property-rights groups, who oppose turning private land into public property, have resisted the idea, and so has the Bush administration. As one official privately notes, Secretary Norton feels that the federal land portfolio "is quite large enough, thank you very much."

AS IT LEAVES the Klamath plateau, the river drops into Northern California's woolly Siskiyou country, home to Bigfoot sightings, marijuana patches, off-the-grid rednecks, and long-toothed hippies. State Highway 96 follows its course past a string of abandoned mines, through former mill towns that now get by on fly-fishing and rafting tours. As the Klamath Mountains segue into the Coast Range, moist Pacific air creeps up the river valley in cottony mists. Moss overcoats wrap around trunks of toyon, the California holly that inspired the name Hollywood. Strengthened by dozens of winter streams, the Klamath throws up rapids and surfable waves that draw kayakers from hundreds of miles away. At the village of Weitchpec (Witch-peck), the green Trinity River joins the muddy Klamath, and the combined channel veers away from the highway. A one-lane road on the Hoopa Indian Reservation continues to shadow the river until finally it, too, gives out and the Klamath rolls on through the woods, for the first time truly wild.

But the water wars don't stop when you leave the river. A few miles from the Klamath-Trinity confluence, I knocked at a house that had a sign posted in its yard. "DYING 4 WATER," it said, over a drawing of a salmon. Duane Sherman Sr., the 33-year-old former Hoopa tribal chairman, invited me in.

"I'm dividing up this deer we killed a week ago," Sherman said, offering me a bite of venison jerky. "You see, Indians are nothing but extended family. This deer feeds my grandmother, my sister, my aunt," he said, nodding to three women chatting at the kitchen table, "and my nephews too." Two boys watched cartoons in the next room. "The salmon's the same way. Those fish aren't just our livelihood. If we don't fish, we don't eat."

Most of the 33,000 fish lost in last September's salmon kill were headed upriver toward the Trinity, where the Hoopa have fishing rights. (Their Yurok neighbors have rights along the lower 40 miles of the Klamath, from the mouth to the Trinity confluence.) Which means there's a lot more deer than salmon on the Sherman family table these days.

For more than a decade the Hoopa, along with the Yurok, state and municipal governments, and landowners up and down the river, have been working to restore the Klamath fishery, once among the most productive on the West Coast. Last year was supposed to be the first season it all paid off. Instead, the salmon kill left local communities with what one study estimates will be about $20 million in losses, roughly as much as the farmers lost during the water crisis of 2001.

"I understand that the irrigators have a contract with the government," Susan Masten, the Yurok tribal chair, tells me over a plate of eggs at Sis' diner, near the mouth of the Klamath. "The government also has a contract with us. And it was the first contract."

This is Masten's second water war. The Yuroks' federal agreements, which date back to the mid-19th century, guarantee them access fish in perpetuity. But during the 1970s, tribal members battled the federal government and commercial fishermen for the right to set their nets. Federal agents with M-16s and riot gear occupied the reservation to keep tribal fishermen out of the river. A popular bumper sticker off the reservation read, "Can an Indian, Save a Salmon."

But over the years, the commercial fishermen have become the Yuroks' strongest allies. "After years of fighting with the tribes, we looked at each other and realized we had fewer and fewer fish to fight over," says Glen Spain, the Northwest regional director of the Pacific Coast Federation of Fishermen's Associations. "It's the same thing with the farmers. We're all workaday people trying to make a living. We understand the market forces driving the farmers down."

Like the farm towns farther upstream, the fishing communities on this stretch of coast are dotted with "For Sale" signs and shuttered family businesses. "From Fort Bragg, California, to Coos Bay, Oregon—the range of Klamath River salmon—we've lost 3,700 jobs and an $80 million-a-year economy," says Spain.

But the fishermen haven't been cast as victims in this crisis—not by the media, and not by the government. Their decline has come slowly, invisibly; they lack the ag industry's national lobbying muscle, or the property-rights movement's clout with the Bush administration. They are no more a political force than the ducks that land in the Tule Lake refuge.

SOME RIVERS ARE SO BIG you can't see them meet the sea. They simply fan out and merge with the tide. The amazing thing about the Klamath is that you can actually watch it pour into the Pacific. In Yurok country the river is a quarter-mile wide and quite; at its mouth it narrows into a frightening rush that cuts between a sandbar and a rocky cliff and then it's gone, lost in the foam. Standing there in the teeth of a January storm, watching a flock of gulls wheel in the cold wind, I thought of the classic closing line in *Chinatown:* "Forget it, Jake, it's Chinatown." Some things, it suggests, are simply too murky to understand.

But the Klamath water war comes down to a single, very simple equation: too many takers, not enough water.

"When I was a boy," John Anderson told me as we watched dusk settle over fields north of Tule Lake, "the ducks and geese came here by the millions. You could hear the flocks roar at night. We've given up a lot of the life that used to be here. Sometimes I question whether it's been a good trade-off."

It wasn't a wholly bad deal. A nation got fed, the West settled, families raised. But the math never did add up, and it's now becoming clear what got shortchanged in the bargain.

Additional reporting by Stephen Baxter.

Based in Seattle, **Bruce Barcott** is a contributing writer for *Outside.* In addition to this story about the battle over water rights on the Klamath River, Barcott covered the Pacific Northwest's salmon-fishing industry in "Aquaculture's Troubled Harvest" (November/December 2001).

A **Human** THIRST

Humans now appropriate more than half of all the freshwater in the world. Rising demands from agriculture, industry, and a growing population have left important habitats around the world high and dry.

by Don Hinrichsen

On March 20, 2000, a group of monkeys, driven mad with thirst, clashed with desperate villagers over drinking water in a small outpost in northern Kenya near the border with Sudan. The Pan African News Agency reported that eight monkeys were killed and 10 villagers injured in what was described as a "fierce two-hour melee." The fight erupted when relief workers arrived and began dispensing water from a tanker truck. Locals claimed that a prolonged drought had forced animals to roam out of their natural habitats to seek life-giving water in human settlements. The monkeys were later identified as generally harmless vervets.

The world's deepening freshwater crisis—currently affecting 2.3 billion people—has already pitted farmers against city dwellers, industry against agriculture, water-rich state against water-poor state, county against county, neighbor against neighbor. Inter-species rivalry over water, such as the incident in northern Kenya, stands to become more commonplace in the near future.

"The water needs of wildlife are often the first to be sacrificed and last to be considered," says Karin Krchnak, population and environment program manager at the National Wildlife Federation (NWF) in Washington, D.C. "We ignore the fact that working to ensure healthy freshwater ecosystems for wildlife would mean healthy waters for all." As more and more water is withdrawn from rivers, streams, lakes and aquifers to feed thirsty fields and the voracious needs of industry and escalating urban demands, there is often little left over for aquatic ecosystems and the wealth of plants and animals they support.

The mounting competition for freshwater resources is undermining development prospects in many areas of the world, while at the same time taking an increasing toll on natural systems, according to Krchnak, who co-authored an NWF report on population, wildlife, and water. In effect, humanity is waging an undeclared water war with nature.

"There will be no winners in this war, only losers," warns Krchnak. By undermining the water needs of wildlife we are not just undermining other species, we are threatening the human prospect as well.

Pulling Apart the Pipes

Currently, humans expropriate 54 percent of all available freshwater from rivers, lakes, streams, and shallow aquifers. During the 20th century water use increased at double the rate of population growth: while the global population tripled, water use per capita increased by six times. Projected levels of population growth in the next 25 years alone are expected to increase the human take of available freshwater to 70 percent, according to water expert Sandra Postel, Director of the Global Water Policy Project in Amherst, Massachusetts. And if per capita water consumption continues to rise at its current rate, by 2025 that share could significantly exceed 70 percent.

As a global average, most freshwater withdrawals—69 percent—are used for agriculture, while industry accounts for 23 percent and municipal use (drinking water, bathing and cleaning, and watering plants and grass) just 8 percent.

The past century of human development—the spread of large-scale agriculture, the rapid growth of industrial development, the construction of tens of thousands of large dams, and the growing sprawl of cities—has profoundly altered the Earth's hydrological cycle. Countless rivers, streams, floodplains, and wetlands have been dammed, diverted, polluted, and filled. These components of the hydrological cycle, which function as the Earth's plumbing system, are being disconnected and plundered, piece by piece. This fragmentation has been so extensive that freshwater ecosystems are perhaps the most severely endangered today.

Left High and Dry

Habitat destruction, water diversions, and pollution are contributing to sharp declines in freshwater biodiversity. One-fifth of all freshwater fish are threatened or extinct. On continents where studies have been done, more than half of amphibians are in decline. And more than 1,000 bird species—many of them aquatic—are threatened.

More than 40,000 large dams bisect waterways around the world, and more than 500,000 kilometers of river have been dredged and channelized for shipping. Deforestation, mining, grazing, industry, agriculture, and urbanization increase pollution and choke freshwater ecosystems with silt and other runoff.

Water diversion for irrigation, industry, and urban use has increased 35-fold in the past 300 years. In some cases, this increased demand has deprived entire ecosystems of water. Sprawl is an increasing concern, as the spread of urban areas is destroying important wetlands, and paved-over area is reducing the amount of water that is able to recharge aquifers.

Consider the plight of wetlands—swamps, marshes, fens, bogs, estuaries, and tidal flats. Globally, the world has lost half of its wetlands, with most of the destruction having taken place over the past half century. The loss of these productive ecosystems is doubly harmful to the environment: wetlands not only store water and transport nutrients, but also act as natural filters, soaking up and diluting pollutants such as nitrogen and phosphorus from agricultural runoff, heavy metals from mining and industrial spills, and raw sewage from human settlements.

In some areas of Europe, such as Germany and France, 80 percent of all wetlands have been destroyed. The United States has lost 50 percent of its wetlands since colonial times. More than 100 million hectares of U.S. wetlands (247 million acres) have been filled, dredged, or channeled—an area greater than the size of California, Nevada, and Oregon combined. In California alone, more than 90 percent of wetlands have been tilled under, paved over, or otherwise destroyed.

Destruction of habitat is the largest cause of biodiversity loss in almost every ecosystem, from wetlands and estuaries to prairies and forests. But biologists have found that the brunt of current plant and animal extinctions have fallen disproportionately on those species dependent on freshwater and related habitats. One fifth of the world's freshwater fish—2,000 of the 10,000 species identified so far—are endangered, vulnerable, or extinct. In North America, the continent most studied, 67 percent of all mussels, 51 percent of crayfish, 40 percent of amphibians, 37 percent of fish, and 75 percent of all freshwater mollusks are rare, imperiled, or already gone.

The global decline in amphibian populations may be the aquatic equivalent of the canary in the coal mine. Data are scarce for many species, but more than half of the amphibians studied in Western Europe, North America, and South America are in a rapid decline.

Around the world, more than 1,000 bird species are close to extinction, and many of these are particularly dependent on wetlands and other aquatic habitats. In Mexico's Sonora Desert, for instance, agriculture has siphoned off 97 percent of the region's water resources, reducing the migratory bird population by more than half, from 233,000 in 1970 to fewer than 100,000 today.

Pollution is also exacting a significant toll on freshwater and marine organisms. For instance, scientists studying beluga whales swimming in the contaminated St. Lawrence Seaway, which connects the Atlantic Ocean to North America's Great Lakes, found that the cetaceans have dangerously high levels of PCBs in their blubber. In fact the contamination is so severe that under Canadian law the whales actually qualify as toxic waste.

Waterways everywhere are used as sewers and waste receptacles. Exactly how much waste ends up in freshwater systems and coastal waters is not known. However, the UN Food and Agriculture Organization (FAO) estimates that every year roughly 450 cubic kilometers (99 million gallons) of wastewater (untreated or only partially treated) is discharged into rivers, lakes, and coastal areas. To dilute and transport this amount of waste requires at least 6,000 cubic kilometers (1.32 billion gallons) of clean water. The FAO estimates that if current trends continue, within 40 years the world's entire stable river flow would be needed just to dilute and transport humanity's wastes.

The Point of No Return?

The competition between people and wildlife for water is intensifying in many of the most biodiverse regions of the world. Of the 25 biodiversity hotspots designated by Conservation International, 10 are located in water-short regions. These regions—including Mexico, Central America, the Caribbean, the western United States, the Mediterranean Basic, southern Africa, and southwestern China—are home to an extremely high number of endemic and threatened species. Population pressures and overuse of resources, combined with critical water shortages, threaten to push these diverse and vital ecosystems over the brink. In a number of cases, the point of no return has already been reached.

China

China, home to 22 percent of the world's population, is already experiencing serious water shortages that threaten both people and wildlife. According to China's former environment minister, Qu Geping, China's freshwater supplies are capable of sustainably supporting no more than 650 million people—half its current population. To compensate for the tremendous shortfall, China is draining its rivers dry and mining ancient aquifers that take thousand of years to recharge.

As a result, the country has completely overwhelmed its freshwater ecosystems. Even in the water-rich Yangtze River Basin, water demands from farms, industry, and a giant population have polluted and degraded freshwater and riparian ecosystems. The Yangtze is one of the longest rivers in Asia, winding

6,300 kilometers on its way to the Yellow Sea. This massive watershed is home to around 400 million people, one-third of the total population of China. But the population density is high, averaging 200 people per square kilometer. As the river, sluggish with sediment and laced with agricultural, industrial, and municipal wastes, nears its wide delta, population densities soar to over 350 people per square kilometer.

The effects of the country's intense water demands, mostly for agriculture, can be seen in the dry lake beds on the Gianghan Plain. In 1950 this ecologically rich area supported over 1,000 lakes. Within three decades, new dams and irrigation canals had siphoned off so much water that only 300 lakes were left.

China's water demands have taken a huge toll on the country's wildlife. Studies carried out in the Yangtze's middle and lower reaches show that in natural lakes and wetlands still connected to the river, the number of fish species averages 100. In lakes and wetlands cut off and marooned from the river because of diversions and drainage, no more than 30 survive. Populations of three of the Yangtze's largest and more productive fisheries—the silver, bighead, and grass carp—have dropped by half since the 1950s.

Mammals and reptiles are in similar straits. The Yangtze's shrinking and polluted waters are home to the most endangered dolphin in the world—the Yangtze River dolphin, or Baiji. There are only around 100 of these very rare freshwater dolphins left in the wild, but biologists predict they will be gone in a decade. And if any survive, their fate will be sealed when the massive Three Gorges Dam is completed in 2013. The dam is expected to decrease water flows downstream, exacerbate the effects of pollution, and reduce the number of prey species that the dolphins eat. Likewise, the Yangtze's Chinese alligators, which live mostly in a small stretch near the river's swollen, silt-laden mouth, are not expected to survive the next 10 years. In recent years, the alligator population has dropped to between 800 and 1,000.

The Aral Sea

The most striking example of human water demands destroying an ecosystem is the nearly complete annihilation of the 64,500 square kilometer Aral Sea, located in Central Asia between Kazakhstan and Uzbekistan. Once the fourth largest inland sea in the world, it has contracted by half its size and lost three-quarters of its volume since the 1960s, when its two feeder rivers—the Amu Darya and the Syr Darya—were diverted to irrigate cotton fields and rice paddies.

The water diversions have also deprived the region's lakes and wetlands of their life source. At the Aral Sea's northern end in Kazakhstan, the lakes of the Syr Darya delta shrank from about 500 square kilometers to 40 square kilometers between 1960 and 1980. By 1995, more than 50 lakes in the Amu Darya delta had dried up and the surrounding wetlands had withered from 550,000 hectares to less than 20,000 hectares.

The unique *tugay* forests—dense thickets of small shrubs, grasses, sedges and reeds—that once covered 13,000 square kilometers around the fringes of the sea have been decimated. By

Alien Invaders

"Rapidly growing populations place heavy demand on freshwater resources and intensify pressures on wildlands," concludes a combined World Resources Institute and Worldwatch report called "Watersheds of the World." But increasingly, the introduction of exotic or alien species is playing a large role in wreaking havoc on freshwater habitats.

The spread of invasive species is a global phenomenon, and is increasingly fostered by the growth of aquaculture, shipping, and commerce. Whether introduced by accident or on purpose, these alien invaders are capable of altering habitats and extirpating native species en masse.

The invasion and insidious spread of the zebra mussel in the U.S. Great Lakes highlights the tremendous costs to ecosystems and species. A native of Eastern Europe, the zebra mussel arrived in the Great Lakes in 1988, released most likely through the discharge of ballast waters from a cargo ship. Once established, it spread rapidly throughout the region.

The mussels have crowded out native species that cannot compete with them for space and food. A study of the mussels in western Lake Erie found that all of the native clams at each of 17 sampling stations had been wiped out. Moreover, the last known population of the winged maple leaf clam, found in the St. Croix River in the upper Mississippi River basin, is now threatened by advancing ranks of the zebra mussel.

1999 less than 1,000 square kilometers of fragmented and isolated forest remained.

The habitat destruction has dramatically reduced the number of mammals that used to flourish around the Aral Sea: of 173 species found in 1960, only 38 remained in 1990. Though the ruined deltas still attract waterfowl and other wetland species, the number of migrant and nesting birds has declined from 500 species to fewer than 285 today.

Plant life has been hard hit by the increase in soil salinity, aridity, and heat. Forty years ago, botanists had identified 1,200 species of flowering plants, including 29 endemic species. Today, the endemics have vanished. The number of plant species that can survive the increasingly harsh climate is a fraction of the original number.

Most experts agree that the sea itself may very well disappear entirely within two decades. But the region's freshwater habitats and related communities of plants and animals have already been consigned to oblivion.

Lake Chad

Lake Chad, too, has shrunk—to one-tenth of its former size. In 1960, with a surface area of 25,000 square kilometers, it was the second-largest lake in Africa. When last surveyed, it was down to only 2,000 square kilometers. And here, too, massive water withdrawals from the watershed to feed irrigated agriculture have reduced the amount of water flowing into the lake to a trickle, especially during the dry season.

Lake Chad is wedged between four nations: populous Nigeria to the southwest, Niger on the northwest shore, Chad to the northeast, and Cameroon on a small section of the south shore. Nigeria has the largest population in Africa, with 130 million inhabitants. Population-growth rates in these countries average 3 percent a year, enough to double human numbers in one generation. And population growth rates in the regions around the lake are even higher than the national averages. People gravitate to this area because the lake and its rivers are the only sources of surface water for agricultural production in an otherwise dry and increasingly desertified region.

Although water has been flowing into the lake from its rivers over the past decade, the lake is still in serious ecological trouble. The lake's fisheries have more or less collapsed from over-exploitation and loss of aquatic habitats as its waters have been drained away. Though some 40 commercially valuable species remain, their populations are too small to be harvested in commercial quantities. Only one species—the mudfish—remains in viable populations.

As the lake has withered, it has been unable to provide suitable habitat for a host of other species. All large carnivores, such as lions and leopards, have been exterminated by hunting and habitat loss. Other large animals, such as rhinos and hippopotamuses, are found in greatly reduced numbers in isolated, small populations. Bird life still thrives around the lake, but the variety and numbers of breeding pairs have dropped significantly over the past 40 years.

A Blue Revolution

As these examples illustrate, the challenge for the world community is to launch a "blue revolution" that will help governments and communities manage water resources on a more sustainable basis for all users. "We not only have to regulate supplies of freshwater better, we need to reduce the demand side of the equation," says Swedish hydrologist Malin Falkenmark, a senior scientist with Sweden's Natural Science Research Council. "We need to ask how much water is available and how best can we use it, not how much do we need and where do we get it." Increasingly, where we get it from is at the expense of aquatic ecosystems.

If blindly meeting demand precipitated, in large measure, the world's current water crisis, reducing demand and matching supplies with end uses will help get us back on track to a more equitable water future for everyone. While serious water initiatives were launched in the wake of the World Summit on Sustainable Development held in Johannesburg, South Africa, not one of them addressed the water needs of ecosystems.

There is an important lesson here: just as animals cannot thrive when disconnected from their habitats, neither can humanity live disconnected from the water cycle and the natural systems that have evolved to maintain it. It is not a matter of "either or" says NWF's Krchnak. "We have no real choices here. Either we as a species live within the limits of the water cycle and utilize it rationally, or we could end up in constant competition with each other and with nature over remaining supplies. Ultimately, if nature loses, we lose."

By allowing natural systems to die, we may be threatening our own future. After all, there is a growing consensus that natural ecosystems have immense, almost incalculable value. Robert Costanza, a resource economist at the University of Maryland, has estimated the global value of freshwater wetlands, including related riverine and lake systems, at close to $5 trillion a year. This figure is based on their value as flood regulators, waste treatment plants, and wildlife habitats, as well as for fisheries production and recreation.

The nightmarish scenarios envisioned for a water-starved not too distant future should be enough to compel action at all levels. The water needs of people and wildlife are inextricably bound together. Unfortunately, it will probably take more incidents like the one in northern Kenya before we learn to share water resources, balancing the needs of nature with the needs of humanity.

Don Hinrichsen *is a UN consultant. He is former editor-in-chief of* Ambio *and was a news correspondent in Europe for 15 years.*

From *World Watch*, January/February 2003. © 2003 by Worldwatch Institute, www.worldwatch.org.

WATER AND HUMAN DESTINY

Our Perilous Dependence on Groundwater

Widespread and widely accessible, aquifers can be a constant and reliable source of water, provided they are not overpumped or contaminated.

Marc J. Defant

Seventeen-year-old Jason Tuskes must have realized his hopeless predicament. Before he was found, drowned, in a water-filled cave in Florida, he had removed the almost empty air tanks from his back and used his knife to scratch a message to his parents and brother on the wall of the cave: "I love you mom, dad, and Christian." Investigators concluded that the boy may have become disoriented in the aquifer's many confining passages after his fins kicked up silt, making visibility almost zero.

Hundreds of springs bubbling from the side of Idaho's Snake River Canyon provide evidence of a big, leaky aquifer, replenished historically by rivers that were completely absorbed into the surrounding plain. Today, farmers' wells drain the aquifer even as seepage from irrigation canals replenishes it.

Aquifers, which provide life-sustaining freshwater to millions of people around the world, can also be deadly when inexperienced divers such as Tuskes explore them. An aquifer is any region of rock that harbors potable water. Florida's water-filled limestone caves lie at one end of a broad spectrum of aquifer types, ranging from those that are shallow, naturally open to interactions with surface water, and flowing through large cavities, to those that are deep, closed to interactions with surface water, and resting in closely packed sand and gravel. Typically, water in aquifers flows slowly through permeable rock (such as sandstone) via small connecting voids. Water-filled caves would not characterize most aquifers.

Where sections of the aquifers are made up of limestone, however, the limestone can be whittled away by slightly acidic water flowing through fractures in the rock. Rainwater in Florida is rather acidic, with an average pH of 4.8 (pH = 7 is neutral), and it has become more so due to the formation of acid rain from the burning of fossil fuels. The acidic water gradually dissolves the limestone, sometimes forming caves.

Groundwater basics

Underground water is common below most of Earth's surface, whether in deserts, plains, or mountains. In some regions, people have been able to access the water for centuries through the use of wells dug by hand. In other regions, the water lies so deep that only modern well-drilling and pumping technology makes it available.

Scientists estimate that the volume of freshwater available underground is 30 times greater than the volume of water contained in all the world's freshwater lakes. Viewed through "groundwater glasses," most rivers and lakes, as well as wetlands, appear as surface manifestations of the groundwater levels.

Florida, with various types of aquifers and water-use issues, provides a good arena for training ourselves to see through groundwater glasses. Geography prescribes that the state be dependent on aquifers for most of its freshwater supplies. Surrounded on three sides by salt water, Florida is like a large island in many respects. Its few rivers are relatively short, and its surface lies over extremely permeable sands and rock that allow the surface water to seep quickly into the ground, recharging underlying aquifers. The end result is a state with a relatively limited supply of surface water and large, easily accessible underground reservoirs of water. Agricultural, industrial, and residential users have all dipped so freely into the aquifers that the state now faces uncertainties about how it will supply its growing population with freshwater in the coming decades.

Floridians tap five aquifers for freshwater. These underground water reservoirs are contained within layers of host

Contamination and Cleanup

Potential sources of groundwater contamination range from the surface runoff of pesticides and fertilizers to injection wells used for disposing of industrial and hazardous wastes deep underground. Other potential sources include septic tank systems, landfills, mining operations, and underground storage tanks. A few examples illustrate the challenges of keeping groundwater pure.

Septic tank systems and injection wells for waste disposal are both based on the idea that wastes can be safely disposed of underground. More than 30 percent of the households around the United States have an on-site sewage disposal system, usually consisting of a septic tank (a big underground tank) coupled through various configurations with a drain field. Inside the septic tank, anaerobic bacteria decompose the wastes into solids (sludge), which sink to the bottom; a scum, which floats; and a relatively clear liquid. The sludge and scum remain in the tank, which needs to be pumped out periodically, while the liquid flows through a pipe to the drain field.

Ideally, the fluid is dispersed into the drain field in such a way that as it percolates through the soil and rock layers, it is completely cleaned by filtration and further bacterial action. In practice, the "cleaned" sewage fluids often reach the underlying aquifer carrying dissolved chemicals such as nitrogen and phosphate. This process leads to groundwater contamination around septic tanks and, particularly in densely populated areas, forms large plumes of contaminated water.

Underground sewage disposal is an even less satisfactory solution if the underlying soil and rock layers do not support filtration and bacterial processing of the sewage fluids released by the septic tanks. Regions underlaid by limestone, for example, with its potential for large caves and caverns, are afforded a very poor natural filtering system underground. Sewage disposal is also extremely problematic in regions positioned over a "perched" water table, which typically exists where clay layers intertongue with gravel or soil. The impermeable clay layer can collect small reservoirs of water above the true water table, or it can collect septic tank effluents.

One of the worst stories of groundwater pollution in the United States centers on leaky underground fuel-storage tanks. Located below every gasoline station in the country is a tank (or tanks) from which gas is pumped to the customer's automobile. Gasoline and other products stored in underground tanks contain toxic and carcinogenic components. By the late 1970s and early '80s, environmentalists were discovering that many of these tanks were slowly leaking their products into the aquifers.

A massive national campaign was initiated to correct the problem in two ways. The first was to require all gasoline stations to install improved, double-walled tanks (electronic pressure gauges were used inside the tanks to monitor leakage). The second focused on cleaning sites around the nation that had been profoundly contaminated. They became known as "superfund" sites because of the large amount of money funneled into their remediation. The environmentally profligate practices responsible for these sites included storing hazardous fluids in leaking tanks, dumping hazardous solid and liquid wastes on the open ground, and storing hazardous liquid wastes in unlined ponds.

Sometimes the groundwater needs to be cleaned after the surface-level sources of contamination have been removed. One common process, air sparging, involves injecting air through deep wells. As the air returns to the surface, it passes through the aquifer collecting volatiles from the contaminants, such as gasoline. The injected air also oxygenates the aquifer, supporting the growth of bacteria that can help eliminate the contaminants. Another practice requires that water be pumped out of the aquifer, cleaned, and then pumped back.

—*M.J.D.*

---

rocks that are generally exposed at their more northerly ends before the layers dip underground. Water from precipitation infiltrates into the aquifers in the areas where the host rock layers are exposed, either at the surface or under a thin layer of soil (weathered rock and decaying organic matter). Four of the state's aquifers are recharged within its borders; the fifth, which in most of Florida lies below the surface under impermeable rock and other aquifers, is recharged in Alabama, Georgia, and South Carolina.

Three of Florida's aquifers are *unconfined*. That is, they exist at or near the surface and are intertwined with the surface waters. They consist of loose sand and gravel or extremely porous limestone. With unconfined aquifers, the groundwater and surface water are so closely connected that the water table (the depth below ground to reach water) and water level in rivers and lakes rise and fall together in response to drought or precipitation. When precipitation dwindles during a drought, large rivers continue to carry water supplied via springs from underground.

Unlike Florida's three surficial and unconfined aquifers, its other two, the intermediate and the Floridan, are *confined* and thus not connected to the water in rivers and lakes. Confined aquifers lie between impermeable layers of rock. In Florida, the confined aquifers outcrop at their higher, northern end, where they are recharged. Their southerly ends discharge through springs into the sea.

The intermediate aquifer, lying below the three surficial aquifers, is the main source of water for sections of Florida's west coast. The huge and ancient Floridan aquifer lies deeper still and supplies more freshwater than any of the other aquifers.

The Floridan provides water for cities ranging from Savannah, Georgia, in the northeast to Tallahassee in the west and Orlando in the south. All are pumping from the same source. In recent years, the four states drinking from the same

Multi-State View of the High Plains Aquifer

KANSAS DEPARTMENT OF AGRICULTURE

Water from the Ogallala aquifer, shown here in dark grey, supports agricultural activity over a swath of the Great Plains.

aquifer have been forced to collaborate in formulating water-use regulations, in an effort to assure that overuse of the aquifer doesn't lead to water deprivation for them all.

Accessing groundwater

Groundwater is pumped from wells to provide water for everything from agriculture to quenching thirst. If the pumping rate is too great, water flow in the aquifer can be altered or even reversed. Springs can go dry or water from a river can be pulled into the well, raising the possibility of either the aquifer being contaminated or the river drying up. Lacking a river to

replenish the aquifer, wells pumped at too great a rate will go dry.

If an aquifer system is along the sea coast, overpumping can lead to salt-water encroachment, in which salt water flows into the aquifer replacing the freshwater lost to pumping. Water in unconfined coastal aquifers eventually reaches the ocean through freshwater springs. During periods in the past when pumping was unregulated, salt-water encroachment became a significant problem for large sections of the world's coastal aquifers, particularly around metropolitan areas. Though coastal aquifer pumping is more closely regulated now, aquifer damage due to salt-water

encroachment remains a major problem in many coastal regions, such as Long Island, New York.

An additional calamity caused by overpumping wells is a reduction in the groundwater pressure supporting the land. The land may subside partially or completely to form a sinkhole. Droughts and overpumping (usually by large industries) in Florida have combined to create considerable damage due to sinkhole activity, leading to vociferous complaints and litigation.

Managing groundwater use

The case of the Edwards aquifer in Texas illustrates some of the complications of managing groundwater use. The Edwards is a confined aquifer; historically, its only discharge has been through two large springs that escaped through cracks in the rocks and sustained the San Antonio River, a ribbon of green in a dry valley. Water feeds into the aquifer in higher mountainous regions, then flows slowly through the rock layers until it lies underground in a band that extends from the valley floor up onto the flanks of the mountainous regions nearby. By merely puncturing the permeable layer overlying the aquifer, farmers and homeowners alike could tap into free water that would flow unceasingly from the well unless they regulated the discharge.

Capitalizing on the beauty of the river, the city of San Antonio built up its enchanting River Walk. The trail weaves along the river through the city center, offering a lush canopy of trees and quaint shops along the river.

As early as the late 1800s, as farmers and local communities sank wells and drained the Edwards aquifer to supply their crops and the region's growing population, the river's flow from the springs was reduced dramatically. In 1910, when the flow became insufficient even to carry away the sewage dumped into the river, public outrage forced the mayor of San Antonio to take drastic action. Wells were established for the purpose of feeding aquifer water directly into the river.

Through the remainder of the twentieth century, San Antonio sustained its River Walk with water from its regional aquifer. Coupled with all the other demands, the community was withdrawing water from the aquifer much faster than it was being replenished in the mountains. Finally, in June 2000, San Antonio stopped pumping aquifer water into its river and now maintains water flow past the River Walk using recycled municipal water. Despite such basic precautions, the region's growing water demands assure that the Edwards aquifer remains in peril.

Surface consequences

If groundwater consumption is unchecked, extensive harmful consequences can occur at the surface. The Southwest provides a good example.

More than 90 percent of Arizona's formerly continually flowing streams and rivers have dried up, due primarily to overpumping of groundwater contained in unconfined aquifers continuous with the streams. Water that previously would have flowed in the rivers is now being naturally drawn underground to replenish aquifers whose water table has sunk below the level of the river. When rivers dry up, streamside vegetation also dies, and animals dependent on the vegetation follow soon thereafter. Loss of vegetative cover leaves the land barren and open to massive erosion during the flash floods that commonly come out of the mountains. The landscape is literally being denuded by groundwater overuse.

Further to the northeast, the Ogallala aquifer supplies large sections of America's bread basket with water for irrigating 30 percent of all the irrigated land in the United States. Although it is presented as one aquifer, in fact, the treat cache of water is extremely variable in depth to the water table and in the saturated thickness of the rocky layers below.

Under the Texas high plains, the aquifer, according to local water managers, has been drained of about half its estimated original water supply, and the pumping continues. In that area, aquifer recharge is minimal in comparison to the amounts withdrawn. The average depth to water is 150–200 feet, but there are great variations. At one extreme, one of the wells, which started with a water table at a depth of 95 feet, now reaches water at 335 feet, and the remaining saturated thickness for that well is only 65 feet.

Nebraska represents perhaps an opposite extreme. It has been blessed with the largest water component of the aquifer and, despite significant draw-down in years past, in some places it still has more than 1,000 feet of saturated rock. Furthermore, in the sand hills region in

the western part of the state, water infiltration continues to recharge the aquifer. The great flood of the Mississippi River in 1993 was an anomaly, as aquifer levels in Nebraska actually rose.

On average, water levels in the aquifer are dropping by a few inches each year, down from drops averaging one or two feet per year that occurred in the decades after high-capacity wells to support intensive irrigation were introduced in the middle of the twentieth century. More efficient irrigation technologies and limits imposed by local water districts on how much water can be used for crops as seen as major factors in lessening the rate of decline in the aquifer's water table.

A parched tomorrow?

For much of the twentieth century, it seemed to a large segment of the U.S. population that water was in endless supply—one of nature's free "gifts." No longer! Several states have constructed desalination plants to add to their freshwater supplies, and all states pay indirectly for their water via the costs of pumping, decontamination, and assuring high standards of water quality. As population and industry grow, the demand for freshwater can only continue to increase. The aquifers can be thought of as temporary storage facilities that, if used properly and kept from pollution, can continue to supply future generations with one of our most crucial commodities. Only time will tell how much future generations will have to invest in order to drink what we still tend to take for granted: potable water.

Further reading
A. F. Randazzo and D. S. Jones, eds., *The Geology of Florida,* University Press of Florida, Gainesville, 1997.

Marc J. Defant is professor of geology at the University of South Florida. He specializes in the study of volcanoes and is the author of Voyage of Discovery: From the Big Bang to the Ice Ages, *a panoramic history of the universe, including our galaxy, solar system, and planet.*

OCEANS
Are on the Critical List

The primary threats to the planet's seas—overfishing, habitat degradation, pollution, introduction of alien species, and climate change—are largely human-induced.

BY ANNE PLATT McGINN

OCEANS FUNCTION as a source of food and fuel, a means of trade and commerce, and a base for cities and tourism. Worldwide, people obtain much of their animal protein from fish. Ocean-based deposits meet one-fourth of the world's annual oil and gas needs, and more than half of world trade travels by ship. More important than these economic figures, however, is the fact that humans depend on oceans for life itself. Harboring a greater variety of animal body types (phyla) than terrestrial systems and supplying more than half of the planet's ecological goods and services, the oceans play a commanding role in the Earth's balance of life.

Due to their large physical volume and density, oceans absorb, store, and transport vast quantities of heat, water, and nutrients. The oceans store about 1,000 times more heat than the atmosphere does, for example. Through processes such as evaporation and photosynthesis, marine systems and species help regulate the climate, maintain a livable atmosphere, convert solar energy into food, and break down natural wastes. The value of these "free" services far surpasses that of

ocean-based industries. Coral reefs alone, for instance, are estimated to be worth $375,000,000,000 annually by providing fish, medicines, tourism revenues, and coastal protection for more than 100 countries.

Despite the importance of healthy oceans to our economy and well-being, we have pushed the world's oceans perilously close to—and in some cases past—their natural limits. The warning signs are clear. The share of overexploited marine fish species jumped from almost none in 1950 to 35% in 1996, with an additional 25% nearing full exploitation. More than half of the world's coastlines and 60% of the coral reefs are threatened by human activities, including intensive coastal development, pollution, and overfishing.

In January, 1998, as the United Nations was launching the Year of the Ocean, more than 1,600 marine scientists, fishery biologists, conservationists, and oceanographers from across the globe issued a joint statement entitled "Troubled Waters." They agreed that the most pressing threats to ocean health are human-induced, in-

cluding species overexploitation, habitat degradation, pollution, introduction of alien species, and climate change. The impacts of these five threats are exacerbated by poorly planned commercial activities and coastal population growth.

Yet, many people still consider the oceans as not only inexhaustible, but immune to human interference. Because scientists just recently have begun to piece together how ocean systems work, society has yet to appreciate—much less protect—the wealth of oceans in its entirety. Indeed, current courses of action are rapidly undermining this wealth.

Nearly 1,000,000,000 people, predominantly in Asia, rely on fish for at least 30% of their animal protein supply. Most of these fish come from oceans, but with increasing frequency they are cultured on farms rather than captured in the wild. Aquaculture, based on the traditional Asian practice of raising fish in ponds, constitutes one of the fastest-growing sectors in world food production.

In addition to harvesting food from the sea, people have traditionally relied on oceans as a transportation route. Sea trade currently is dominated by multinational companies that are more influenced by the rise and fall of stock prices than by the tides and trade winds. Modern fishing trawlers, oil tankers, aircraft carriers, and container ships follow a path set by electronic beams, satellites, and computers.

Society derives a substantial portion of energy and fuel from the sea—a trend that was virtually unthinkable a century ago. In an age of falling trade barriers and mounting pressures on land-based resources, new ocean-based industries such as tidal and thermal energy production promise to become even more vital to the workings of the world economy. Having increased six-fold between 1955 and 1995, the volume of international trade is expected to triple again by 2020, according to the U.S. National Oceanographic and Atmospheric Administration, and 90% of it is expected to move by ocean.

In contrast to familiar fishing grounds and sea passageways, the depths of the ocean were long believed to be a vast wasteland that was inhospitable, if not completely devoid of life. Since the first deployment of submersibles in the 1930s and more advanced underwater acoustics and pressure chambers in the 1960s, scientific and commercial exploration has helped illuminate life in the deep sea and the geological history of the ancient ocean. Mining for sand, gravel, coral, and minerals (including sulfur and, most recently, petroleum) has taken place in shallow waters and continental shelves for decades, although offshore mining is severely restricted in some national waters.

Isolated, but highly concentrated, deep-sea deposits of manganese, gold, nickel, and copper, first discovered in the late 1970s, continue to tempt investors. These valuable nodules have proved technologically difficult and expensive to extract, given the extreme pressures and depths of their location. An international compromise on the deep seabed mining provisions of the Law of the Sea in 1994 has opened the way to some mining in international waters, but it appears unlikely to lead to much as long as mineral prices remain low, demand is largely met from the land, and the cost of underwater operations remains prohibitively high.

Perhaps more valuable than the mineral wealth in oceans are still-undiscovered living resources—new forms of life, potential medicines, and genetic material. For example, in 1997, medical researchers stumbled across a compound in dogfish that

stops the spread of cancer by cutting off the blood supply to tumors. The promise of life-saving cures from marine species is gradually becoming a commercial reality for bioprospectors and pharmaceutical companies as anti-inflammatory and cancer drugs have been discovered and other leads are being pursued.

Tinkering with the ocean for the sake of shortsighted commercial development, whether for mineral wealth or medicine, warrants close scrutiny, however. Given how little we know—a mere 1.5% of the deep sea has been explored, let alone adequately inventoried—any development could be potentially irreversible in these unique environments. Although seabed mining is subjected to some degree of international oversight, prospecting for living biological resources is completely unregulated.

During the past 100 years, scientists who work both underwater and among marine fossils found high in mountains have shown that the tree of life has its evolutionary roots in the sea. For about 3,200,000,000 years, all life on Earth was marine. A complex and diverse food web slowly evolved from a fortuitous mix there of single-celled algae, bacteria, and several million trips around the sun. Life remained sea-bound until 245,000,000 years ago, when the atmosphere became oxygen-rich.

Thanks to several billion years' worth of trial and error, the oceans today are home to a variety of species that have no descendants on land. Thirty-two out of 33 animal life forms are represented in marine habitats. (Only insects are missing.) Fifteen of these are exclusively marine phyla, including those of comb jellies, peanut worms, and starfish. Five phyla, including that of sponges, live predominantly in salt water. Although, on an individual basis, marine species account for just nine percent of the 1,800,000 species described for the entire planet, there may be as many as 10,000,000 additional species in the sea that as yet have not been classified.

In addition to hosting a vast array of biological diversity, the marine environment performs such vital functions as oxygen production, nutrient recycling, storm protection, and climate regulation—services that often are taken for granted.

Marine biological activity is concentrated along the world's coastlines (where sunlit surface waters receive nutrients and sediments from land-based runoff, river deltas, and rainfall) and in upwelling systems (where cold, nutrient-rich, deep-water currents run up against continental

margins). It provides 25% of the planet's primary biological productivity and an estimated 80–90% of the global commercial fish catch. It is estimated that coastal environments account for 38% of the goods and services provided by the Earth's ecosystems, while open oceans contribute an additional 25%. The value of all marine goods and services is estimated at 21 trillion dollars annually, 70% more than terrestrial systems come to.

Oceans are vital to both the chemical and biological balance of life. The same mechanism that created the present atmosphere—photosynthesis—continues to feed the marine food chain. Phytoplankton—tiny microscopic plants—take carbon dioxide (CO_2) from the atmosphere and convert it into oxygen and simple sugars, a form of carbon that can be consumed by marine animals. Other types of phytoplankton process nitrogen and sulfur, thereby helping the oceans function as a biological pump.

Although most organic carbon is consumed in the marine food web and eventually returned to the atmosphere via respiration, the unused balance rains down to the deep waters that make up the bulk of the ocean, where it is stored temporarily. Over the course of millions of years, these deposits have accumulated to the point that most of the world's organic carbon, approximately 15,000,000 gigatons (a gigaton equals 1,000,000,000 tons), is sequestered in marine sediments, compared with 4,000 gigatons in land-based reserves. On an annual basis, about one-third of the world's carbon emissions—around two gigatons—is taken up by oceans, an amount roughly equal to the uptake by land-based resources. If deforestation continues to diminish the ability of forests to absorb carbon dioxide, oceans are expected to play a more important role in regulating the planet's CO_2 budget in the future as human-induced emissions keep rising.

Perhaps no other example so vividly illustrates the connections between the oceans and the atmosphere than El Niño. Named after the Christ child because it usually appears in December, the El Niño Southern Oscillation takes place when trade winds and ocean surface currents in the eastern and central Pacific Ocean reverse direction. Scientists do not know what triggers the shift, but the aftermath is clear: Warm surface waters essentially pile up in the eastern Pacific and block deep, cold waters from upwelling, while a low pressure system hovers over South America, collecting heat and moisture that would otherwise be distributed at sea. This produces severe weather conditions in many parts of the world—increased precipitation, heavy flooding, drought, fire, and deep freezes—which, in turn, have enormous economic impact. During the 1997–98 El Niño, for instance, Argentina lost more than $3,000,000,000 in agricultural products due to these ocean-climate reactions, and Peru reported a 90% drop in anchovy harvests compared with the previous year.

A sea of problems

As noted earlier, the primary threats to oceans are largely human-induced and synergistic. Fishing, for example, has drastically altered the marine food web and underwater habitat areas. Meanwhile, the ocean's front line of defense—the coastal zone—is crumbling from years of degradation and fragmentation, while its waters have been treated as a waste receptacle for generations. The combination of overexploitation, the loss of buffer areas, and a rising tide of pollution has suffocated marine life and the livelihoods based on it in some areas. Upsetting the marine ecosystem in these ways has, in turn, given the upper hand to invasive species and changes in climate.

Overfishing poses a serious biological threat to ocean health. The resulting reductions in the genetic diversity of the spawning populations make it more difficult for the species to adapt to future environmental changes. The orange roughy, for instance, may have been fished down to the point where future recovery is impossible. Moreover, declines in one species can alter predator-prey relations and leave ecosystems vulnerable to invasive species. The overharvesting of triggerfish and pufferfish for souvenirs on coral reefs in the Caribbean has sapped the health of the entire reef chain. As these fish declined, populations of their prey—sea urchins—exploded, damaging the coral by grazing on the protective layers of algae and hurting the local reef-diving industry.

These trends have enormous social consequences as well. The welfare of more than 200,000,000 people around the world who depend on fishing for their income and food security is severely threatened. As the fish disappear, so do the coastal communities that depend on fishing for their way of life. Subsistence and small-scale fishers, who catch nearly half of the world's fish, suffer the greatest losses as they cannot afford to compete with large-scale vessels or changing technology. Furthermore, the health of more than 1,000,000,000 poor consumers who depend on minimal quantities of fish to constitute their diets is at risk as an ever-growing share of fish—83% by value—continues to be exported to industrial countries each year.

Despite a steadily growing human appetite for fish, large quantities are wasted each year because the fish are undersized or a nonmarketable sex or species, or because a fisher does not have a permit to catch them and must therefore throw them out. The United Nations' Food and Agricultural Organization estimates that discards of fish alone—not counting marine mammals, seabirds, and turtles—total 20,000,000 tons, equivalent to one-fourth of the annual marine catch. Many of these fish do not survive the process of getting entangled in gear, being brought on board, and then tossed back to sea.

Another threat to habitat areas stems from trawling, with nets and chains dragged across vast areas of mud, rocks, gravel, and sand, essentially sweeping everything in the vicinity. By recent estimates, all the ocean's continental shelves are trawled by fishers at least once every two years, with some areas hit several times a season. Considered a major cause of habitat degradation, trawling disturbs bottom-dwelling communities as well as localized species diversity and food supplies.

The conditions that make coastal areas so productive for fish—proximity to nutrient flows and tidal mixing and their place at the crossroads between land and water—also make them vulnerable to human assault. Today, nearly 40% of the world lives within 60 miles of a coastline. As more people move to coastal areas and further stress the seams between land and sea, coastal ecosystems are losing ground.

Human activities on land cause a large portion of offshore contamination. An estimated 44% of marine pollution comes from land-based pathways, flowing down rivers into tidal estuaries, where it bleeds out to sea; an additional 33% is airborne pollution that is carried by winds and deposited far off shore. From nutrient-rich sediments, fertilizers, and human waste to toxic heavy metals and synthetic chemicals, the outfall from human society ends up circulating in the fluid and turbulent seas.

Excessive nutrient loading has left some coastal systems looking visibly sick. Seen from an airplane, the surface waters of Manila Bay in the Philippines resemble green soup due to dense carpets of algae. Nitrogen and phosphorus are necessary for life and, in limited quantities, can help boost plant productivity, but too much of a good thing can be bad. Excessive nutrients build up and create conditions that are conducive to outbreaks of dense algae blooms, also known as "red tides." The blooms block sunlight, absorb dissolved oxygen, and disrupt food-web dynamics. Large portions of the Gulf of Mexico are now considered a biological "dead zone" due to algal blooms.

The frequency and severity of red tides has increased in the past couple of decades. Some experts link the recent outbreaks to increasing loads of nitrogen and phosphorus from nutrient-rich wastewater and agricultural runoff in poorly flushed waters.

Organochlorines, a fairly recent addition to the marine environment, are proving to have pernicious effects. Synthetic organic compounds such as chlordane, DDT, and PCBs are used for everything from electrical wiring to pesticides. Indeed, one reason they are so difficult to control is that they are ubiquitous. The organic form of tin (tributyltin), for example, is used in most of the world's marine paints to keep barnacles, seaweed, and other organisms from clinging to ships. Once the paint is dissolved in the water, it accumulates in mollusks, scallops, and rock crabs, which are consumed by fish and marine mammals.

As part of a larger group of chemicals known collectively as persistent organic pollutants (POPs), these compounds are difficult to control because they do not degrade easily. Highly volatile in warm temperatures, POPs tend to circulate toward colder environments where the conditions are more stable, such as the Arctic Circle. Moreover, they do not dissolve in water, but are lipid-soluble, meaning that they accumulate in the fat tissues of fish that are then consumed by predators at a more concentrated level.

POPs have been implicated in a wide range of animal and human health problems—from suppression of immune systems, leading to higher risk of illness and infection, to disruption of the endocrine system, which is linked to birth defects and infertility. Their continued use in many parts of the world poses a threat to marine life and fish consumers everywhere.

Because marine species are extremely sensitive to fluctuations in temperature, changes in climate and atmospheric conditions pose high risks to them. Recent evidence shows that the thinning ozone layer above Antarctica has allowed more ultraviolet-B radiation to penetrate the waters. This has affected photosynthesis and the growth of phytoplankton and macroalgae. The effects are not limited to the base of the food chain. By striking aquatic species during their most vulnerable stages of life and reducing their food supply at the same time, increases in UV-B could have devastating impacts on world fisheries production.

Because higher temperatures cause water to expand, a warming world may trigger more frequent and damaging storms. Ironically, the coastal barriers, seawalls, jetties, and levees that are designed to protect human settlements from storm surges likely exacerbate the problem of coastal erosion and instability, as they create deeper inshore troughs that boost wave intensity and sustain winds.

Depending on the rate and extent of warming, global sea levels may rise as much as three feet by 2100—up to five times as much as during the last century. Such a rise would flood most of New York City, including the entire subway system and all three major airports. Economic damages and losses could cost the global economy as much as $970,000,000,000 in 2100, according to the Organisation for Economic Co-operation and Development. The human costs would be unimaginable, especially in the low-lying, densely populated river deltas of Bangladesh, China, Egypt, and Nigeria.

These damages could be just the tip of the iceberg. Warmer temperatures would likely accelerate polar ice cap melting and could boost this rising wave by several feet. Just four years after a large portion of Antarctica melted, another large ice sheet fell off into the Southern Sea in February, 1998, rekindling fears that global warming could ignite a massive thaw that would flood coastal areas worldwide. Because oceans play such a vital role in regulating the Earth's climate and maintaining a healthy planet, minor changes in ocean circulation or in its temperature or chemical balance could have repercussions many orders of magnitude larger than the sum of human-induced wounds.

While understanding past climatic fluctuations and predicting future developments are an ongoing challenge for scientists, there is clear and growing evidence of the overuse—indeed abuse—that many marine ecosystems and species are suffering from direct human actions. The situation is probably much worse, for many sources of danger are still unknown or poorly monitored. The need to take preventive and decisive action on behalf of oceans is more important than ever.

Saving the oceans

Scientists' calls for precaution and protective measures are largely ignored by policymakers, who focus on enhancing commerce, trade, and market supply and look to extract as much from the sea as possible, with little regard for the effects on marine species or habitats. Overcoming the interest groups that favor the status quo will require engaging all potential stakeholders and reformulating the governance equation to incorporate the stewardship obligations that come with the privilege of use.

Fortunately for the planet, a new sea ethic is emerging. From tighter dumping regulations to recent international agreements, policymakers have made initial progress toward the goal of cleaning up humans' act. Still, much more is needed in the way of public education to build political support for marine conservation.

To boost ongoing efforts, two key principles are important. First, any dividing up of the waters should be based on equity, fairness, and need as determined by dependence on the resource and the best available scientific knowledge, not simply on economic might and political pressure. In a similar vein, resource users should be responsible for their actions, with decision-making and accountability shared by stakeholders and government officials. Second, given the uncertainty in scientific knowledge and management capabilities, it is necessary to err on the side of caution and take a precautionary approach.

Replanting mangroves and constructing artificial reefs are two concrete steps that help some fish stocks rebound quickly while letting people witness firsthand the results of their labors. Once they see the immediate payoff of their work, they are more likely to stay involved in longer-term protection efforts, such as marine sanctuaries, which involve removing an area from use entirely.

Marine protected areas are an important tool to help marine scientists and resource planners incorporate an ecologically based approach to oceans protection. By limiting accessibility and easing pressures on the resource, these areas allow stocks to rebound and profits to return. Globally, more than 1,300 marine and coastal sites have some form of protection, but most lack effective on-the-ground management.

Meanwhile, efforts to establish marine refuges and parks lag far behind similar efforts on land. The World Heritage Convention, which identifies and protects areas of special significance to mankind, identifies just 31 sites that include either a marine or a coastal component, out of a total of 522. John Waugh, Senior Program Officer of the World Conservation Union-U.S., and others argue that the World Heritage List could be extended to a number of marine hotspots and should include representative areas of the continental shelf, the deep sea, and the open ocean. Setting these and other areas aside as off-limits to commercial development can help advance scientific understanding of marine systems and provide refuge for threatened species.

To address the need for better data, coral reef scientists have enlisted the help of recreational scuba divers. Sport divers who volunteer to collect data are given basic training to identify and survey fish and coral species and conduct rudimentary site assessments. The data then are compiled and put into a global inventory that policymakers use to monitor trends and to target intervention. More efforts like these— that engage the help of concerned individuals and volunteers—could help overcome funding and data deficiencies and build greater public awareness of the problems plaguing the world's oceans.

Promoting sustainable ocean use also means shifting demand away from environmentally damaging products and extraction techniques. To this end, market forces, such as charging consumers more for particular fish and introducing industry codes of conduct, can be helpful. In April, 1996, the World Wide Fund for Nature teamed up with one of the world's largest manufacturers of seafood products, Anglo-Dutch Unilever, to create economic incentives for sustainable fishing. Implemented through an independent Marine Stewardship Council, fisheries products that are harvested in a sustainable manner will qualify for an ecolabel. Similar efforts could help convince industries to curb wasteful practices and generate greater consumer awareness of the need to choose products carefully.

Away from public oversight, companies engaged in shipping, oil and gas extraction, deep-sea mining, bioprospecting, and tidal and thermal energy represent a coalition of special interests whose activities help determine the fate of the oceans. It is crucial to get representatives of these industries engaged in implementing a new ocean charter that supports sustainable use.

Their practices not only affect the health of oceans, they help decide the pace of a transition toward a more sustainable energy economy, which, in turn, affects the balance between climate and oceans.

Making trade data and industry information publicly available is an important way to build industry credibility and ensure some degree of public oversight. While regulations are an important component of environmental protection, pressure from consumers, watchdog groups, and conscientious business leaders can help develop voluntary codes of action and standard industry practices that can move industrial sectors toward cleaner and greener operations. Economic incentives targeted to particular industries, such as low-interest loans for thermal projects, can aid companies in making a quicker transition to sustainable practices.

The fact that oceans are so central to the global economy and to human and planetary health may be the strongest motivation for protective action. Although the range of assaults and threats to ocean health are broad, the benefits that oceans provide are invaluable and shared by all. These huge bodies of water represent an enormous opportunity to forge a new system of cooperative, international governance based on shared resources and common interests. Achieving these far-reaching goals, however, begins with the technically simple, but politically daunting, task of overcoming several thousand years' worth of ingrained behavior. It requires seeing oceans not as an economic frontier for exploitation, but as a scientific frontier for exploration and a biological frontier for careful use.

For generations, oceans have drawn people to their shores for a glimpse of the horizon, a sense of scale, and awe at nature's might. Today, oceans offer careful observers a different kind of awe—a warning that humans' impacts on the Earth are exceeding natural bounds and in danger of disrupting life. Protection efforts already lag far behind what is needed. How humans choose to react will determine the future of the planet. Oceans are not simply one more system under pressure—they are critical to man's survival. As Carl Safina writes in *The Song for the Blue Ocean*, "we need the oceans more than they need us."

Anne Platt McGinn *is a senior researcher, Worldwatch Institute, Washington, D.C.*

From *USA Today Magazine*, January 2000, pp. 32–35. © 2000 by the Society for the Advancement of Education. Reprinted by permission.

UNIT 6

The Hazards of Growth: Polution and Climate Change

Unit Selections

Key Points to Consider

- Why has the U.S. federal government refused to acknowledge carbon dioxide as a pollutant and therefore subject to federal regulation? Is present and pending legislation sufficient to stem the flow of new greenhouse gases from U.S. industries?

- How has the Clean Water Act of 1972 moved the United States ahead of so many other countries in cleaning up surface waters? Compare water quality issues in the United States with those of countries in Latin America, Africa, and Asia in terms of both environmental and economic issues.

- How are human activities altering the role played in the global atmosphere by particulate matter? Explain both physical and chemical atmospheric change that results from such things as fossil fuel combustion.

- Describe the debate between advocates of global climate change as a consequence of human activity and those who believe that climate change is simply a natural process in which humans play no role. Where do you come down on this issue and why?

 **Links: www.dushkin.com/online/**
These sites are annotated in the World Wide Web pages.

IISDnet
http://www.iisd.org/default.asp

Persistent Organic Pollutants (POP)
http://www.chem.unep.ch/pops/

School of Labor and Industrial Relations (SLIR): Hot Links
http://www.lir.msu.edu/hotlinks/

Space Research Institute
http://arc.iki.rssi.ru/Welcome.html

Worldwatch Institute
http://www.worldwatch.org

Of all the massive technological changes that have combined to create our modern industrial society, perhaps none has been as significant for the environment as the chemical revolution. The largest single threat to environmental stability is the proliferation of chemical compounds for a nearly infinite variety of purposes, including the universal use of organic chemicals (fossil fuels) as the prime source of the world's energy systems. The problem is not just that thousands of new chemical compounds are being discovered or created each year, but that their long-term environmental effects are often not known until an environmental disaster involving humans or other living organisms occurs. The problem is exacerbated by the time lag that exists between the recognition of potentially harmful chemical contamination and the cleanup activities that are ultimately required.

A critical part of the process of dealing with chemical pollutants is the identification of toxic and hazardous materials, a problem that is intensified by the myriad ways in which a vast number of such materials, natural and man-made, can enter environmental systems. Governmental legislation and controls are important in correcting the damages produced by toxic and hazardous materials such as DDT, PCBs, or CFCs; in limiting fossil fuel burning; or in preventing the spread of living organic hazards such as pests and disease-causing agents. Unfortunately, as evidenced by most of the articles in this unit, we are losing the battle against harmful substances regardless of

legislation, and chemical pollution of the environment is probably getting worse rather than better.

The first article in this unit deals with one of the newest legislative approaches to several forms of pollution resulting from the chemical revolution: the increases in sulfur dioxide, nitrogen oxides, and mercury in soil, water, and air as a consequence of industrial activities, transportation, and heavy applications of artificial fertilizer. In "Three Pollutants and an Emission," author Dallas Burtraw, a senior fellow at Resources for the Future, describes how the U.S. Congress may be poised to enact the first major pollution legislation in more than a decade—but is still failing to recognize carbon dioxide as a pollutant, calling it an "emission" instead. Such a designation is problematic for the future of international agreements regarding the reduction in greenhouse gases in an attempt to retard global warming. In the second article in this unit, an emphasis is also placed on pollution, in this instance related to what may be humanity's most important environmental problem: the quality of the supply of freshwater. In "The Quest for Clean Water," civil engineer Joseph Orlins and attorney Anne Wehrly address the need for implementing new techniques to assess, prevent, and mitigate water pollution. The authors take account of the tremendous strides that have been made in cleaning up the nation's water supply since the passage of the Clean Water Act of 1972 but note that there is still much to be done regarding water pollution—nearly 40 percent of the inland and coastal waters in the United States are unfit for fishing, swimming, or drinking. Part of the problem is that the clean water legislation dealt only with pollution from point-sources: factories, power plants, wastewater treatment facilities, and so on. But most water pollution in the United States and elsewhere is attributable to nonpoint sources, such as runoff from crop and animal agriculture, urban areas, roads, and forests. By their nature, nonpoint sources are much more difficult to control. The authors conclude that, while the United States has done a great deal to clean up its water supplies, worldwide nearly 1/5 of the world's population lacks access to safe water supplies.

The unit's third and fourth selections move to the theme of atmospheric pollution and its consequences. In "Solving Hazy Mysteries," science writer Sid Perkins introduces the peculiar atmospheric component of "aerosols"—bits of particulate matter that range from the purely natural (soot and ash from volcanic eruptions, for example) to the human-produced particles that are emitted by coal-burning factories or electrical power generating facilities. There is a difference between the natural haze produced over Tennessee's Great Smoky Mountains and the smog of Los Angeles. Aerosols have always been an important part of the atmosphere and play vital physical and chemical roles in atmospheric interactions on a fairly small scale. But when the penchant of humans to burn fossil fuels and add significant numbers of aerosols to the atmosphere is added to the equation, then the role of aerosols is magnified to a crucial global-scale role in climate and atmospheric chemistry. Some scientists have predicted that increasing aerosols may, in fact, balance out the global warming problem by screening out incoming solar radiation. The latest research, however, suggests that this is less likely to be the case and that aerosol-derived cooling effects may not be nearly as great as hoped. As the authors point out, this is "cold comfort" for a world with increasing amounts of greenhouse gases in its atmosphere.

The final article in the unit addresses the issue of these greenhouse gases and their potential for climate change. In "Feeling the Heat," a special report from *Time* magazine, journalist Michael D. Lemonick and a team of investigative reporters discuss the social, political, economic, and scientific aspects of the global warming debate and conclude that, in spite of scientific and nonscientific rhetoric and argument about global warming, the fact that the Earth is becoming warmer is indisputable. It is equally indisputable that human activities are playing some role in the process. While a small segment of the scientific community still does not see human influence in short-term climate trends, the majority believes in a greenhouse effect enhanced by human activity. Interestingly, some of the strongest proponents of action to curb human-induced global warming are some of the biggest energy companies.

The pollution problem might appear nearly impossible to solve. But solutions do exist: massive cleanup campaigns to remove existing harmful chemicals from the environment and to severely restrict their future use; strict regulation of the production, distribution, use, and disposal of potentially hazardous chemicals; the development of sound biological techniques to replace existing uses of chemicals for such purposes as pest control; the adoption of energy and material resource conservation policies; and the use of more conservative and protective agricultural and construction practices. We now possess the knowledge and the tools to ensure that environmental cleanup is carried through. It will not be an easy task, and it will be terribly expensive. It will also demand a new way of thinking about humankind's role in the environmental systems upon which all life forms depend. If we do not complete the task, however, the support capacity of the environment may be damaged or diminished beyond our capacities to repair it. The consequences would be fatal for all who inhabit this planet.

Three Pollutants and an Emission

A Playbill for the Multipollutant Legislative Debate

By Dallas Burtraw

For the first time since 1990, Congress may be poised to enact major clean air legislation. Proposals now before Congress would affect electricity generation and large industrial facilities only and, as such, fall short of a full Clean Air Act reauthorization. But in other ways, the proposals are dramatic. They address multiple pollutants, broadly applying market-based approaches and aggregate emission caps, and would mandate deep emission cuts.

All the Congressional proposals and that of the Bush administration would address emissions of three pollutants—sulfur dioxide, nitrogen oxides, and mercury—from large stationary sources. The affected plants account for two-thirds of total U.S. emissions of sulfur dioxide, a quarter of nitrogen oxides, and a third of mercury. The proposals differ in whether they address carbon dioxide, even though the sources in question give off 40 percent of all U.S. emissions of carbon dioxide, a key greenhouse gas. Although carbon dioxide regulation is destined one day to play the leading role in clean air policy and thereby shape controls on the other three pollutants, for now it is backstage. Paradoxically, however, even from backstage carbon dioxide plays a critical role in the proposals' prospects for enactment.

The stage was set in spring of last year, when President Bush used previous clean air legislation to distinguish the three conventional pollutants from carbon dioxide. Because carbon dioxide was not identified as a pollutant in the Clean Air Act, the president termed it an "emission," not a "pollutant" and thus attempted to keep carbon dioxide out of the debate. The president's Clean Skies Initiative, announced this past Valentine's Day, excludes mandatory controls on carbon dioxide until at least 2012. Instead, the proposal embraces the goal of simply maintaining the modest rate of reduction in the intensity of carbon emissions per dollar of gross domestic product that was achieved in the last decade. Champions in Congress are divided mostly along partisan lines. Some proposals, including most notably that of Senator James Jeffords (I-VT), have kept carbon dioxide in the script.

All the proposals would make important—and costly—cuts in emissions of sulfur dioxide, nitrogen oxides, and mercury. But costs of the proposals that include carbon dioxide go far higher—so high that they would pretty well swamp the costs of controlling the other pollutants and thereby determine the direction of investments and technological choices for the future. Including carbon dioxide in the upcoming legislation would lead industry to begin to change its choice of fuel for power production. But if carbon dioxide emissions continue unregulated, coal—which now accounts for about 51 percent of electricity production—will continue to fuel power plants at about the same level. As recent studies by the Energy Information Administration and Environmental Protection Agency show, three-pollutant legislation that excluded carbon dioxide would necessitate retrofitting coal-fired facilities with ambitious controls for the conventional pollutants, but would have little effect on coal use and carbon dioxide emissions.

Legislation that also mandates big cuts in carbon dioxide emissions, however, would require many plants to switch from coal to natural gas. The switch to natural gas, which burns more cleanly than coal, would also mean sizable cuts in the other pollutants.

Will the Bush administration's proposal to leave out carbon dioxide and target only the three conventional pollutants resolve the future of carbon dioxide regulation, or will the issue be revisited? Power producers who invest hundreds of millions of dollars to comply with cuts mandated by three-pollutant legislation will not expect to see major new carbon dioxide controls just around the corner. Passing three-pollutant legislation would thus seem implicitly to commit the nation to a long-term delay in carbon dioxide policy and to an energy policy with a large coal component for many years. Indeed, for a future Congress to consider regulating carbon dioxide right on the heels of passing three-pollutant legislation could be perceived by power producers as highly unfair. Anticipating that reaction, environmental advocates may have little incentive to compromise over some pretty dicey issues involved with controls on the conventional pollutants.

A key goal of the proposed multipollutant legislation is to improve on the current regulatory framework by giving industry greater regulatory predictability and certainty about compliance requirements. Paradoxically, however, achieving such certainty, as well as taking into account the interests of industry, may require giving carbon dioxide cuts at least a small role in the multipollutant strategy.

Reducing Multipollutant Levels

Debate over the legislative proposals involves three main questions. The first is what level of control to impose for each of the three pollutants. The stakes here are high, for the leading proposals would impose large costs—several billion dollars—and reap large benefits as well. But the costs are spread roughly equally over the three pollutants, while the benefits are not.

Reducing sulfur dioxide emissions will likely yield the lion's share of quantifiable benefits. Sulfur dioxide as a gas is recognized as a potent health threat, but it is more noteworthy in economic terms for its role in forming secondary particulates and acidifying soil and water.

In the public policy world, sulfur dioxide hit marquee status with passage of the 1990 Clean Air Act Amendments, which mandated a 50 percent cut in aggregate emissions from electricity generation. At the time, Paul Portney, of Resources for the Future (RFF)—the only analyst bold enough to offer an estimate of the benefits and costs of the amendments—predicted that benefits of sulfur dioxide cuts would about justify the costs. Subsequent analysis by the EPA and RFF suggests that two unexpected developments caused benefits to exceed costs by far. First, since 1990, a cascade of new research—most recently, a study published in the *Journal of the American Medical Association* early in March—has linked secondary particulates from sulfur dioxide emissions with premature mortality, raising the quantifiable economic benefits of sulfur dioxide cuts many times over what was expected. Second, cutting sulfur dioxide proved less costly than anticipated.

In 1990, although the primary benefits were ultimately found to stem from improvements in health, arguably the main justification for the sulfur dioxide cuts was acid rain. Since 1990, sulfur deposition in sensitive ecological areas has fallen dramatically, but recovery has been slower than expected, leading many advocates in the northeast to call for further sulfur dioxide cuts of as much as 75 percent. Meeting that target will mean more than simply switching to low-sulfur coal at more facilities. Compliance will require the widespread installation of scrubbers—and will impose large capital costs on industry. Even so, from a benefit-cost perspective, the proposed sulfur dioxide cut is an easy pill to swallow.

Nitrogen Oxides

Pending multipollutant proposals would reduce nitrogen oxides from electricity generation to some 50–75 percent of current baseline emissions. The cuts have multifaceted and important benefits but are harder to justify on a cost-benefit basis than the cuts in sulfur dioxide.

Nitrogen oxides have effects similar to those of sulfur dioxide. They contribute to secondary particulates but are usually thought to be less potent in affecting health. They are also instrumental in acidification, but again most analyses attribute to them a lesser role. Both pollutants impair visibility. Nitrogen oxides also contribute to forming ground-level ozone, another pollutant identified in the Clean Air Act, but health scientists and economists assign lower benefits to reducing ozone than to reducing particulates.

The 1990 Clean Air Act Amendments first imposed modest cuts in nitrogen oxides emissions at power plants. In complying with the 1990 law, plants have already made the least expensive cuts. Dramatic further reductions will require widespread installation of post-combustion controls (mostly, selective catalytic reduction) and will be quite costly.

In sum, viewed from the basis of each ton reduced, the benefits of cutting nitrogen oxides appear to be equal to (or perhaps a little less than) those of cutting sulfur dioxide. But the cost per ton of the nitrogen oxide reductions is greater than the cost per ton of cutting sulfur dioxide. Thus, benefit-cost considerations would argue for greater relative reductions in sulfur dioxide—at least as long as they continue to yield the biggest bang for the buck. But current proposals would reduce the emissions of both by equal percentages.

Mercury Levels

Proposed mercury cuts are even larger, in percentage terms, than cuts in the other two pollutants. Though highly toxic and with potentially profound ecological and human health effects, mercury is emitted in much smaller quantities. And the benefits of reducing mercury emissions are hard to quantify. The Clean Air Act would normally require maximum achievable emission reduction for mercury as a hazardous air pollutant, leading to more than 90 percent removal, but the 1990 amendments gave mercury a temporary special exception. Mercury may, however, come under strict control in the next few years within the existing regulatory process unless exempted by legislation or blocked by litigation. Whichever ending is played out—implementation or exemption—will determine the baseline against which the multipollutant legislative proposals are measured. Environmental advocates seem to expect a baseline that reflects tightening controls on mercury. Realistically, however, long delays in implementing mercury controls on electricity generation are likely unless multipollutant legislation is passed.

Cost-effectiveness is an important issue. Byron Swift of the Environmental Law Institute notes that cutting use of mercury from various small sources through pollution prevention—for example, by placing a declining cap on the use of mercury in products and process—would be far less costly than controlling mercury emitted in power production. That would argue for pursuing other, less costly options before pushing far in the power sector.

A key question in the multipollutant legislative debate is what role to give to limiting carbon dioxide emissions.

Furthermore, installing post-combustion controls for sulfur dioxide and nitrogen oxides would yield sizable ancillary reductions in mercury. Selective catalytic reduction controls for nitrogen oxides oxidize much of the available mercury, and sulfur dioxide scrubbers capture the oxidized mercury. Although technical experience is limited and prediction is thus uncertain, simultaneous controls for sulfur dioxide and nitrogen oxides probably cut more than half of mercury emissions. Making greater cuts would require installing new technologies, such as activated carbon injection, that are as yet unproven in widescale application. Their cost is especially high when viewed in marginal cost terms, because the incremental reductions (those over and above the ancillary benefits of sulfur dioxide and nitrogen oxides controls) are small.

In sum, moderate mercury emissions cuts are available as part of a multipollutant strategy that targets sulfur dioxide and nitrogen oxides comprehensively, and these cuts are reflected in the administration's proposal. Making further modest cuts would encourage development of new technologies, but the dramatic reduction targets for mercury controls embedded in Jeffords' legislative proposals come at a high cost.

New Source Review

A second question in the debate over multipollutant legislation is whether to revise requirements for the New Source Review (NSR) program, originally established by the 1977 Clean Air Act and now under scrutiny by the Bush administration. The program requires sources that add or modify equipment that could generate new emissions to use new technology to cut emissions. New source performance standards characterize allowable emission rates for either modified or new sources anywhere in the nation. Sources that increase emissions in areas of the country that do not meet federal air quality standards are held to even stricter emission levels and must cut emissions in other facilities to offset new emissions. To some,

NSR appears to be a bargaining chip in the debate; to others it is the central issue.

The NSR program's biggest problem is uncertainty. A firm that modifies a source does not know whether it will trigger NSR requirements and what those requirements might be until a determination is made by the EPA. The technologies that are acceptable in areas that do not meet federal standards are subject to change.

Differences in emission standards between facilities under NSR and those not under NSR lead to inefficient spending on pollution control. In addition, firms under NSR may try to avoid expense and uncertainty by delaying investments to modernize facilities. And new sources in areas that do not require emission offsets increase emissions in the aggregate.

Advocates vigorously defend NSR, arguing that it has led to dramatic emission cuts and that it identifies new technologies and innovations and brings them to application. But a key goal of the legislative debate must be to streamline and increase the predictability of the NSR process as it applies to electricity generation.

Design

During the mid-1990s Washington sharply revised its view of market-based approaches to environmental regulation. After years of suspicion, the default question is now: why not use market-based approaches? All the pending multipollutant proposals would regulate sulfur dioxide and nitrogen oxides within an emission allowance trading program based on the trading program established in the 1990 Clean Air Act Amendments. But how the trading programs are designed and whether mercury is included are major issues still in play.

The existing regional cap and trading program for nitrogen oxides operates during the five-month summer ozone season in 11 northeastern states. It will be expanded to a 19-state program in 2004. Current three-pollutant proposals would make the program national and year-round. Recent research suggests that operating the program year-round in the 19-state region would improve its cost-effectiveness dramatically. But expanding it nationwide is harder to justify because population density is much less, and consequently the aggregate benefits of reducing exposure to pollution are much less. It would also lead to a divergence in the distribution of benefits and costs across the nation that could be its undoing in Congress.

It remains to be seen whether mercury would be regulated through trading or through inflexible technology standards. Textbook treatment of pollution implies that the marginal costs of control among facilities converge as emission reduction targets approach 100 percent, because essentially every facility has to eliminate the emission. But Alex Farrell of Carnegie Mellon University has demonstrated that the marginal costs of mercury control among facilities fan out widely as controls approach 100 percent because of differences in coal types and in the other pollution controls that are in place. That argues for a trading program, which would be a departure from the usual controls on hazardous air pollutants.

One other design issue is a sleeper, but its importance will surface soon enough. How tradable emission allowances are allocated portends potentially large transfers of wealth and affects total costs as well. Because of expanded competition in the electricity industry, the value of the emission allowances may far outstrip the cost of compliance for power producers when allowances are allocated at zero cost. And prices for consumers may go higher than is justified by producer costs. When a similar scenario unfolded as industry complied with the 1987 Montreal Protocol to protect stratospheric ozone, Congress captured the "windfall profits" by taxing them. The same thing may happen here.

Looking Ahead

Compromise would enhance the cost-effectiveness of the multipollutant legislation and, at least from the view in the balcony seats, increase its chance of passage. Sulfur dioxide reductions of the size being proposed have apparent justification, but the nitrogen oxide reductions proposed for the entire nation are hard to justify economically and in any case face political opposition in the western states. A compelling alternative would be to transform the seasonal program in the 19 eastern states into an annual one and to impose somewhat less stringent controls in the other states. The case for maximum reductions in mercury emissions is the weakest. A better strategy would be to claim the mercury reductions that are achieved ancillary to imposed reductions on sulfur dioxides and nitrogen oxides as a victory. In fact, this approach may provide a justification for the cost-effectiveness of nationwide nitrogen oxide controls that is missing if those controls come on top of controls on mercury (and vice versa). One could also question whether the NSR program needs to exist at all if stringent caps are placed on emissions. But unless those caps decline continuously over time, the NSR program will be essential from the perspective of environmentalists.

The administration's proposal embodies many of these suggestions with respect to levels of control (72 percent reduction for sulfur dioxide and 67 percent reduction for nitrogen oxides). It goes most of the way to meeting the

Jeffords proposal with respect to sulfur dioxide and nitrogen oxides (75 percent reductions for both pollutants), but calls for mercury reductions that would be mostly ancillary to these (69 percent) rather than the maximum achievable (90 percent). The big difference between the proposals is timing. The Jeffords proposal would achieve these reductions a decade sooner than does the administration plan. Given the maturity of control technologies, the shorter timeline is practical and preferable.

The most important consideration, however, is what role is given to limiting carbon dioxide emissions. Large cuts do not appear plausible, but zero reductions do nothing to resolve the most important environmental uncertainty facing the electricity industry. A policy of modest cuts, beginning now and implemented through market-based policies, would stabilize the setting for new investment, provide incentive for innovation in technologies and institutions, and provide the opportunity to learn about the costs of climate change policy.

Every negotiation is shaped by expectations about what will happen if it breaks down. If the multipollutant debate collapses, what is expected is a return to the status quo—the so-called emission baseline. But like so much in this debate, the baseline itself is uncertain—so much so that the parties may find it more useful to compromise over legislation than to continue the familiar battles over a shifting baseline. Environmental advocates feel that regulatory inertia favors continuing reductions in conventional emissions. Regulatory calendars stretching to 2015 foresee the implementation of new particulate standards, mercury controls, and regional haze rules that will affect the various emissions individually. But advocates of multipollutant legislation have to feel their possible advances are at risk under the Bush administration.

The bellwether may be how the administration completes its ongoing review of NSR. If it backs away from proposed changes that would give power plants more freedom to expand or modernize without coming under NSR requirements, industry may feel it has the most to gain from legislation. If the administration review leads to substantial revisions or elimination of NSR, the baseline will be changed. Either fur may fly in Congress, or the stage may be set for an important compromise around multipollutant legislation. Or maybe both.

Dallas Burtraw is a senior fellow at Resources for the Future.

The Quest for Clean Water

As water pollution threatens our health and environment, we need to implement an expanding array of techniques for its assessment, prevention, and remediation.

Joseph Orlins and Anne Wehrly

In the 1890s, entrepreneur William Love sought to establish a model industrial community in the La Salle district of Niagara Falls, New York. The plan included building a canal that tapped water from the Niagara River for a navigable waterway and a hydroelectric power plant. Although work on the canal was begun, a nationwide economic depression and other factors forced abandonment of the project.

By 1920, the land adjacent to the canal was sold and used as a landfill for municipal and industrial wastes. Later purchased by Hooker Chemicals and Plastics Corp., the landfill became a dumping ground for nearly 21,000 tons of mixed chemical wastes before being closed and covered over in the early 1950s. Shortly thereafter, the property was acquired by the Niagara Falls Board of Education, and schools and residences were built on and around the site.

In the ensuing decades, groundwater levels in the area rose, parts of the landfill subsided, large metal drums of waste were uncovered, and toxic chemicals oozed out. All this led to the contamination of surface waters, oily residues in residential basements, corrosion of sump pumps, and noxious odors. Residents began to question if these problems were at the root of an apparent prevalence of birth defects and miscarriages in the neighborhood.

Eventually, in 1978, the area was declared unsafe by the New York State Department of Health, and President Jimmy Carter approved emergency federal assistance. The school located on the landfill site was closed and nearby houses were condemned. State and federal agencies worked together to relocate hundreds of residents and contain or destroy the chemical wastes.

That was the bitter story of Love Canal. Although not the worst environmental disaster in U.S. history, it illustrates the tragic consequences of water pollution.

Water quality standards

In addition to toxic chemical wastes, water pollutants occur in many other forms, including pathogenic microbes (harmful bacteria and viruses), excess fertilizers (containing compounds of phosphorus and nitrogen), and trash floating on streams, lakes, and beaches. Water pollution can also take the form of sediment eroded from stream banks, large blooms of algae, low levels of dissolved oxygen, or abnormally high temperatures (from the discharge of coolant water at power plants).

The United States has seen a growing concern about water pollution since the middle of the twentieth century, as the public recognized that pollutants were adversely affecting human health and rendering lakes unswimmable, streams unfishable, and rivers flammable. In response, in 1972, Congress passed the Federal Water Pollution Control Act Amendments, later modified and referred to as the Clean Water Act. Its purpose was to "restore and maintain the chemical, physical, and biological integrity of the nation's waters."

The Clean Water Act set the ambitious national goal of completely eliminating the discharge of pollutants into navigable waters by 1985, as well as the interim goal of making water clean enough to sustain fish and wildlife, while being safe for swimming and boating. To achieve these goals, certain standards for water quality were established.

The "designated uses" of every body of water subject to the act must first be identified. Is it a source for drinking water? Is it used for recreation, such as swimming? Does it supply agriculture or industry? Is it a significant habitat for fish and other aquatic life? Thereafter, the water must be tested for pollutants. If it fails to meet the minimum standards for its designated uses, then steps must be taken to limit pollutants entering it, so that it becomes suitable for those uses.

On the global level, the fundamental importance of clean water has come into the spotlight. In November 2002, the UN Committee on Economic, Cultural and Social Rights declared access to clean

Tragedy at Minamata Bay

The Chisso chemical factory, located on the Japanese island of Kyushu, is believed to have discharged between 70 and 150 tons of methylmercury (an organic form of mercury) into Minamata Bay between 1932 and 1968. The factory, a dominant presence in the region, used the chemical to manufacture acetic acid and vinyl chloride.

Methylmercury is easily absorbed upon ingestion, causing widespread damage to the central nervous system. Symptoms include numbing and unsteadiness of extremities, failure of muscular coordination, and impairment of speech, hearing and vision. Exposure to high levels of the substance can be fatal. In addition, the effects are magnified for infants exposed to methylmercury through their mothers, both before birth and while nursing.

In the 1960s and '70s, it was revealed that thousands of Minamata Bay residents had been exposed to methylmercury. The chemical had been taken up from the bay's waters by its fish and then made its way into the birds, cats, and people who ate the fish. Consequently, methylmercury poisoning came to be called Minamata disease.

Remediation, which took as long as 14 years, involved removing the mercury-filled sediments and containing them on reclaimed land in Minamata Bay. Fish in the bay had such high levels of methylmercury that they had to be prevented from leaving the bay by a huge net, which was in place from 1947 to 1997.

Mercury poisoning has recently appeared in the Amazon basin, where deforestation has led to uncontrolled runoff of natural accumulations of mercury from the soil into rivers and streams. In the United States, testing has revealed that predator fish such as bass and walleye in certain lakes and rivers contain enough mercury to justify warnings against consuming them in large amounts.

—*J.O. and A.W.*

Minamata Bay residents who were exposed to methylmercury have been suffering from such problems as loss of muscular control, numbing of extremities, and impairment of speech, hearing, and vision.

water a human right. Moreover, the United Nations has designated 2003 to be the International Year of Freshwater, with the aim of encouraging sustainable use of freshwater and integrated water resources management.

Here, there, and everywhere

Implementing the Clean Water Act requires clarifying the sources of pollutants. They are divided into two groups: "point sources" and "nonpoint sources." Point sources correspond to discrete, identifiable locations from which pollutants are emitted. They include factories, wastewater treatment plants, landfills, and underground storage tanks. Water pollution that originates at point sources is usually what is associated with headline-grabbing stories such as those about Love Canal.

Nonpoint sources of pollution are diffuse and therefore harder to control. For instance, rain washes oil, grease, and solid pollutants from streets and parking lots into storm drains that carry them into bays and rivers. Likewise, irrigation and rainwater leach fertilizers, herbicides, and insecticides from farms and lawns and into streams and lakes.

In the United States, the Clean Water Act requires that industrial wastes be neutralized or broken down before being released into rivers and lakes.

The direct discharge of wastes from point sources into lakes, rivers, and streams is regulated by a permit program known as the National Pollutant Discharge Elimination System (NPDES). This program, established through the Clean Water Act, is administered by the Environmental Protection Agency (EPA) and authorized states. By regulating the wastes discharged, NPDES has helped reduce point-source pollution dramatically. On the other hand, water pollution in the United States is now mainly from nonpoint sources, as reported by the EPA.

In 1991, the U.S. Geological Survey (USGS, part of the Department of the Interior) began a systematic, long-term program to monitor watersheds. The National Water-Quality Assessment Program (NAWQA), established to help manage surface and groundwater supplies, has involved the collection and analysis of water quality data in over 50 major river basins and aquifer systems in nearly all 50 states.

The program has encompassed three principal categories of investigation: (1) the current conditions of surface water and groundwater; (2) changes in those conditions over time; and (3) major factors—such as climate, geography, and land use—that affect water quality. For each of these categories, the water and sediment have been tested for such pollutants as pesticides, plant nutrients, volatile organic compounds, and heavy metals.

The NAWQA findings were disturbing. Water quality is most affected in watersheds with highest population density and urban development. In agricultural areas, 95 percent of tested streams and 60 percent of shallow wells contained herbicides, insecticides, or both. In urban areas, 99 percent of tested streams and 50 percent of shallow wells had herbicides, especially those used on lawns and golf courses. Insecticides were found more frequently in urban streams than in agricultural ones.

The study also found large amounts of plant nutrients in water supplies. For instance, 80 percent of agricultural streams and 70 percent of urban streams were found to contain phosphorus at concentrations that exceeded EPA guidelines.

Moreover, in agricultural areas, one out of five well-water samples had nitrate concentrations higher than EPA standards for drinking water. Nitrate contamination can result from nitrogen fertilizers or material from defective septic systems leaching into the groundwater, or it may reflect defects in the wells.

Effects of pollution

According to the UN World Water Assessment Programme, about 2.3 billion people suffer from diseases associated with polluted water, and more than 5 million people die from these illnesses each year. Dysentery, typhoid, cholera, and hepatitis A are some of the ailments that result from ingesting water contaminated with harmful microbes. Other illnesses—such as malaria, filariasis, yellow fever, and sleeping sickness—are transmitted by vector organisms (such as mosquitoes and tsetse flies) that breed in or live near stagnant, unclean water.

A number of chemical contaminants—including DDT, dioxins, polychlorinated biphenyls (PCBs), and heavy metals—are associated with conditions ranging from skin rashes to various cancers and birth defects. Excess nitrate in an infant's drinking water can lead to the "blue baby syndrome" (methemoglobinemia)—a condition in which the child's digestive system cannot process the nitrate, diminishing the blood's ability to carry adequate concentrations of oxygen.

Besides affecting human health, water pollution has adverse effects on ecosystems. For instance, while moderate amounts of nutrients in surface water are generally not problematic, large quantities of phosphorus and nitrogen compounds can lead to excessive growth of algae and other nuisance species. Known as *eutrophication*, this phenomenon reduces the penetration of sunlight through the water; when the plants die and decompose, the body of water is left with odors, bad taste, and reduced levels of dissolved oxygen.

Low levels of dissolved oxygen can kill fish and shellfish. In addition, aquatic weeds can interfere with recreational activities (such as boating and swimming) and can clog intake by industry and municipal systems.

Some pollutants settle to the bottom of streams, lakes, and harbors, where they may remain for many years. For instance, although DDT and PCBs were banned years ago, they are still found in sediments in many urban and rural streams. They occur at levels harmful to wildlife at more than two-thirds of the urban sites tested.

Prevention and remediation

As the old saying goes, an ounce of prevention is worth a pound of cure. This is especially true when it comes to controlling water pollution. Several important steps taken since the passage of the Clean Water Act have made surface waters today cleaner in many ways than they were 30 years ago.

For example, industrial wastes are mandated to be neutralized or broken down before being discharged to streams, lakes, and harbors. Moreover, the U.S. government has banned the production and use of certain dangerous pollutants such as DDT and PCBs.

In addition, two major changes have been introduced in the handling of sewage. First, smaller, less efficient sewage treatment plants are being replaced with modern, regional plants that include biological treatment, in which microorganisms are used to break down organic matter in the sewage. The newer plants are releasing much cleaner discharges into the receiving bodies of water (rivers, lakes, and ocean).

Second, many jurisdictions throughout the United States are building separate sewer lines for storm water and sanitary wastes. These upgrades are needed because excess water in the older, "combined" sewer systems would simply bypass the treatment process, and untreated sewage would be discharged directly into receiving bodies of water.

To minimize pollutants from nonpoint sources, the EPA is requiring all municipalities to address the problem of runoff from roads and parking lots. At the same time, the use of fertilizers and pesticides needs to be reduced. Toward this end, county extension agents are educating farmers and homeowners about their proper application and the availability of nutrient testing.

To curtail the use of expensive and potentially harmful pesticides, the approach known as *integrated pest management* can be implemented [see "Safer Modes of Pest Control," THE WORLD & I, May 2000, p. 164]. It involves the identification of specific pest problems and the use of nontoxic chemicals and chemical-free alternatives whenever possible. For instance, aphids can be held in check by ladybug beetles and caterpillars can be controlled by applying neem oil to the leaves on which they feed.

Moreover, new urban development projects in many areas are required to implement storm-water management practices. They include such features as: oil and grease traps in storm drains; swales to slow down runoff, allowing it to infiltrate back into groundwater; "wet" detention basins (essentially artificial ponds) that allow solids to settle out of runoff; and artificial wetlands that help break down contaminants in runoff. While such additions may be costly, they significantly improve water quality. They are of course much more expensive to install after those areas have been developed.

Once a waterway is polluted, cleanup is often expensive and time consuming. For instance, to increase the concentration of dissolved oxygen in a lake that has undergone eutrophication, fountains and aerators may be necessary. Specially designed boats may be needed to harvest nuisance weeds.

At times, it is costly just to identify the source of a problem. For example, if a body of water contains high levels of coliform bacteria, expensive DNA testing may be needed to determine whether the bacteria came from leakage of human sewage, pet waste, or the feces of waterfowl or other wildlife.

> **Testing the effect of simulated rainfall on fresh manure, scientists with the Agricultural Research Service have found that grass strips are highly effective at preventing manure-borne microbes from being washed down a slope and contaminating surface waters.**

Contaminated sediments are sometimes difficult to treat. Available techniques range from dredging the sediments to "capping" them in place, to limit their potential exposure. Given that they act as reservoirs of pollutants, it is often best to remove the sediments and burn off the contaminants. Alternatively, the extracted sediments may be placed in confined disposal areas that prevent the pollutants from leaching back into groundwater. Dredging, however, may create additional problems by releasing pollutants back into the water column when the sediment is stirred up.

The future of clean water

The EPA reports that as a result of the Clean Water Act, millions of tons of sewage and industrial waste are being treated before they are discharged into U.S. coastal waters. In addition, the majority of lakes and rivers now meet mandated water quality goals.

Yet the future of federal regulation under the Clean Water Act is unclear. In 2001, a Supreme Court decision (*Solid Waste Agency of Northern Cook County v. United States Army Corps of Engineers, et al.*) brought into question the power of federal agencies to regulate activities affecting water quality in smaller, nonnavigable bodies of water. This and related court decisions have set the stage for the EPA and other federal agencies to redefine which bodies of water can be protected from unregulated dumping and discharges under the Clean Water Act. As a result, individual states may soon be faced with much greater responsibility for the protection of water resources.

Worldwide, more than one billion people presently lack access to clean water sources, and over two billion live without basic sanitation facilities. A large proportion of those who die from water-related diseases are infants. We would hope that by raising awareness of these issues on an international level, the newly recognized right to clean water will become a reality for a much larger percentage of the world's population.

Joseph Orlins, professor of civil engineering at Rowan University in Glassboro, New Jersey, specializes in water resources and environmental engineering. Anne Wehrly is an attorney and freelance writer.

On the Internet

2003: INTERNATIONAL YEAR OF FRESHWATER (UN)
www.wateryear2003.org

CLEAN WATER ACT (EPA)
www.epa.gov/watertrain/cwa

DO'S AND DON'T'S AROUND THE HOME (EPA)
www.epa.gov/owow/nps/dosdont.html

LOVE CANAL COLLECTION (SUNY BUFFALO)
ublib.buffalo.edu/libraries/projects/lovecanal

NATIONAL WATER-QUALITY ASSESSMENT PROGRAM (USGS)
water.usgs.gov/nawqa

OFFICE OF WATER (EPA)
www.epa.gov/ow

THE WORLD'S WATER
www.worldwater.org

This article appeared in the May 2003 issue and is reprinted with permission from *The World & I,* a publication of The Washington Times Corporation, © 2003.

SOLVING HAZY MYSTERIES

New research plumbs the chemical evolution and climatic effect of aerosols

BY SID PERKINS

The picturesque hazes of Tennessee's Smoky Mountains appear when volatile organic chemicals released by trees react with other gases in the atmosphere. And every time a raindrop falls into the ocean, microscopic droplets of salt water splatter upward into the atmosphere. Both mountain haze and fine ocean spray are nature's aerosols. People's penchant for burning fossil fuels is now creating a rival source of aerosol formation.

Liquid droplets and solid particles small enough to remain suspended in air can scatter light in ways that make for hazy mountain vistas, stunning sunrises and sunsets, and urban smog. Scientists, however, are finding that these aerosols also play critical global-scale roles in climate and atmospheric chemistry. The researchers have been examining lab-made and natural aerosols in an effort to discover how their constituent particles form, what effects they have on the atmosphere, and whether something needs to be done about them.

IN A FOG Clouds, smoke, soot, smog, sea spray, sneezes, suspended powders, and plumes of sulfuric acid droplets from active volcanoes—aerosols are everywhere.

Although microscopic, aerosol grains or droplets of a certain size scatter light back into space. This produces a cooling effect that counteracts global warming. That's why climatologists are so keen on measuring the phenomenon.

Aerosols play a big role in atmospheric chemistry, too. Droplets of liquid aerosols often serve as tiny wafting beakers in which dissolved substances interact. Also, solid particles suspended in the atmosphere can react with surrounding gases or act as catalytic sites where other substances can conveniently intermingle and react.

Now, scientists at the University of North Carolina (UNC) at Chapel Hill suggest why conditions in some areas generate much higher aerosol concentrations than are found elsewhere.

One city that pumps out a lot of pollution is Houston. Scientists noticed that on days when the atmosphere there contained particularly high concentrations of organic chemicals and sulfur dioxide, the air over oil refineries was especially thick with haze, says Myoseon Jang, an environmental scientist at UNC. When sulfur dioxide—which reacts with water vapor in the air to form sulfuric acid—wasn't abundant over Houston, the aerosols weren't so prevalent.

To study this phenomenon, Jang and her colleagues conducted laboratory tests in which they spewed extremely small particles into Teflon-coated chambers, some with air volumes equivalent to that of a medium-size room. The particles served as seeds around which aerosol droplets could grow. Then, the researchers introduced various chemicals that form when hydrocarbon vapors undergo reactions with other gases in the atmosphere.

In experiments in which the particles had been coated with small amounts of sulfuric acid—about the same concentrations found in diesel soot—the mixture produced up to 10 times as much aerosol as was generated in chambers holding uncoated particles. The sulfuric acid apparently serves as a catalyst and boosts the mist-creating reactions, says Jang. She and the team report their findings in the Oct. 25 *Science*.

"This finding opens up a new area of research," says Edward O. Edney, an atmospheric chemist with the Environmental Protection Agency in Research Triangle Park, N.C. Jang's team is the first to propose that aerosols result from acid-catalyzed reactions, Edney notes. "If it can happen in a lab, it may happen in the atmosphere," he says.

Jang and other chemists employ both the traditional enclosed vessels and newer, flow-through chambers. By fine-tuning the temperature, humidity, and other aspects of a mixture of flowing gases in such chambers, researchers can analyze how reactions might progress under a variety of weather conditions or at different times of day.

"We don't know enough about how [aerosols] form," says James G. Hudson of the Desert Research Institute in Reno, Nev. For example, there are thousands of chemicals that can be produced by such reactions, he says. Edney agrees: "The atmosphere is chemically complicated." The challenge, he notes, is in sorting through that complexity and striking a balance between simulations that are simple enough to be manageable but realistic enough to be useful.

Jang and her colleagues are now using their test results to develop mathematical models of aerosol production. If these formulas accurately portray actual atmospheric conditions, scientists could predict, for example, variations in aerosol production resulting from changes in emissions of hydrocarbons and other organic chemicals from cars or industrial sources.

MARINE MISTS Although substantial amounts of aerosols stem from the organic substances emitted by human activity, the largest source of aerosols is the oceans. Scientists with the Geneva-based Intergovernmental Panel on Climate Change estimate that in the year 2000, as many as 3.3 billion metric tons of salt spray entered the atmosphere.

Those saline droplets form in a variety of ways. Breaking waves toss them skyward, and drops of rain or bursting bubbles splash them into the air. Particle diameters range from 100 nanometers to several micrometers, but the predominant size is 1 to 2 μm across, says Murray V. Johnston, an atmospheric chemist at the University of Delaware in Newark. All of these dimensions are smaller than a typical biological cell. If they stay airborne, the droplets' chemical composition almost immediately begins to evolve.

As the droplet's water evaporates into the surrounding atmosphere, the tiny globule's salt concentration goes up, says Johnston. When the airborne droplet becomes saturated, the dissolved salt begins to crystallize. The air may contain nitric acid vapor formed when nitrogen oxides from cars and other sources combined with water vapor. That nitric acid can react with the sodium chloride in a droplet to form sodium nitrate and hydrochloric acid vapor. The acid vapor then reacts with any ozone that's available and destroys it. In some situations—for example, if hydrocarbon vapors are present—the acid vapor may instead create ozone.

In either case, says Johnston, the nitrate that remains in the droplet eventually ends up fertilizing the ocean or the land onto which it falls, perhaps as much as 160 kilometers inland. That can be bad or good, leading to either harmful algal blooms or greener fields.

In some circumstances, the droplets flung into the air carry more than just salt water. Scraps of dead marine microorganisms may hitch a ride heavenward on the spray.

Researchers have long suspected that ocean-spawned aerosols could be coated with organic chemicals, but they had only indirect evidence to support the idea. The long-chain, carbon-based molecules that make up cell membranes have one end that attracts water and one end that repels it. That repulsion could drive the molecules to the surface of the ocean, where they would form a thin layer—a biological oil slick. Droplets splashed from the surface could carry a coating of this slick.

Direct evidence for this scenario came to light when Heikki Tervahattu, an atmospheric scientist at the University of Helsinki collected particles over Helsinki early in 1998. When he and his colleagues looked at those aerosols with a scanning electron microscope, they noticed that many of the droplets pulsated, enlarging in one part while shrinking in another. This odd behavior, which suggested that a film may have coated these aerosols, spurred the researchers to perform chemical analyses of the droplets. The team reported their observations online April 11 in the *Journal of Geophysical Research (Atmospheres)*.

The results indicate that the droplets began as salt spray over the North Atlantic Ocean and accumulated industrial pollutants as they passed over France, Germany, and southern Scandinavia. Tervahattu and his colleagues also detected long-chain, carbon-based molecules associated with the aerosols.

During subsequent analyses, the scientists blasted the droplets with an ion beam. Those tests, reported online Aug. 31 in the same journal, showed a crystal of sea salt at the heart of each aerosol droplet and a fatty acid coating no more than a molecule or two thick. The scientists determined that the film is primarily palmitic acid, which is produced when a microorganism's cell membrane disintegrates.

Though thin, the film significantly affects a droplet's physical, chemical, and optical properties, says Tervahattu. For example, the water-repellent end of each fatty acid molecule points toward the outside of the aerosol particle. As a result, once completely coated by a layer of fatty acids, the droplet would have been much less likely to grow by attracting more water vapor, Tervahattu suggests. Because droplets must grow above a certain size to fall as rain, organic films on a cloud's particles might squelch precipitation. At this early stage in the research, the film's global impact on rainfall remains uncertain.

COOL IT, OR NOT New research suggests that the aerosols produced by human activity may not cool Earth's climate as much as some scientists have predicted.

Ulrike Lohmann and Glen Lesins of Dalhousie University in Halifax, Nova Scotia, recently used satellite observations to validate a climate model that estimates the indirect effects of human-generated aerosols on various properties of the most familiar aerosol of all—clouds. Sulfate- and carbon-rich emissions from automobiles and industrial activity have substantially increased the number of aerosol particles now present in the atmosphere, as compared with preindustrial times, say the researchers.

Because the individual droplets are smaller, clouds tainted with industrial aerosols tend to scatter radiation back into space more effectively than pristine clouds do. There's less precipitation because the water droplets don't grow to a size at which they'd fall out of the cloud. That, in turn, means that the clouds last longer than they would if the artificial aerosols weren't present. These indirect effects, when combined, add up to a net global cooling.

When Lohmann and Lesins removed the human-produced aerosols from the mathematical model they were testing, the simulation's results for cloud reflectivity and cloud-droplet size didn't match those estimated from satellite observations. By including those aerosols in the model, results better matched reality, the researchers note. They report their findings in the Nov. 1 *Science*.

During their analyses, Lohmann and Lesins discovered that pollutant aerosols influenced the clouds in the simulations more strongly than they did in the satellite observations. So, the researchers recalibrated their model. The revised simulations suggest that human-generated aerosols block 40 percent less of the incoming solar radiation than the previous model had indicated. The new data indicate that aerosol-derived cooling effects may not be as large as many might hope.

That's cold comfort for a world full of industries and automobiles that spew a seemingly ever-increasing amount of planet-warming greenhouse gases.

SPECIAL REPORT • GLOBAL WARMING

FEELING THE HEAT
LIFE IN THE GREENHOUSE

Except for nuclear war or a collision with an asteroid, no force has more potential to damage our planet's web of life than global warming. It's a "serious" issue, the White House admits, but nonetheless George W. Bush has decided to abandon the 1997 Kyoto treaty to combat climate change—an agreement the U.S. signed but the new President believes is fatally flawed. His dismissal last week of almost nine years of international negotiations sparked protests around the world and a face-to-face disagreement with German Chancellor Gerhard Schröder. Our special report examines the signs of global warming that are already apparent, the possible consequences for our future, what we can do about the threat and why we have failed to take action so far.

By MICHAEL D. LEMONICK

There is no such thing as normal weather. The average daytime high temperature for New York City this week should be 57°F, but on any given day the mercury will almost certainly fall short of that mark or overshoot it, perhaps by a lot. Manhattan thermometers can reach 65° in January every so often and plunge to 50° in July. And seasons are rarely normal. Winter snowfall and summer heat waves beat the average some years and fail to reach it in others. It's tough to pick out overall changes in climate in the face of these natural fluctuations. An unusually warm year, for example, or

even three in a row don't necessarily signal a general trend.

Yet the earth's climate does change. Ice ages have frosted the planet for tens of thousands of years at a stretch, and periods of warmth have pushed the tropics well into what is now the temperate zone. But given the normal year-to-year variations, the only reliable signal that such changes may be in the works is a long-term shift in worldwide temperature.

And that is precisely what's happening. A decade ago, the idea that the planet was warming up as a result of human activity was largely

theoretical. We knew that since the Industrial Revolution began in the 18th century, factories and power plants and automobiles and farms have been loading the atmosphere with heat-trapping gases, including carbon dioxide and methane. But evidence that the climate was actually getting hotter was still murky.

Not anymore. As an authoritative report issued a few weeks ago by the United Nations-sponsored Intergovernmental Panel on Climate Change makes plain, the trend toward a warmer world has unquestionably begun. Worldwide temperatures have climbed more than 1°F over the past

MAKING THE CASE THAT OUR CLIMATE IS CHANGING

From melting glaciers to rising oceans, the signs are everywhere. Global warming can't be blamed for any particular heat wave, drought or deluge, but scientists say a hotter world will make such extreme weather more frequent—and deadly.

EXHIBIT A

Thinning Ice

ANTARCTICA, home to these Adélie penguins, is heating up. The annual melt season has increased up to three weeks in 20 years.

MOUNT KILIMANJARO has lost 75% of its ice cap since 1912. The ice on Africa's tallest peak could vanish entirely within 15 years.

LAKE BAIKAL in eastern Siberia now feezes for the winter 1.1 days later than it did a century ago.

VENEZUELAN mountaintops had six glaciers in 1972. Today only two remain.

EXHIBIT B

Hotter Times

TEMPERATURES SIZZLED from Kansas to New England last May.

CROPS WITHERED and Dallas temperatures topped 100°F for 29 daysstraight in a Texas hot spell that struck during the summer of 1998.

INDIA'S WORST heat shock in 50 years killed more than 2,500 people in May 1998.

CHERRY BLOSSOMS in Washington bloom seven days earlier in the spring than they did in 1970.

EXHIBIT C

Wild Weather

HEAVY RAINS in England and Wales made last fall Britain's wettest three-month period on record.

FIRES due to dry conditions and record-breaking heat consumed 20% of Samos Island, Greece, last July.

FLOODS along the Ohio River in March 1997 caused 30 deaths and at least $500 million in property damage.

HURRICAN FLOYD brought flooding rains and 130-m.p.h. winds through the Atlantic seabord in September1999, killing 77 people and leaving thousands homeless.

EXHIBIT D

Nature's Pain

PACIFIC SALMON populations fell sharply in 1997 and 1998, when local ocean temperatures rose 6°F.

POLAR BEARS in Hudson Bay are having fewer cubs, possibly as a result of earlier spring ice breakup.CORAL REEFS suffer from the loss of algae that color and nourish them. The process, called bleaching, is caused by warmer oceans.DISEASES like dengue fever are expanding their reach northward in the U.S.

BUTTERFLIES are relocating to higher latitudes. The Edith's Checkerspot butterfly of western North America has moved almost 60 miles north in 100 years.

EXHIBIT E

Rising Sea Levels

CAPE HATTERAS Lighthouse was 1,500 ft. from the North Carolina shoreline when it was built in 1870. By the late 1980s teh ocean had crept to within 160 ft., and the lighthouse had to be moved to avoid collapse.

JAPANESE FORTIFICATIONS were built on Kosrae Island in the southwest Pacific Ocean during World War II to guard against U.S. Marines' invading the beach. Today the fortifications are awash at high tide.

FLORIDA FARMLAND up to 1,000 ft. inland from Biscayne Bay is being infiltrated by salt water, rendering the land too toxic for crops. Salt water is also nibbling at the edges of farms on Maryland's Eastern Shore.

BRAZILIAN SHORELINE in the region of Recife receded more than 6 ft. a year from 1915 to 1950 and more than 8 ft. a year from 1985 to 1995.

century, and the 1990s were the hottest decade on record. After analyzing data going back at least two decades on everything from air and ocean temperatures to the spread and retreat of wildlife, the IPCC asserts that this slow but steady warming has had an impact on no fewer than 420 physical processes and animal and plant species on all continents.

Glaciers, including the legendary snows of Kilimanjaro, are disappearing from mountaintops around the globe. Coral reefs are dying off as the seas get too warm for comfort. Drought is the norm in parts of Asia and Africa. El Niño events, which trigger devastating weather in the eastern Pacific, are more frequent. The Arctic permafrost is starting to melt. Lakes and rivers in colder climates are freezing later and thawing earlier each year. Plants and animals are shifting their ranges poleward and to higher altitudes, and migration patterns for animals as diverse

as polar bears, butterflies and beluga whales are being disrupted.

Faced with these hard facts, scientists no longer doubt that global warming is happening, and almost nobody questions the fact that humans are at least partly responsible. Nor are the changes over. Already, humans have increased the concentration of carbon dioxide, the most abundant heat-trapping gas in the atmosphere, to 30% above pre-industrial levels—and each year the

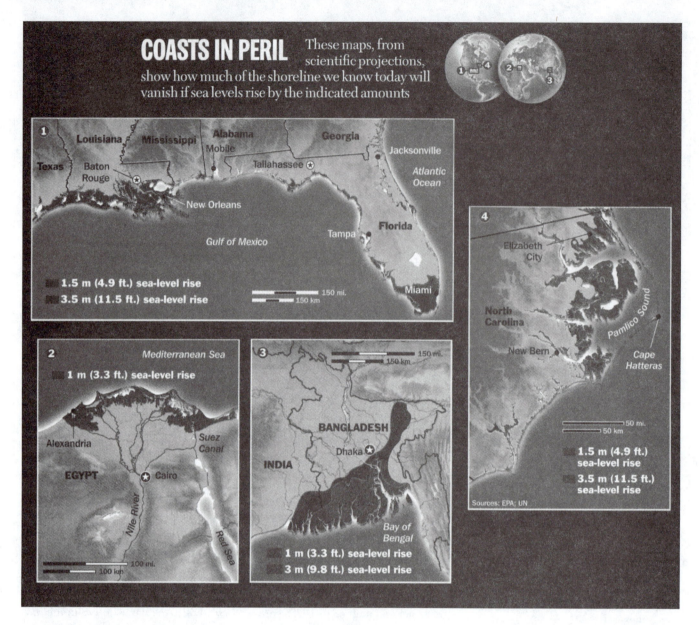

COASTS IN PERIL

These maps, from scientific projections, show how much of the shoreline we know today will vanish if sea levels rise by the indicated amounts

1
Texas · Louisiana · Mississippi · Alabama · Georgia
Baton Rouge · Mobile · Tallahassee · Jacksonville
New Orleans · Atlantic Ocean
Gulf of Mexico · Tampa · Florida · Miami

- 1.5 m (4.9 ft.) sea-level rise
- 3.5 m (11.5 ft.) sea-level rise

150 mi.
150 km

2
Mediterranean Sea
- 1 m (3.3 ft.) sea-level rise
Alexandria · Suez Canal
EGYPT · Cairo · Nile River · Red Sea
100 mi.
100 km

3
150 mi.
150 km
BANGLADESH
Dhaka
INDIA
Bay of Bengal
- 1 m (3.3 ft.) sea-level rise
- 3 m (9.8 ft.) sea-level rise

4
Elizabeth City
North Carolina · Pamlico Sound
New Bern · Cape Hatteras
50 mi.
50 km
- 1.5 m (4.9 ft.) sea-level rise
- 3.5 m (11.5 ft.) sea-level rise
Sources: EPA; UN

rate of increase gets faster. The obvious conclusion: temperatures will keep going up.

Unfortunately, they may be rising faster and heading higher than anyone expected. By 2100, says the IPCC, average temperatures will increase between 2.5°F and 10.4°F—more than 50% higher than predictions of just a half-decade ago. That may not seem like much, but consider that it took only a 9°F shift to end the last ice age. Even at the low end, the changes could be problematic enough, with storms getting more frequent and intense, droughts more pronounced,

coastal areas ever more severely eroded by rising seas, rainfall scarcer on agricultural land and ecosystems thrown out of balance.

But if the rise is significantly larger, the result could be disastrous. With seas rising as much as 3 ft., enormous areas of densely populated land—coastal Florida, much of Louisiana, the Nile Delta, the Maldives, Bangladesh—would become uninhabitable. Entire climatic zones might shift dramatically, making central Canada look more like central Illinois, Georgia more like Guatemala. Agriculture would be

thrown into turmoil. Hundreds of millions of people would have to migrate out of unlivable regions.

Public health could suffer. Rising seas would contaminate water supplies with salt. Higher levels of urban ozone, the result of stronger sunlight and warmer temperatures, could worsen respiratory illnesses. More frequent hot spells could lead to a rise in heat-related deaths. Warmer temperatures could widen the range of disease-carrying rodents and bugs, such as mosquitoes and ticks, increasing the incidence of dengue fever, malaria, encephalitis,

Lyme disease and other afflictions. Worst of all, this increase in temperatures is happening at a pace that outstrips anything the earth has seen in the past 100 million years. Humans will have a hard enough time adjusting, especially in poorer countries, but for wildlife, the changes could be devastating.

Like any other area of science, the case for human-induced global warming has uncertainties—and like many pro-business lobbyists, President Bush has proclaimed those uncertainties a reason to study the problem further rather than act. But while the evidence is circumstantial, it is powerful, thanks to the IPCC's painstaking research. The U.N.-sponsored group was organized in the late 1980s. Its mission: to sift through climate-related studies from a dozen different fields and integrate them into a coherent picture. "It isn't just the work of a few green people," says Sir John Houghton, one of the early leaders who at the time ran the British Meteorological Office. "The IPCC scientists come from a wide range of backgrounds and countries."

Measuring the warming that has already taken place is relatively simple; the trick is unraveling the causes and projecting what will happen over the next century. To do that, IPCC scientists fed a wide range of scenarios involving varying estimates of population and economic growth, changes in technology and other factors into computers. That process gave them about 35 estimates, ranging from 6 billion to 35 billion tons, of how much excess carbon dioxide will enter the atmosphere.

Then they loaded those estimates into the even larger, more powerful computer programs that attempt to model the planet's climate. Because no one climate model is considered definitive, they used seven different versions, which yielded 235 independent predictions of global temperature increase. That's where the range of 2.5°F to 10.4°F (1.4°C to 5.8°C) comes from.

The computer models were criticized in the past largely because the climate is so complex that the limited hardware and software of even a half-decade ago couldn't do an adequate simulation. Today's climate models, however, are able to take into account the heat-trapping effects not just of CO_2 but also of other greenhouse gases, including methane. They can also factor in natural variations in the sun's energy and the effect of substances like dust from volcanic eruptions and particulate matter spewed from smokestacks.

That is one reason the latest IPCC predictions for temperature increase are higher than they were five years ago. Back in the mid-1990s, climate models didn't include the effects of the El Chichon and Mount Pinatubo volcanic eruptions, which threw enough dust into the air to block out some sunlight and slow down the rate of warming. That effect has dissipated, and the heating should start to accelerate. Moreover, the IPCC noted, many countries have begun to reduce their emissions of sulfur dioxide in order to fight acid rain. But sulfur dioxide particles, too, reflect sunlight; without this shield, temperatures should go up even faster.

The models still aren't perfect. One major flaw, agree critics and champions alike, is that they don't adequately account for clouds. In a warmer world, more water will evaporate from the oceans and presumably form more clouds. If they are billowy cumulus clouds, they will tend to shade the planet and slow down warming; if they are high, feathery cirrus clouds, they will trap even more heat.

Research by M.I.T. atmospheric scientist Richard Lindzen suggests that warming will tend to make cirrus clouds go away. Another critic, John Christy of the University of Alabama in Huntsville, says that while the models reproduce the current climate in a general way, they fail to get right the amount of warming at different levels in the atmosphere. Neither Lindzen nor Christy (both IPCC

authors) doubts, however, that humans are influencing the climate. But they question how much—and how high temperatures will go. Both scientists are distressed that only the most extreme scenarios, based on huge population growth and the maximum use of dirty fuels like coal, have made headlines.

It won't take the greatest extremes of warming to make life uncomfortable for large numbers of people. Even slightly higher temperatures in regions that are already drought- or flood-prone would exacerbate those conditions. In temperate zones, warmth and increased CO_2 would make some crops flourish—at first. But beyond 3° of warming, says Bill Easterling, a professor of geography and agronomy at Penn State and a lead author of the IPCC report, "there would be a dramatic turning point. U.S. crop yields would start to decline rapidly." In the tropics, where crops are already at the limit of their temperature range, the decrease would start right away.

Even if temperatures rise only moderately, some scientists fear, the climate would reach a "tipping point"—a point at which even a tiny additional increase would throw the system into violent change. If peat bogs and Arctic permafrost warm enough to start releasing the methane stored within them, for example, that potent greenhouse gas would suddenly accelerate the heat-trapping process.

By contrast, if melting ice caps dilute the salt content of the sea, major ocean currents like the Gulf Stream could slow or even stop, and so would their warming effects on northern regions. More snowfall reflecting more sunlight back into space could actually cause a net cooling. Global warming could, paradoxically, throw the planet into another ice age.

Even if such a tipping point doesn't materialize, the more drastic effects of global warming might be only postponed rather than avoided. The IPCC's calculations end with the year 2100, but the warming won't.

World Bank chief scientist, Robert Watson, currently serving as IPCC chair, points out that the CO_2 entering the atmosphere today will be there for a century. Says Watson: "If we stabilize (CO_2 emissions) now, the concentration will continue to go up for hundreds of years. Temperatures will rise over that time."

That could be truly catastrophic. The ongoing disruption of ecosystems and weather patterns would be bad enough. But if temperatures reach the IPCC's worst-case levels and stay there for as long as 1,000 years, says Michael Oppenheimer, chief scientist at Environmental Defense, vast ice sheets in Greenland and Antarctica could melt, raising sea level more than 30 ft. Florida would be history, and every city on the U.S. Eastern seaboard would be inundated.

In the short run, there's not much chance of halting global warming, not even if every nation in the world ratifies the Kyoto Protocol tomorrow. The treaty doesn't require reductions in carbon dioxide emissions until 2008. By that time, a great deal of damage will already have been done. But we can slow things down. If action today can keep the climate from eventually reaching an unstable tipping point or can finally begin to reverse the warming trend a century from now, the effort would hardly be futile. Humanity embarked unknowingly on the dangerous experiment of tinkering with the climate of our planet. Now that we know what we're doing, it would be utterly foolish to continue.

Reported by David Bjerklie,
Robert H. Boyle and
Andrea Dorfman/New York and
Dick Thompson/Washington

Environmental Information Retrieval

ON FINDING OUT MORE

There is probably more printed information on environmental issues, regulations, and concerns than on any other major topic. So much is available from such a wide and diverse group of sources, that the first effort at finding information seems an intimidating and even impossible task. Attempting to ferret out what agencies are responsible for what concerns, what organizations to contact for specific environmental information, and who is in charge of what becomes increasingly more difficult.

To list all of the governmental agencies private and public organizations, and journals devoted primarily to environmental issues is, of course, beyond the scope of this current volume. However, we feel that a short primer on environmental information retrieval should be included in order to serve as a springboard for further involvement; for it is through informed involvement that issues, such as those presented, will eventually be corrected.

I. SELECTED OFFICES WITHIN FEDERAL AGENCIES AND FEDERAL-STATE AGENCIES FOR ENVIRONMENTAL INFORMATION RETRIEVAL

Appalachian Regional Commission
1666 Connecticut Avenue, NW, Suite 700, Washington, DC 20009 (202) 884-7799
http://www.arc.gov

Council on Environmental Quality
722 Jackson Place NW, Washington, DC 20503 (202) 395-5750
http://www.whitehouse.gov/CEQ/

Delaware River Basin Commission
25 State Police Drive, P.O. Box 7360, West Trenton, NJ 08628-0360 (609) 883-9500
http://www.state.nj.us/drbc/drbc.htm

Department of Agriculture
1400 Independence Avenue, SW, Washington, DC 20250 (202) 720-3631
http://www.usda.gov

Department of the Army (Corps of Engineers)
441 G. Street NW, Washington, DC 20314-1000 (202) 761-0008
http://www.usace.army.mil

Department of Commerce
1401 Constitution Ave. NW, Washington, DC 20230 (202) 482-4883
http://www.doc.gov

Department of Defense
Public Affairs, 1400 Defense Pentagon, Room 3A750, Washington, DC 20301-1400 (703) 428-0711
http://www.defenselink.mil/index.html

Department of Health and Human Services
200 Independence Avenue, SW, Washington, DC 20201 (202) 619-0257   (877) 696-6775
http://www.os.dhhs.gov

Department of the Interior
1849 C Street, NW, Washington, DC 20240-0001 (202) 208-3100
http://www.doi.gov
- Bureau of Indian Affairs (202) 208-37110
- Bureau of Land Management (202) 452-5125
- National Park Service (202) 208-6843
- United States Fish and Wildlife Service (202) 208-4131

Department of State, Bureau of Oceans and International Environmental and Scientific Affairs
2201 C Street, NW, Washington, DC 20520 (202) 647-2492
http://www.state.gov/www/global/oes/index.html

Department of the Treasury, U.S. Customs Service
1300 Pennsylvania Avenue, NW, Washington, DC 20229 (202) 927-1000
http://www.customs.ustreas.gov

Environmental Protection Agency (EPA)
1200 Pennsylvania Avenue, Washington, DC 20460 (202) 260-2090, http://www.epa.gov
- Region 1, One Congress Street, Suite 1100, Boston, MA 02114-2023 (617) 918-1111 (888) 372-7341
  http://www.epa.gov/region01/ (Connecticut, Maine, Massachusetts, New Hampshire, Rhode Island, Vermont)
- Region 2, 290 Broadway, New York, NY 10007-1866 (212) 637-5000
  http://www.epa.gov/region02/ (New Jersey, New York, Puerto Rico, Virgin Islands)
- Region 3, 1650 Arch Street, Philadelphia, PA 19103-2029 (215) 814-5000 (800) 438-2474
  http://www.epa.gov/region03/ (Delaware, District of Columbia, Maryland, Pennsylvania, Virginia, West Virginia)
- Region 4, 61 Forsyth Street, SW, Atlanta, GA 30303-3104 (404) 562-9900
  http://www.epa.gov/region04/ (Alabama, Florida, Georgia, Kentucky, Mississippi, North Carolina, South Carolina, Tennessee)
- Region 5, 77 West Jackson Blvd., Chicago, IL 60604-3507 (312) 353-2000
  http://www.epa.gov/region5/ (Illinois, Indiana, Michigan, Minnesota, Ohio, Wisconsin)
- Region 6, 1445 Ross Avenue, Suite 1200, Dallas, TX 75202 (214) 665-2200 (800) 887-6063
  http://www.epa.gov/region06/ (Arkansas, Louisiana, New Mexico, Oklahoma, Texas)
- Region 7, 901 North 5th Street, Kansas City, KS 66101 (913) 551-7003 (800) 223-0425
  http://www.epa.gov/region07/ (Iowa, Kansas, Missouri, Nebraska)
- Region 8, 999 18th Street, Suite 500, Denver, CO 80202-2466 (303) 312-6339 (800) 227-8917
  http://www.epa.gov/region08/ (Colorado, Montana, North Dakota, South Dakota, Utah, Wyoming)

- Region 9, 75 Hawthorne Street, San Francisco, CA 94105 (415) 744-1305
  http://www.epa.gov/ region09/ (Arizona, California, Hawaii, Nevada, American Samoa, Guam, Trust Territories of Pacific Islands, Wake Island)
- Region 10, 1200 Sixth Avenue, Seattle, WA 98101 (206) 553-1200 (800) 424-4372
  http://www.epa.gov/region10/ (Alaska, Idaho, Oregon, Washington)

Federal Energy Regulatory Commission
888 First Street, NE, Washington, DC 20426 (202) 502-6088   (866) 208-0372
http://www.ferc.fed.us

Interstate Commission on the Potomac River Basin
6110 Executive Boulevard, Suite 300, Rockville, MD 20852-3903 (301) 984-1908
http://www.potomacriver.org

Nuclear Regulatory Commission
One White Flint North, 11555 Rockville Pike, Rockville, MD 20852-2738 (301) 415-7000
http://www.nrc.gov

Susquehanna River Basin Commission
1721 North Front Street, Harrisburg, PA 17102 (717) 238-0423
http://www.srbc.net

Tennessee Valley Authority
400 West Summit Hill Drive, Knoxville, TN 37902-1499 (865) 632-2101
http://www.tva.gov

II. SELECTED STATE, TERRITORIAL, AND CITIZENS' ORGANIZATIONS FOR ENVIRONMENTAL INFORMATION RETRIEVAL

A. Government Agencies

Alabama:
Department of Conservation and Natural Resources
64 North Union Street, Montgomery, AL 36130-1456 (334) 242-3486   (800) 262-3151
http://www.dcnr.state.al.us

Alaska:
Department of Environmental Conservation
410 Willoughby Avenue, Suite 303, Juneau, AL 99801-1795 (907) 465-5010
http://www.state.ak.us/dec/

Arizona:
Department of Water Resources
500 North 3rd Street, Phoenix, AZ 85004-3226 (602) 417-2400
http://www.adwr.state.az.us/dec/

Natural Resources Division
1616 West Adams Street, Phoenix, AZ 85007 (602) 542-4625

Arkansas:
Department of Environmental Quality
8001 National Drive, Little Rock, AR 72209 (501) 682-0744
http://www.adeq.state.ar.us

Energy Office
1 Capitol Mall, Little Rock, AR 72201-1012 (501) 682-7319

California:
Department of Conservation
801 K Street, MS24-01, Sacramento, CA 95814 (916) 322-1080
http://www.consrv.ca.gov

Environmental Protection Agency
P.O. Box 2815, 1001 I Street, Sacramento, CA 95814 (916) 445-3846
http://www.calepa.ca.gov

Colorado:
Department of Natural Resources
1313 Sherman Street, Room 718, Denver, CO 80203-2239 (303) 866-3311
http://www.dnr.state.co.us

Connecticut:
Department of Environmental Protection
State Office Building, 79 Elm Street, Hartford, CT 06106-5127 (860) 424-3000
http://www.dep.state.ct.us

Delaware:
Natural Resources and Environmental Control Department
89 Kings Highway, Dover, DE 19901 (302) 739-4506
http://www.dnrec.state.de.us

District of Columbia:
Environmental Regulation Administration
2100 Martin Luther King Jr. Avenue, SE, Suite 203, Washington, DC 20020 (202) 404-1167
http://clean.rti.org/state.cfm?id=dc

Florida:
Department of Environmental Protection
3900 Commonwealth Blvd., M.S.49, Tallahassee, FL 32399-3000 (850) 245-2118
http://www.dep.state.fl.us

Georgia:
Department of Natural Resources
2 Martin Luther King Drive, SE, Suite 1252 East Tower, Atlanta, GA 30334-4100 (404) 656-3500
http://www.dnr.state.ga.us

Guam:
Environmental Protection Agency
17-3304 Mariner Avenue, Tiyan, Guam 96913 (671) 475-1658
http://www.quamepa.govguam.net

Hawaii:
Department of Land and Natural Resources
1151 Punchbowl Street, Room 130, Honolulu, HI 96813 (808) 587-0330
http://www.hawaii.gov/dlnr/Welcome.html

Idaho:
Department of Lands
954 West Jefferson, Boise, ID 83720-0050 (208) 334-0200
http://www2.state.id.us/lands/index.htm
Department of Water Resources
1301 North Orchard Street, Boise, ID 83706 (208) 327-7900
http://www.idwr.state.id.us

Illinois:

Department of Natural Resources

One Natural Resources Way, Springfield, IL 62702-1271 (217) 782-6302

http://dnr.state.il.us

Indiana:

Department of Natural Resources

402 W. Washington St., Indianapolis, IN 46204-2212 (317) 232-4020

http://www.state.in.us/dnr/

Iowa:

Department of Natural Resources

502 East 9th Street, Des Moines, IA 50319-0034 (515) 281-5385

http://www.state.ia.us/dnr/

Kansas:

Department of Environment-1367

1000 SW Jackson, Suite 420, Topeka, KS 66612-1367 (785) 296-1535

http://www.state.ks.us/public/kdhe/

Kentucky:

Natural Resources and Environmental Protection Cabinet

Capital Plaza Tower, Frankfort, KY 40601 (502) 564-5525

http://www.nr.state.ky.us/nrhome.htm

Louisiana:

Department of Environmental Quality

P.O. Box 4301, Baton Rouge, LA 70821-4301 (225) 765-0741

http://www.deq.state.la.us

Department of Natural Resources

617 North 3rd Street, P.O. Box 94396, Baton Rouge, LA 70804 (225) 342-4500

http://www.dnr.state.la.us/index.ssi

Maine:

Department of Environmental Protection

17 State House Station, Augusta, ME 04333-0017 (207) 287-7688 (800) 452-1942

http://www.state.me.us/dep/

Maryland:

Department of Natural Resources

580 Taylor Avenue, Tawes State Office Bldg., Annapolis, MD 21401 (410) 260-8100

http://www.dnr.state.md.us

Massachusetts:

Department of Environmental Management

251 Causeway Street, Suite 600, Boston, MA 02114-2104 (617) 626-1250

http://www.state.ma.us/dem/dem.htm

Michigan:

Department of Natural Resources

Box 30028, Lansing, MI 48909 (517) 373-2329

http://www.michigan.gov/dnr

Minnesota:

Department of Natural Resources

500 Lafayette Road, St. Paul, MN 55155-4040 (651) 296-6157

http://www.dnr.state.mn.us

Mississippi:

Department of Environmental Quality

P.O. Box 20305, Jackson, MS 39289 (601) 961-5053

http://www.deq.state.ms.us/

Missouri:

Department of Natural Resources

P.O. Box 176, Jefferson City, MO 65102 (573) 751-3443 (800) 361-4827

http://www.dnr.state.mo.us/homednr.htm

Montana:

Department of Natural Resources and Conservation

1625 11th Avenue, P.O. Box 201601, Helena, MT 59620-1601 (406) 444-2074

http://www.dnrc.state.mt.us

Nebraska:

Department of Environmental Quality

P.O. Box 98922, Lincoln, NE 68509 (402) 471-2186

http://www.deq.state.ne.us

Nevada:

Department of Conservation and Natural Resources

123 W. Nye Lane, Room 230, Carson City, NV 89706-0818 (775) 687-4360

http://www.dcnr.nv.gov

New Hampshire:

Department of Environmental Services

6 Hazen Drive, P.O. Box 95, Concord, NH 03302-0095 (603) 271-3503

http://www.state.nh.us/des/

Department of Resources and Economic Development

P.O. Box 541, Concord, NH 03302-1856 (603) 271-2411

http://www.dred.state.nh.us

New Jersey:

Department of Environmental Protection

401 E. State Street, P.O. Box 402, Trenton, NJ 08625-0402 (609) 292-2885

http://www.state.nj.us/dep/

New Mexico:

Environmental Department

1190 Saint Francis Drive, Santa Fe, NM 87502-0110 (505) 827-2855 (800) 219-6157

http://www.nmenv.state.nm.us

New York:

Department of Environmental Conservation

625 Broadway, Albany, NY 12233-1010 (518) 402-8540

http://www.dec.state.ny.us

North Carolina:

Department of Environment and Natural Resources

1601 Mail Service Center, Raleigh, NC 27699-1601 (919) 733-4984

http://www.enr.state.nc.us/

North Dakota:
Game & Fish Department
    100 North Bismarck Expressway, Bismarck, ND 58501-5095 (701) 328-6300
    http://www.state.nd.us/gnf/

Ohio:
Department of Natural Resources
    1952 Belcher Drive, Building C-1, Columbus, OH 43224-1387 (614) 265-6811
    http://www.dnr.state.oh.us
Environmental Protection Agency
    P.O. Box 1049, Columbus, OH 43216-1049 (614) 644-3020
    http://www.epa.state.oh.us

Oklahoma:
Conservation Commission
    2800 North Lincoln Boulevard, Suite 160, Oklahoma City, OK 73105-4210 (405) 521-2384
    http://www.okcc.state.ok.us
Department of Environmental Quality
    707 North Robinson, Oklahoma City, OK 73102 (405) 702-1000
    http://www.deq.state.ok.us

Oregon:
Department of Environmental Quality
    811 S.W. 6th Avenue, Portland, OR 97204-1390 (503) 229-5696
    http://www.deq.state.or.us

Pennsylvania:
Department of Environmental Resources
    400 Market Street, Harrisburg, PA 17105 (717) 787-2814
    http://www.dep.state.pa.us

Puerto Rico:
Department of Natural Resources and Environment
    P.O. Box 5887, San Juan, PR 00906 (787) 723-3090

Rhode Island:
Department of Environmental Management
    235 Promenade Street, Providence, RI 02908 (401) 222-6800
    http://www.state.ri.us

South Carolina:
Department of Health and Environmental Control
    2600 Bull Street, Columbia, SC 29201 (803) 898-3432
    http://www.myscgov.com
Department of Natural Resources
    1000 Assembly Street, Columbia 29201 (803) 734-3888
    http://water.dnr.state.sc.us

South Dakota:
Department of Environment and Natural Resources
    Joe Foss Bldg., 523 East Capitol, Pierre, SD 57501 (605) 773-3151
    http://www.state.sd.us/denr/denr.html

Tennessee:
Department of Environment and Conservation
    401 Church St., 21st Floor, Nashville, TN 37243 (888) 891-8332
    http://www.state.tn.us/environment/

Texas:
Commission on Environmental Quality
    P.O. Box 13087, Austin, TX 78711 (512) 239-1000
    http://www.tceq.state.tx.us

Utah:
Department of Environmental Quality
    168 North 1950 West, Box 144810, Salt Lake City, UT 84114 (801) 536-4400
    http://www.deq.utah.gov

Vermont:
Agency of Natural Resources
    103 South Main Street, Waterbury, VT 05671-0301 (802) 241-3600
    http://www.anr.state.vt.us

Virgin Islands:
Department of Planning & Natural Resources
    396-1 Annas Retreat, Foster Bldg., Charlotte Amalie, U.S. Virgin Islands 00802 (340) 774-3320
    http://www.gov.vi/pnr/

Virginia:
Secretary of Natural Resources
    Ninth Street Office Bldg. 7th floor 202 N. Ninth St., Richmond, VA 23219 (804) 786-0044
    http://snr.vipnet.org

Washington:
Department of Ecology
    P.O. Box 47600, Olympia, WA 98504-7600 (360) 407-6000
    http://www.ecg.wa.gov/
Department of Natural Resources
    P.O. Box 47000, Olympia, WA 98504-7000 (360) 902-1000
    http://www.dnrwa.gov

West Virginia:
Division of Natural Resources
    1900 Kanawha Blvd. East, Charleston, WV 25305 (304) 558-2791
    http://www.dnr.state.wv.us

Wisconsin:
Department of Natural Resources
    Box 7921, Madison, WI 53707 (608) 266-2621
    http://www.dnr.state.wi.us

Wyoming:
Department of Environmental Quality
    122 West 25th Street, Herschler Bldg., Cheyenne, WY 82002 (307) 777-7937
    http://deq.state.wy.us

B. Citizens' Organizations
Advancement of Earth & Environmental Sciences
    International Association for, Northeastern Illinois University, Geography and Environmental Studies Department 5500 North St. Louis Avenue, Chicago, Illinois 60625 (312) 794-2628
Air Pollution Control Association
    1 Gateway Center, 3rd Floor, Pittsburgh, PA 15222 (412) 232-3444
American Association for the Advancement of Science
    1200 New York Avenue, NW, Washington, DC 20005 (202) 326-6400
    http://www.aaas.org

American Chemical Society
1155 16th Street, NW, Washington, DC 20036 (800) 227-5558
http://chemistry.org

American Farm Bureau Federation
225 Touhy Avenue, Park Ridge, IL 60068 (847) 685-8600
http://www.fb.com

American Fisheries Society,
5410 Grosvenor Lane, Suite 110, Bethesda, MD 20814 (301) 897-8616
http://www.fisheries.org

American Forest & Paper Association
1111 19th Street, NW, Suite 800, Washington, DC 20036 (800) 878-8878
http://www.afandpa.org

American Forests
P.O. Box 2000, Washington, DC 20013 (202) 955-4500
http://www.amfor.org

American Institute of Biological Sciences
1444 Eye Street NW, Washington, DC 20005 (202) 628-1500
http://www.aibs.org

American Museum of Natural History
Central Park West at 79th Street, New York, NY 10024-5192 (212) 769-5100
http://www.amnh.org

American Petroleum Institute
1220 L Street, NW, Washington, DC 20005-4070 (202) 682-8000
http://www.api.org

American Rivers
1025 Vermont Avenue NW, Suite 720, Washington, DC 20005 (202) 347-7550
http://www.amrivers.org

Association for Conservation Information
P.O. Box 12559, Charleston, SC 29412 (803) 762-5032
http://www.aci-net.org

Boone and Crockett Club
250 Station Dr., Missoula, MT 59801 (406) 542-1888
http://www.boone-crockett.org

The Ocean Conservancy
1725 DeSales Street, NW, Suite 600, Washington, DC 20036 (202) 429-5609
http://www.cmc-ocean.org

Citizens for a Better Environment
1845 N. Farwell Ave., Suite 220, Milwaukee, WI 53202 (414) 271-7280
http://www.cbemw.org

Texas Coastal Conservation Association
6919 Port West, Suite 100, Houston, TX 77024 (713) 626-4222   (800) 626-4222
http://www.ccatexas.org

Conservation Foundation
103404 Knoch Knolls Road, Naperville, IL 60565 (630) 428-4500
www.theconservationfoundation.org

Conservation Fund
1800 North Kent Street, Suite 1120, Arlington, VA 22209-2156 (703) 525-6300
http://www.conservationfund.org

Conservation International
1919 M Street, NW, Suite 600, Washington, DC 20036 (202) 912-1000 (800) 406-2306
http://www.conservation.org

Defenders of Wildlife
1130 17th Street, NW, #1400, Washington, DC, 20030 (202) 682-9400
http://www.defenders.org

Ducks Unlimited, Inc.
One Waterfowl Way, Memphis, TN 38120 (901) 758-3825
http://www.ducks.org

Earthwatch International
3 Clock Tower Place, Suite 100, Maynard, MA 01754 (978) 461-0081
http://www.earthwatch.org

Environmental Action Foundation, Inc.
333 John Caryle St. Suite 209, Alexandria, VA, 22314 (703) 837-5335
http://www.egc.org/environmentalinfo/EAF.asp

Food and Agriculture Organization of the United Nations (FAO)
Via delle Terme di Caracalla, 00100, Rome, Italy (396) 57051
http://www.fao.org

Friends of the Earth
1025 Vermont Avenue, NW, Suite 300, Washington, DC 20005 (202) 783-7400
http://foe.co.uk

Greenpeace U.S.A.
702 H Street, NW, Washington, DC 20019 (202) 462-1177
http://greenpeaceusa.org

International Association of Fish and Wildlife Agencies
444 North Capitol Street, NW, Suite 544, Washington, DC 20001 (202) 624-7890
http://www.teaming.com/iafwa.htm

International Fund for Agricultural Development (IFAD)
Via del Serafico, 107, 00142, Rome, Italy (3960) 54591
http://www.ifad.org

Keep America Beautiful, Inc.
1010 Washington Blvd., Stamford, CT 06901 (203) 323-8987
http://www.kab.org

National Association of Conservation Districts
509 Capitol Court, NE, Washington, DC 20002 (202) 547-6223
http://www.nacdnet.org

National Audubon Society
700 Broadway, New York, NY 10003 (212) 979-3000
http://www.audubon.org

National Environmental Health Association
720 South Colorado Boulevard, 970 South Tower, Denver, CO 80246 (303) 756-9090
http://www.neha.org

National Fisheries Institute
1901 N. Fort Myer Dr., Suite 700, Arlington, VA 22209 (703) 524-8880
http://www.nfi.org

National Geographic Society
1145 17th Street, NW, Washington, DC 20036 (800) 647-5463
http://www.nationalgeographic.com

National Parks and Conservation Association
1300 19th Street, NW, Suite 300, Washington, DC 20036 (800) 628-7275
http://www.npca.org

National Wildlife Federation
11100 Wildlife Center Drive, Reston, VA 20190 (703)438-6000
http://www.nwf.org

Natural Resources Council of America
1025 Thomas Jefferson Street, NW, Suite 109, Washington, DC 20007 (202) 333-0411
http://www.naturalresourcescouncil.org

The Nature Conservancy
4245 Fairfax Drive, Suite 100, Arlington, VA 22203 (703) 841-5300
http://www.tnc.org

Population Association of America
8630 Fenton Street, Suite 722, Silver Spring, MD 20910 (301) 565-6710
http://www.popassoc.org

Population Connection
1400 16th Street, NW, Suite 320, Washington, DC 20036 (202) 332-2200
http://www.populationconnection.org

Rainforest Alliance
665 Broadway, Suite 500, New York, NY 10012 (212) 677-1900
http://www.rainforest-alliance.org

Save-the-Redwoods League
114 Sansome Street, Room 1200, San Francisco, CA 94104 (415) 362-2352
http://www.savetheredwoods.org

Sierra Club
85 Second Street, 2nd Floor, San Francisco, CA 94105 (415) 977-5500
http://www.sierraclub.org

Smithsonian Institution
P.O. Box 37012, SI Building, Room 153, MNC 010, Washington, DC 20013 (202) 357-2700
http://smithsonian.org

Society of American Foresters
5400 Grosvenor Lane, Bethesda, MD 20814-2198 (301) 897-8720
http://www.safnet.org

Sport Fishing Institute
1010 Massachusetts Avenue, NW, Suite 320, Washington, DC 20001 (202) 898-0770

United Nations Educational, Scientific, and Cultural Organization (UNESCO)
UNESCO House, 7, Place de Fontenoy, 75352 Paris 07 SP France, (313) 45 68 10 00
http://www.unesco.org

United Nations Environment Programme/Industry & Environment Centre
Tour Mirabeau 39-43, quai André Citröen 75739 Paris Cedex 15, France (313) 44 37 14 41
http://www. unepie.org

Wilderness Society
1615 M Street, NW, Washington, DC 20036 (800) The Wild
http://www.wilderness.org

World Wildlife Fund
1250 24th Street, NW, Washington, DC 20037 (800) 225-5993
http://www.wwf.org

III. CANADIAN AGENCIES AND CITIZENS' ORGANIZATIONS

A: Government Agencies

Alberta:

Alberta Environment
MainFloor, Petroleum Plaza, South Tower, 9915-108 Street, Edmonton, AB T5K 268 (780) 427-2391
http://www.gov.ab.ca/env.html

British Columbia:

Ministry of Water, Land and Air Protection
P.O. Box 9339 Stn. Prov. Govt., Victoria, BC V8W 9M1 (250) 387-1161
http://www.env.gov.bc.ca/w\ap/

Manitoba:

Manitoba Conservation
200 Saulteaux Crescent, Winnipeg, MB R3J 3W3 (204) 945-6784
http://www.gov.mb.ca/conservation/index.html

New Brunswick:

Department of Environment and Local Government
P.O. Box 6000, Fredericton, NB E3B SHI (506) 453-2690
http://www.gnb.ca/0009/index-e.asp

Department of Natural Resources and Energy
P.O. Box 6000, Fredericton, NB E3B 5H1 (506) 453-3826
http://www.qnb.ca/0078/index-e.asp

Newfoundland and Labrador:

Department of Environment
Confederation Bldg., PO Box 8700, St. John's NL, Canada, A1B 4J6 (709) 729-2664
http://www.govt.nf.ca/env/

Northwest Territories:

Department of Resources, Wildlife, and Economic Development, P.O. Box 1320, Yellowknife, NT 2L9 X1A Canada, (867) 873-7817
http://www.rwed.gov.nt.ca

Nova Scotia:

Department of Natural Resources, P.O. Box 698, Halifax, NS B3J 2T9 (902) 424-5935
http://www.gov.ns.ca/natr/

Department of Environment and Labour
P.O. Box 697, Halifax, NS B3J 2T8 (902) 424-5300
http://www.gov.ns.ca/enla/

Ontario:

Ministry of Natural Resources
300 Water Street, P.O. Box 7000, Peterborough, ON K9J 8M5 (705) 755-2000 (800) 667-1940
http://www.mnr.gov. on.ca/mnr/

Prince Edward Island:

Department of Environment and Energy
Jones Bldg., 11 Kent Street, 4th Floor, P.O. Box 2000, Charlottetown, PEI C1A 7N8 (902) 368-5000
http://www.gov.pe.ca/te/index.asp

Quebec:
Ministère de l'Environnement et de la Faune
Edifice Marie-Guyart, 675, boulevard René-Lévesque, Est, Québec, PC G1R 5V7 (418) 521-3830 (800) 561-1616
http://www.menv.gouv.qc.ca
Ministère des Ressources Naturelles
#B-302, 5700, 4 Avenue Ouest, Charlesbourg, PQ G1H 6R1 (418) 627-8600
http://www.mrn.gouv.qc.ca

Saskatchewan:
Saskatchewan Environment and Resource Management
3211 Albert Street, Regina, SK S4S 5W6 (306) 787-2700
http://www.se.gov.sk.ca

Yukon Territory:
Department of Environment
P.O. Box 2703, Whitehorse, YT Y1A 2C6 (867) 667-5652, (800) 661-0408
http://www.gov.yk.ca/main/index.shtml

B. Citizens' Groups
Alberta Wilderness Association
Box 6398, Station D, Calgary, AB T2P 2E1 (403) 283-2025
http://www.web.net/~awa/
BC Environmental Network (BCEN)
1672 East 10th Avenue, Vancouver, BC V5N 1X5 (604) 879-2272
http://www.bcen.bc.ca
Canadian EarthCare Society
Box 66 PBC, Kelowna, BC V1Y 793 (250) 861-4788
http://www.earthcares.org
Ducks Unlimited Canada
Oak Hammock Marsh, Stonewall, P.O. Box 1160, Oak Hammock Marsh, MB R0C 2Z0 (204) 467-3000 (800) 665-3825
http://www.ducks.ca
Federation of Ontario Naturalists
355 Lesmill Road, Toronto, ON M3B 2W8 (416) 444-8419
http://www.ontarionature.org
L'Association des Entrepreneurs de Service en Environnement du Quebec (AESEQ)
911 Jean-Talon, Est 220, Montreal, H2R 1V5 (514) 270-7110
New Brunswick Environment Industry Association
P.O. Box 637, Stn. A, Fredericton, NB E3B 5B3 (506) 455-0212
http://www.nbeia.nb.ca
Prince Edward Island Environmental Network (PEIEN)
126 Richmond Street, Charleston, PEI C1A 1H9 (902) 566-4170
http://www.isn.net/~network/
Yukon Conservation Society (YCS)
302 Hawkins Street, Whitehorse, YT Y1A 1X6 (867) 668-5678
http://www.yukonconservation.org

IV. SELECTED JOURNALS AND PERIODICALS OF ENVIRONMENTAL INTEREST
American Forests
P.O. Box 2000, Washington, DC 20013 (202) 955-4500 (800) 368-5748
http://www.amfor.org

American Scientist
Scientific Research Society, P.O. Box 13975, Research Triangle Park, NC 27709-3975 (919) 549-0097
http://www.amsci.org/amsci/amsci.html
Audubon
National Audubon Society, 700 Broadway, New York, NY 10003 (212) 979-3000
http://magazine.audubon.org
BioScience
American Institute of Biological Sciences, 1444 I St. NW, Suite 200, Washington, DC 20005 (202) 628-1500
http://www.aibs.org
The Canadian Field-Naturalist
Box 35069 Westgate, Ottawa, ON, Canada K1Z 1A2 (613) 722-3050
http://www.achilles.net/ofnc/cfn.htm
Conservation Directory
National Wildlife Federation, 8925 Leesburg Pike, Vienna, VA 22184 (703) 790-4000
http://www.nwf.org/conservationdirectory
Earth First! Journal
P.O. Box 3023, Tucson, AZ 97440-1415 (520) 620-6900
http://www.envirolink.org
E: The Environmental Magazine
Earth Action Network, P.O. Box 5098, Westport, CT 06881 (203) 854-5559
http://www.emagazine.com
Environment
Heldref Publications, 1319 18th Street, NW, Washington, DC 20036-1802 (202) 296-6267
http://www.heldref.org
Environment Reporter
Bureau of National Affairs, Inc. 1231 25th Street, NW, Washington, DC 2037 (202) 452-4200
http://www.bna.com/prodcatalog/desc/ER.html
Environmental Action Magazine
Environment Action, Inc. 6930 Carroll Ave., Suite 600, Takoma Park, MD 20912-4414 (800) 372-1033
Environmental Science and Technology
American Chemical Society Publications Support Services, 1155 16th Street, NW, Washington, DC 20036 (202) 872-4582 (800) 227-5558
http://pubs.acs.org/journals/esthag/
Focus (bimonthly newsletter)
World Wildlife Fund, 1250 24th Street, NW, Washington, DC 20037 (202) 293-4800
http://www.worldwildlife.org/news/focus.htm
The Futurist
World Future Society, 7910 Woodmont Avenue, Suite 450, Bethesda, MD 20814 (301) 656-8274 (800) 989-8274
http://www.wfs.org/wfs/
Greenpeace Magazine
Greenpeace USA, 702 H Street, NW, Washington, DC 20001 (202) 462-1177 (800) 326-0959
http://www.greenpeaceusa.org
InterEnvironmental
California Institute of Public Affairs, Box 189040, Sacramento, CA 95818 (916) 442-2472
http://www.interenvironment.org

Journal of Soil and Water Conservation
  Soil and Water Conservation Society, 7515 Northeast Ankeny Road, Ankeny, IA 50021-9764 (515) 289-2331
  http://www.swcs.org

Journal of Wildlife Management
  The Wildlife Society, 5410 Grosvenor Lane, Suite 200, Bethesda, MD 20814-2197 (301) 897-9770
  http://www.wildlife.org/journal.html

Mother Earth News
  Ogden Publications, Inc., 1503 SW 42nd Street, Topeka, KS, 66609, (785) 274-4300
  http://www.MotherEarthNews.Com

National Wildlife
  National Wildlife Federation, 11100 Wildlife Center Drive, Reston, VA 20190 (703) 438-6000
  http://www.nwf.org/nationalwildlife/

Natural Resources Journal
  University of New Mexico, School of Law, MSC11 6070, 1 University of New Mexico, Albuquerque, NM 87131 (505) 277-4910
  http://lawschool.unm.edu/Nrj/

Nature
  529 14th Street NW, Washington, DC 20045-1938 (202) 737-2355
  http://www.nature.com/nature/

Nature Canada
  Canadian Nature Federation, One Nicholas Street, Suite 606, Ottawa, ON, Canada K1N 787 (613) 562-3447 (800) 267-40880
  http://www. cnf.ca/nc_main.html

Nature Conservancy Magazine
  4245 North Fairfax Drive, Suite 100, Arlington, VA 22203-1606 (703) 841-5300 (800) 628-6860
  http://www.tnc.org

Omni
  Omni Publications International Ltd., 277 Park Ave., New York, NY 10172 (212) 702-6000
  http://www.omnimag.com

Pollution Abstracts
  Cambridge Scientific Abstracts, 7200 Wisconsin Avenue, Suite 601, Bethesda, MD 20814 (301) 961-6700 (800) 843-7751
  http://www.csa.com

Science
  American Association for the Advancement of Science, 1200 New York Avenue NW, Washington, DC 20005 (202) 326-6501
  http://www.sciencemag.org

Sierra Magazine
  Sierra Club, 85 2nd Street, 2nd Floor, San Francisco, CA 94105-3441 (415) 977-5750
  http://www.sierraclub.org/sierra/

Smithsonian
  900 Jefferson Drive, Washington, DC 20560 (202) 786-2900
  http://smithsonianmag.com

Technology Review
  1 Main Street, Cambridge, MA 02142 (617) 475-8000
  http://www.techreview.com

U.S. News and World Report
  1050 Thomas Jefferson Street, NW, Washington, DC 20037 (202) 955-2000
  http://www.usnews.com/usnews/home.htm

The World & I
  New World Communications,3600 New York Avenue, NE, Washington, DC 20002 (202) 636-3334
  http://www.worldandi.com

SOURCES used to compile this list: Canadian Almanac Directory 1997; Carroll's Federal Directory, April 1997; Carroll's State Directory, February 1997; Congressional Quarterly's Washington Information Directory 1997–1998; Encyclopedia of Associations, 32nd Edition, 1997; Gale Directory of Publications and Broadcast Media, 131st edition; The World Almanac, 1999, Web search engines: Google, Metacrawler.

Glossary

This glossary of environmental terms is included to provide you with a convenient and ready reference as you encounter general terms in your study of environment that are unfamiliar or require a review. It is not intended to be comprehensive, but taken together with the many definitions included in the articles themselves, it should prove to be quite useful.

A

Abiotic   Without life; any system characterized by a lack of living organisms.

Absorption   Incorporation of a substance into a solid or liquid body.

Acid   Any compound capable of reacting with a base to form a salt; a substance containing a high hydrogen ion concentration (low pH).

Acid Rain   Precipitation containing a high concentration of acid.

Adaptation   Adjustment of an organism to the conditions of its environment, enabling reproduction and survival.

Additive   A substance added to another in order to impart or improve desirable properties or suppress undesirable ones.

Adsorption   Surface retention of solid, liquid, or gas molecules, atoms, or ions by a solid or liquid.

Aerobic   Environmental conditions where oxygen is present; aerobic organisms require oxygen in order to survive.

Aerosols   Tiny mineral particles in the atmosphere onto which water droplets, crystals, and other chemical compounds may adhere.

Air Quality Standard   A prescribed level of a pollutant in the air that should not be exceeded.

Alcohol Fuels   The processing of sugary or starchy products (such as sugar cane, corn, or potatoes) into fuel.

Allergens   Substances that activate the immune system and cause an allergic response.

Alpha Particle   A positively charged particle given off from the nucleus of some radioactive substances; it is identical to a helium atom that has lost its electrons.

Ammonia   A colorless gas comprised of one atom of nitrogen and three atoms of hydrogen; liquefied ammonia is used as a fertilizer.

Anthropocentric   Considering humans to be the central or most important part of the universe.

Aquaculture   Propagation and/or rearing of any aquatic organism in artificial "wetlands" and/or ponds.

Aquifers   Porous, water-saturated layers of sand, gravel, or bedrock that can yield significant amounts of water economically.

Atom   The smallest particle of an element, composed of electrons moving around an inner core (nucleus) of protons and neutrons. Atoms of elements combine to form molecules and chemical compounds.

Atomic Reactor   A structure fueled by radioactive materials that generates energy usually in the form of electricity; reactors are also utilized for medical and biological research.

Autotrophs   Organisms capable of using chemical elements in the synthesis of larger compounds; green plants are autotrophs.

B

Background Radiation   The normal radioactivity present; coming principally from outer space and naturally occurring radioactive substances on Earth.

Bacteria   One-celled microscopic organisms found in the air, water, and soil. Bacteria cause many diseases of plants and animals; they also are beneficial in agriculture, decay of dead matter, and food and chemical industries.

Benthos   Organisms living on the bottom of bodies of water.

Biocentrism   Belief that all creatures have rights and values and that humans are not superior to other species.

Biochemical Oxygen Demand (BOD)   The oxygen utilized in meeting the metabolic needs of aquatic organisms.

Biodegradable   Capable of being reduced to simple compounds through the action of biological processes.

Biodiversity   Biological diversity in an environment as indicated by numbers of different species of plants and animals.

Biogeochemical Cycles   The cyclical series of transformations of an element through the organisms in a community and their physical environment.

Biological Control   The suppression of reproduction of a pest organism utilizing other organisms rather than chemical means.

Biomass   The weight of all living tissue in a sample.

Biome   A major climax community type covering a specific area on Earth.

Biosphere   The overall ecosystem of Earth. It consists of parts of the atmosphere (troposphere), hydrosphere (surface and ground water), and lithosphere (soil, surface rocks, ocean sediments, and other bodies of water).

Biota   The flora and fauna in a given region.

Biotic   Biological; relating to living elements of an ecosystem.

Biotic Potential   Maximum possible growth rate of living systems under ideal conditions.

Birthrate   Number of live births in one year per 1,000 midyear population.

Breeder Reactor   A nuclear reactor in which the production of fissionable material occurs.

C

Cancer   Invasive, out-of-control cell growth that results in malignant tumors.

Carbon Cycle   Process by which carbon is incorporated into living systems, released to the atmosphere, and returned to living organisms.

Carbon Monoxide (CO)   A gas, poisonous to most living systems, formed when incomplete combustion of fuel occurs.

Carcinogens   Substances capable of producing cancer.

Carrying Capacity   The population that an area will support without deteriorating.

Glossary

Chlorinated Hydrocarbon Insecticide   Synthetic organic poisons containing hydrogen, carbon, and chlorine. Because they are fat-soluble, they tend to be recycled through food chains, eventually affecting nontarget systems. Damage is normally done to the organism's nervous system. Examples include DDT, Aldrin, Deildrin, and Chlordane.

Chlorofluorocarbons (CFCs)   Any of several simple gaseous compounds that contain carbon, chlorine, fluorine, and sometimes hydrogen; they are suspected of being a major cause of stratospheric ozone depletion.

Circle of Poisons   Importation of food contaminated with pesticides banned for use in this country but made here and sold abroad.

Clear-Cutting   The practice of removing all trees in a specific area.

Climate   Description of the long-term pattern of weather in any particular area.

Climax Community   Terminal state of ecological succession in an area; the redwoods are a climax community.

Coal Gasification   Process of converting coal to gas; the resultant gas, if used for fuel, sharply reduces sulfur oxide emissions and particulates that result from coal burning.

Commensalism   Symbiotic relationship between two different species in which one benefits while the other is neither harmed nor benefited.

Community Ecology   Study of interactions of all organisms existing in a specific region.

Competitive Exclusion   Resulting from competition; one species forced out of part of an available habitat by a more efficient species.

Conservation   The planned management of a natural resource to prevent overexploitation, destruction, or neglect.

Conventional Pollutants   Seven substances (sulfur dioxide, carbon monoxide, particulates, hydrocarbons, nitrogen oxides, photochemical oxidants, and lead) that make up the largest volume of air quality degradation, as identified by the Clean Air Act.

Core   Dense, intensely hot molten metal mass, thousands of kilometers in diameter, at Earth's center.

Cornucopian Theory   The belief that nature is limitless in its abundance and that perpetual growth is both possible and essential.

Corridor   Connecting strip of natural habitat that allows migration of organisms from one place to another.

Crankcase Smog Devices (PCV System)   A system, used principally in automobiles, designed to prevent discharge of combustion emissions into the external environment.

Critical Factor   The environmental factor closest to a tolerance limit for a species at a specific time.

Cultural Eutrophication   Increase in biological productivity and ecosystem succession resulting from human activities.

D

Death Rate   Number of deaths in one year per 1,000 midyear population.

Decarbonization   To remove carbon dioxide or carbonic acid from a substance.

Decomposer   Any organism that causes the decay of organic matter; bacteria and fungi are two examples.

Deforestation   The action or process of clearing forests without adequate replanting.

Degradation (of water resource)   Deterioration in water quality caused by contamination or pollution that makes water unsuitable for many purposes.

Demography   The statistical study of principally human populations.

Desert   An arid biome characterized by little rainfall, high daily temperatures, and low diversity of animal and plant life.

Desertification   Converting arid or semiarid lands into deserts by inappropriate farming practices or overgrazing.

Detergent   A synthetic soap-like material that emulsifies fats and oils and holds dirt in suspension; some detergents have caused pollution problems because of certain chemicals used in their formulation.

Detrivores   Organisms that consume organic litter, debris, and dung.

Dioxin   Any of a family of compounds known chemically as dibenzo-p-dioxins. Concern about them arises from their potential toxicity as contaminants in commercial products. Tests on laboratory animals indicate that it is one of the more toxic anthropogenic (man-made) compounds.

Diversity   Number of species present in a community (species richness), as well as the relative abundance of each species.

DNA (Deoxyribonucleic Acid)   One of two principal nucleic acids, the other being RNA (Ribonucleic Acid). DNA contains information used for the control of a living cell. Specific segments of DNA are now recognized as genes, those agents controlling evolutionary and hereditary processes.

Dominant Species   Any species of plant or animal that is particularly abundant or controls a major portion of the energy flow in a community.

Drip Irrigation   Pipe or perforated tubing used to deliver water a drop at a time directly to soil around each plant. Conserves water and reduces soil waterlogging and salinization.

E

Ecological Density   The number of a singular species in a geographical area, including the highest concentration points within the defined boundaries.

Ecological Succession   Process in which organisms occupy a site and gradually change environmental conditions so that other species can replace the original inhabitants.

Ecology   Study of the interrelationships between organisms and their environments.

Ecosystem   The organisms of a specific area, together with their functionally related environments; considered as a definitive unit.

Ecotourism   Wildlife tourism that could damage ecosystems and disrupt species if strict guidelines governing tours to sensitive areas are not enforced.

Edge Effects   Change in ecological factors at the boundary between two ecosystems. Some organisms flourish here; others are harmed.

Effluent   A liquid discharged as waste.

El Niño   Climatic change marked by shifting of a large warm water pool from the western Pacific Ocean toward the East.

Electron   Small, negatively charged particle; normally found in orbit around the nucleus of an atom.

Eminent Domain   Superior dominion exerted by a governmental state over all property within its boundaries that authorizes it to appropriate all or any part thereof to a necessary public use, with reasonable compensation being made.

Endangered Species   Species considered to be in imminent danger of extinction.

Endemic Species   Plants or animals that belong or are native to a particular ecosystem.

Environment   Physical and biological aspects of a specific area.

Environmental Impact Statement (EIS)   A study of the probable environmental impact of a development project before federal funding is provided (required by the National Environmental Policy Act of 1968).

Environmental Protection Agency (EPA)   Federal agency responsible for control of air and water pollution, radiation and pesticide problems, ecological research, and solid waste disposal.

Erosion   Progressive destruction or impairment of a geographical area; wind and water are the principal agents involved.

Estuary   Water passage where an ocean tide meets a river current.

Eutrophic   Well nourished; refers to aquatic areas rich in dissolved nutrients.

Evolution   A change in the gene frequency within a population, sometimes involving a visible change in the population's characteristics.

Exhaustible Resources   Earth's geologic endowment of minerals, nonmineral resources, fossil fuels, and other materials present in fixed amounts.

Extinction   Irrevocable elimination of species due to either normal processes of the natural world or through changing environmental conditions.

F

Fallow   Cropland that is plowed but not replanted and is left idle in order to restore productivity mainly through water accumulation, weed control, and buildup of soil nutrients.

Fauna   The animal life of a specified area.

Feral   Refers to animals or plants that have reverted to a non-cultivated or wild state.

Fission   The splitting of an atom into smaller parts.

Floodplain   Level land that may be submerged by floodwaters; a plain built up by stream deposition.

Flora   The plant life of an area.

Flyway   Geographic migration route for birds that includes the breeding and wintering areas that it connects.

Food Additive   Substance added to food usually to improve color, flavor, or shelf life.

Food Chain   The sequence of organisms in a community, each of which uses the lower source as its energy supply. Green plants are the ultimate basis for the entire sequence.

Fossil Fuels   Coal, oil, natural gas, and/or lignite; those fuels derived from former living systems; usually called nonrenewable fuels.

Fuel Cell   Manufactured chemical systems capable of producing electrical energy; they usually derive their capabilities via complex reactions involving the sun as the driving energy source.

Fusion   The formation of a heavier atomic complex brought about by the addition of atomic nuclei; during the process there is an attendant release of energy.

G

Gaia Hypothesis   Theory that Earth's biosphere is a living system whose complex interactions between its living organisms and nonliving processes regulate environmental conditions over millions of years so that life continues.

Gamma Ray   A ray given off by the nucleus of some radioactive elements. A form of energy similar to X rays.

Gene   Unit of heredity; segment of DNA nucleus of the cell containing information for the synthesis of a specific protein.

Gene Banks   Storage of seed varieties for future breeding experiments.

Genetic Diversity   Infinite variation of possible genetic combinations among individuals; what enables a species to adapt to ecological change.

Geothermal Energy   Heat derived from the Earth's interior. It is the thermal energy contained in the rock and fluid (that fills the fractures and pores within the rock) in the Earth's crust.

Germ Plasm   Genetic material that may be preserved for future use (plant seeds, animal eggs, sperm, and embryos).

Global Warming   An increase in the near surface temperature of the Earth. Global warming has occurred in the distant past as the result of natural influences, but the term is most often used to refer to the warming predicted to occur as a result of increased emissions of greenhouse gases. Scientists generally agree that the Earth's surface has warmed by about 1 degree Fahrenheit in the past 140 years.

Green Revolution   The great increase in production of food grains (as in rice and wheat) due to the introduction of high-yielding varieties, to the use of pesticides, and to better management techniques.

Greenhouse Effect   The effect noticed in greenhouses when shortwave solar radiation penetrates glass, is converted to longer wavelengths, and is blocked from escaping by the windows. It results in a temperature increase. Earth's atmosphere acts in a similar manner.

Gross National Product (GNP)   The total value of the goods and services produced by the residents of a nation during a specified period (such as a year).

Groundwater   Water found in porous rock and soil below the soil moisture zone and, generally, below the root zone of plants. Groundwater that saturates rock is separated from an unsaturated zone by the water table.

H

Habitat   The natural environment of a plant or animal.

Habitat Fragmentation   Process by which a natural habitat/landscape is broken up into small sections of natural ecosystems, isolated from each other by sections of land dominated by human activities.

Hazardous Waste   Waste that poses a risk to human or ecological health and thus requires special disposal techniques.

Herbicide   Any substance used to kill plants.

Heterotroph   Organism that cannot synthesize its own food and must feed on organic compounds produced by other organisms.

Glossary

Hydrocarbons   Organic compounds containing hydrogen, oxygen, and carbon. Commonly found in petroleum, natural gas, and coal.

Hydrogen   Lightest-known gas; major element found in all living systems.

Hydrogen Sulfide   Compound of hydrogen and sulfur; a toxic air contaminant that smells like rotten eggs.

Hydropower   Electrical energy produced by flowing or falling water.

I

Infiltration   Process of water percolation into soil and pores and hollows of permeable rocks.

Intangible Resources   Open space, beauty, serenity, genius, information, diversity, and satisfaction are a few of these abstract commodities.

Integrated Pest Management (IPM)   Designed to avoid economic loss from pests, this program's methods of pest control strive to minimize the use of environmentally hazardous, synthetic chemicals.

Invasive   Refers to those species that have moved into an area and reproduced so aggressively that they have replaced some of the native species.

Ion   An atom or group of atoms, possessing a charge; brought about by the loss or gain of electrons.

Ionizing Radiation   Energy in the form of rays or particles that have the capacity to dislodge electrons and/or other atomic particles from matter that is irradiated.

Irradiation   Exposure to any form of radiation.

Isotopes   Two or more forms of an element having the same number of protons in the nucleus of each atom but different numbers of neutrons.

K

Keystone Species   Species that are essential to the functioning of many other organisms in an ecosystem.

Kilowatt   Unit of power equal to 1,000 watts.

L

Leaching   Dissolving out of soluble materials by water percolating through soil.

Limnologist   Individual who studies the physical, chemical, and biological conditions of aquatic systems.

M

Malnutrition   Faulty or inadequate nutrition.

Malthusian Theory   The theory that populations tend to increase by geometric progression (1, 2, 4, 8, 16, etc.) while food supplies increase by arithmetic means (1, 2, 3, 4, 5, etc.).

Metabolism   The chemical processes in living tissue through which energy is provided for continuation of the system.

Methane   Often called marsh gas (CH^4); an odorless, flammable gas that is the major constituent of natural gas. In nature it develops from decomposing organic matter.

Migration   Periodic departure and return of organisms to and from a population area.

Monoculture   Cultivation of a single crop, such as wheat or corn, to the exclusion of other land uses.

Mutation   Change in genetic material (gene) that determines species characteristics; can be caused by a number of agents, including radiation and chemicals, called mutagens.

N

Natural Selection   The agent of evolutionary change by which organisms possessing advantageous adaptations leave more offspring than those lacking such adaptations.

Niche   The unique occupation or way of life of a plant or animal species; where it lives and what it does in the community.

Nitrate   A salt of nitric acid. Nitrates are the major source of nitrogen for higher plants. Sodium nitrate and potassium nitrate are used as fertilizers.

Nitrite   Highly toxic compound; salt of nitrous acid.

Nitrogen Oxides   Common air pollutants. Formed by the combination of nitrogen and oxygen; often the products of petroleum combustion in automobiles.

Nonrenewable Resource   Any natural resource that cannot be replaced, regenerated, or brought back to its original state once it has been extracted, for example, coal or crude oil.

Nutrient   Any nutritive substance that an organism must take in from its environment because it cannot produce it as fast as it needs it or, more likely, at all.

O

Oil Shale   Rock impregnated with oil. Regarded as a potential source of future petroleum products.

Oligotrophic   Most often refers to those lakes with a low concentration of organic matter. Usually contain considerable oxygen; Lakes Tahoe and Baikal are examples.

Organic   Living or once living material; compounds containing carbon formed by living organisms.

Organophosphates   A large group of nonpersistent synthetic poisons used in the pesticide industry; include parathion and malathion.

Ozone   Molecule of oxygen containing three oxygen atoms; shields much of Earth from ultraviolet radiation.

P

Particulate   Existing in the form of small separate particles; various atmospheric pollutants are industrially produced particulates.

Peroxyacyl Nitrate (PAN)   Compound making up part of photochemical smog and the major plant toxicant of smog-type injury; levels as low as 0.01 ppm can injure sensitive plants. Also causes eye irritation in people.

Pesticide   Any material used to kill rats, mice, bacteria, fungi, or other pests of humans.

Pesticide Treadmill   A situation in which the cost of using pesticides increases while the effectiveness decreases (because pest species develop genetic resistance to the pesticides).

Petrochemicals   Chemicals derived from petroleum bases.

pH   Scale used to designate the degree of acidity or alkalinity; ranges from 1 to 14; a neutral solution has a pH of 7; low pHs are acid in nature, while pHs above 7 are alkaline.

Phosphate   A phosphorous compound; used in medicine and as fertilizers.

Photochemical Smog   Type of air pollution; results from sunlight acting with hydrocarbons and oxides of nitrogen in the atmosphere.

Photosynthesis   Formation of carbohydrates from carbon dioxide and hydrogen in plants exposed to sunlight; involves a release of oxygen through the decomposition of water.

Photovoltaic Cells   An energy-conversion device that captures solar energy and directly converts it to electrical current.

Physical Half-Life   Time required for half of the atoms of a radioactive substance present at some beginning to become disintegrated and transformed.

Phytoplankton   That portion of the plankton community comprised of tiny plants, e.g., algae, diatoms.

Pioneer Species   Hardy species that are the first to colonize a site in the beginning stage of ecological succession.

Plankton   Microscopic organisms that occupy the upper water layers in both freshwater and marine ecosystems.

Plutonium   Highly toxic, heavy, radioactive, manmade, metallic element. Possesses a very long physical half-life.

Pollution   The process of contaminating air, water, or soil with materials that reduce the quality of the medium.

Polychlorinated Biphenyls (PCBs)   Poisonous compounds similar in chemical structure to DDT. PCBs are found in a wide variety of products ranging from lubricants, waxes, asphalt, and transformers to inks and insecticides. Known to cause liver, spleen, kidney, and heart damage.

Population   All members of a particular species occupying a specific area.

Predator   Any organism that consumes all or part of another system; usually responsible for death of the prey.

Primary Production   The energy accumulated and stored by plants through photosynthesis.

R

Rad (Radiation Absorbed Dose)   Measurement unit relative to the amount of radiation absorbed by a particular target, biotic or abiotic.

Radioactive Waste   Any radioactive by-product of nuclear reactors or nuclear processes.

Radioactivity   The emission of electrons, protons (atomic nuclei), and/or rays from elements capable of emitting radiation.

Rain Forest   Forest with high humidity, small temperature range, and abundant precipitation; can be tropical or temperate.

Recycle   To reuse; usually involves manufactured items, such as aluminum cans, being restructured after use and utilized again.

Red Tide   Population explosion or bloom of minute single-celled marine organisms (dinoflagellates), which can accumulate in protected bays and poison other marine life.

Renewable Resources   Resources normally replaced or replenished by natural processes; not depleted by moderate use.

Riparian Water Right   Legal right of an owner of land bordering a natural lake or stream to remove water from that aquatic system.

S

Salinization   An accumulation of salts in the soil that could eventually make the soil too salty for the growth of plants.

Sanitary Landfill   Land waste disposal site in which solid waste is spread, compacted, and covered.

Scrubber   Antipollution system that uses liquid sprays in removing particulate pollutants from an airstream.

Sediment   Soil particles moved from land into aquatic systems as a result of human activities or natural events, such as material deposited by water or wind.

Seepage   Movement of water through soil.

Selection   The process, either natural or artificial, of selecting or removing the best or less desirable members of a population.

Selective Breeding   Process of selecting and breeding organisms containing traits considered most desirable.

Selective Harvesting   Process of taking specific individuals from a population; the removal of trees in a specific age class would be an example.

Sewage   Any waste material coming from domestic and industrial origins.

Smog   A mixture of smoke and air; now applies to any type of air pollution.

Soil Erosion   Detachment and movement of soil by the action of wind and moving water.

Solid Waste   Unwanted solid materials usually resulting from industrial processes.

Species   A population of morphologically similar organisms, capable of interbreeding and producing viable offspring.

Species Diversity   The number and relative abundance of species present in a community. An ecosystem is said to be more diverse if species present have equal population sizes and less diverse if many species are rare and some are very common.

Strip Mining   Mining in which Earth's surface is removed in order to obtain subsurface materials.

Strontium-90   Radioactive isotope of strontium; it results from nuclear explosions and is dangerous, especially for vertebrates, because it is taken up in the construction of bone.

Succession   Change in the structure and function of an ecosystem; replacement of one system with another through time.

Sulfur Dioxide (SO^2)   Gas produced by burning coal and as a by-product of smelting and other industrial processes. Very toxic to plants.

Sulfur Oxides (SO^x)   Oxides of sulfur produced by the burning of oils and coal that contain small amounts of sulfur. Common air pollutants.

Sulfuric Acid ($H2 SO^4$)   Very corrosive acid produced from sulfur dioxide and found as a component of acid rain.

Sustainability   Ability of an ecosystem to maintain ecological processes, functions, biodiversity, and productivity over time.

Sustainable Agriculture   Agriculture that maintains the integrity of soil and water resources so that it can continue indefinitely.

T

Technology   Applied science; the application of knowledge for practical use.

Tetraethyl Lead   Major source of lead found in living tissue; it is produced to reduce engine knock in automobiles.

Thermal Inversion   A layer of dense, cool air that is trapped under a layer of less dense warm air (prevents upward flowing air currents from developing).

Glossary

Thermal Pollution   Unwanted heat, the result of ejection of heat from various sources into the environment.

Thermocline   The layer of water in a body of water that separates an upper warm layer from a deeper, colder zone.

Threshold Effect   The situation in which no effect is noticed, physiologically or psychologically, until a certain level or concentration is reached.

Tolerance Limit   The point at which resistance to a poison or drug breaks down.

Total Fertility Rate (TFR)   An estimate of the average number of children that would be born alive to a woman during her reproductive years.

Toxic   Poisonous; capable of producing harm to a living system.

Tragedy of the Commons   Degradation or depletion of a resource to which people have free and unmanaged access.

Trophic   Relating to nutrition; often expressed in trophic pyramids in which organisms feeding on other systems are said to be at a higher trophic level; an example would be carnivores feeding on herbivores, which, in turn, feed on vegetation.

Turbidity   Usually refers to the amount of sediment suspended in an aquatic system.

U

Uranium 235   An isotope of uranium that when bombarded with neutrons undergoes fission, resulting in radiation and energy. Used in atomic reactors for electrical generation.

Z

Zero Population Growth   The condition of a population in which birthrates equal death rates; it results in no growth of the population.

Index

Index

Test Your Knowledge Form

We encourage you to photocopy and use this page as a tool to assess how the articles in *Annual Editions* expand on the information in your textbook. By reflecting on the articles you will gain enhanced text information. You can also access this useful form on a product's book support Web site at *http://www.dushkin.com/online/*.

NAME: _____ DATE: _____

TITLE AND NUMBER OF ARTICLE:
_____

BRIEFLY STATE THE MAIN IDEA OF THIS ARTICLE:
_____

LIST THREE IMPORTANT FACTS THAT THE AUTHOR USES TO SUPPORT THE MAIN IDEA:
_____

WHAT INFORMATION OR IDEAS DISCUSSED IN THIS ARTICLE ARE ALSO DISCUSSED IN YOUR TEXTBOOK OR OTHER READINGS THAT YOU HAVE DONE? LIST THE TEXTBOOK CHAPTERS AND PAGE NUMBERS:
_____

LIST ANY EXAMPLES OF BIAS OR FAULTY REASONING THAT YOU FOUND IN THE ARTICLE:
_____

LIST ANY NEW TERMS/CONCEPTS THAT WERE DISCUSSED IN THE ARTICLE, AND WRITE A SHORT DEFINITION:

We Want Your Advice

ANNUAL EDITIONS revisions depend on two major opinion sources: one is our Advisory Board, listed in the front of this volume, which works with us in scanning the thousands of articles published in the public press each year; the other is you—the person actually using the book. Please help us and the users of the next edition by completing the prepaid article rating form on this page and returning it to us. Thank you for your help!

ANNUAL EDITIONS: Environment 04/05

ARTICLE RATING FORM

Here is an opportunity for you to have direct input into the next revision of this volume.
We would like you to rate each of the articles listed below, using the following scale:

1. **Excellent: should definitely be retained**
2. **Above average: should probably be retained**
3. **Below average: should probably be deleted**
4. **Poor: should definitely be deleted**

Your ratings will play a vital part in the next revision.
Please mail this prepaid form to us as soon as possible.
Thanks for your help!

RATING	ARTICLE	RATING	ARTICLE
_____	1. How Many Planets? A Survey of the Global Environment		
_____	2. Forget Nature. Even Eden Is Engineered		
_____	3. Crimes of (a) Global Nature		
_____	4. Toward a Sustainability Transition: The International Consensus		
_____	5. Making the Global Local: Responding to Climate Change Concerns From the Ground Up		
_____	6. Population and Consumption: What We Know, What We Need to Know		
_____	7. An Economy for the Earth		
_____	8. Factory Farming in the Developing World		
_____	9. Common Ground for Farmers and Forests		
_____	10. Where the Sidewalks End		
_____	11. Beyond Oil: The Future of Energy		
_____	12. Powder Keg		
_____	13. Living Without Oil		
_____	14. Renewable Energy: A Viable Choice		
_____	15. Fossil Fuels and Energy Independence		
_____	16. What Is Nature Worth?		
_____	17. Where Wildlife Rules		
_____	18. Invasive Species: Pathogens of Globalization		
_____	19. On the Termination of Species		
_____	20. Where Have All the Farmers Gone?		
_____	21. What's a River For?		
_____	22. A Human Thirst		
_____	23. Our Perilous Dependence on Groundwater		
_____	24. Oceans Are on the Critical List		
_____	25. Three Pollutants and an Emission		
_____	26. The Quest for Clean Water		
_____	27. Solving Hazy Mysteries		
_____	28. Feeling the Heat: Life in the Greenhouse		

(Continued on next page)

NO POSTAGE
NECESSARY
IF MAILED
IN THE
UNITED STATES

BUSINESS REPLY MAIL
FIRST CLASS MAIL PERMIT NO. 551 DUBUQUE IA

POSTAGE WILL BE PAID BY ADDRESEE

McGraw-Hill/Dushkin
2460 KERPER BLVD
DUBUQUE, IA 52001-9902

ABOUT YOU

Name _____ Date _____

Are you a teacher? ☐ A student? ☐
Your school's name _____

Department _____

Address _____ City _____ State _____ Zip _____

School telephone # _____

YOUR COMMENTS ARE IMPORTANT TO US!

Please fill in the following information:
For which course did you use this book?
_____

Did you use a text with this ANNUAL EDITION? ☐ yes ☐ no
What was the title of the text?
_____

What are your general reactions to the *Annual Editions* concept?
_____

Have you read any pertinent articles recently that you think should be included in the next edition? Explain.
_____

Are there any articles that you feel should be replaced in the next edition? Why?
_____

Are there any World Wide Web sites that you feel should be included in the next edition? Please annotate.
_____

May we contact you for editorial input? ☐ yes ☐ no
May we quote your comments? ☐ yes ☐ no